Lecture Notes in Applied and Computational Mechanics

Volume 70

Series Editors

For further volumes:
http://www.springer.com/series/4623

Wolfgang Grellmann · Gert Heinrich
Michael Kaliske · Manfred Klüppel
Konrad Schneider · Thomas Vilgis
Editors

Fracture Mechanics and Statistical Mechanics of Reinforced Elastomeric Blends

Springer

Editors

Wolfgang Grellmann
Zentrum für Ingenieurwissenschaften
Martin-Luther-Universität Halle-Wittenberg
Halle (Saale)
Germany

Manfred Klüppel
Deutsches Institut für Kautschuktechnologie e.V.
Hannover
Germany

Gert Heinrich
Leibniz-Institut für Polymerforschung Dresden e.V.
Dresden
Germany

Konrad Schneider
Leibniz-Institut für Polymerforschung Dresden e.V.
Dresden
Germany

Michael Kaliske
Institute for Structural Analysis
Technische Universität Dresden
Dresden
Germany

Thomas Vilgis
Max-Planck-Institute for Polymer Research
Mainz
Germany

ISSN 1613-7736 ISSN 1860-0816 (electronic)
ISBN 978-3-642-37909-3 ISBN 978-3-642-37910-9 (eBook)
DOI 10.1007/978-3-642-37910-9
Springer Heidelberg New York Dordrecht London

Library of Congress Control Number: 2013935633

Printed on acid-free paper

Springer is part of Springer Science+Business Media (www.springer.com)

Preface

Fracture mechanics and fatigue behaviour of elastomers are of great technical importance for rubber parts. Durability is essential for the development of tailor made rubber materials for high performance elastomeric products like tires, conveyor belts, seals, air springs, dampers and other components. In spite of recent progress, more scientific and technological potential for further development of fracture mechanical models and testing concepts of reinforced elastomers can be identified, especially if considering the strongly nonlinear viscoelastic material behaviour.

For many years, effort for a quantitative characterization and physical understanding of crack propagation and tear resistance in viscoelastic solids has been undertaken. Most of the work is based on the fundamental approaches of Rivlin and Thomas. The main focus of our work lies in a better understanding of fatigue crack propagation under dynamical loading conditions and a more fundamental, physically motivated prediction of service life properties of rubber parts, which is based on micro-mechanical mechanisms of crack initiation and propagation in elastomeric materials. At present, it is still not exactly known how these processes start and how they proceed under dynamic loading conditions.

The present book gives some new insights into the field of fracture mechanics of rubber and approaches the rupture phenomenon by different methods ranging from molecular to macroscopic length and time scales. In particular, a detailed analysis of structure-property relationships of filled elastomers and of multi-phase elastomeric blends is presented with respect to dynamic crack propagation, stress-strain behaviour and micro-mechanical mechanisms of reinforcement by fillers like carbon black. By applying suitable testing, analytical and numerical methods, the morphologic structure of the filler network and the phase morphology of heterogeneous elastomeric blends are characterized on microscopic and mesoscopic length scales. Fracture mechanical investigations on crack propagation under harmonic pulsed deformation are carried out by means of an automatic crack propagation measurement device – the Tear Fatigue Analyser (TFA). Further, the influence of "confinement" on the interfacial glass dynamics of polymer chains within nano-scaled filler structures is investigated in order to obtain a deeper understanding of the polymer-filler and filler-filler interaction. On the structural level, numerical investigations based on physical properties have been carried out using newly developed methods.

In Chapter 1, some elementary statements based on statistical physics approaches are presented. When a material breaks at the crack tip, individual molecules rupture. This phenomenon is easily understood when crack tips are considered in detail. Right at the tip, polymer chains become highly stretched and oriented before they rupture. In consequence, their local high deformation ratio implicates a complete change of their polymer dynamics. In contrast to low stretching ratios, when the polymer motion is easily described by linear Rouse dynamics, it changes to highly nonlinear characteristics which need to be the starting point for more refined theories of individual chain rupturing. The strong nonlinear character of the theory implies involved numerical simulations, which are able to explain the limits of elementary linearized theories also proposed in this chapter. The ideas are discussed at higher dimensional networks, mainly two-dimensional cross-linked polymer systems, which allow a detailed analysis of the rupture phenomena as a function of the deformation modes.

In Chapter 2, synchrotron wide-angle X-ray diffraction is described which was employed to follow the strain-induced crystallization in natural rubber on previously inaccessibly short time scales during dynamic cyclic loading and during step-like loading. Strain-induced crystallization is a fascinating phenomenon that enhances the mechanical properties of some elastomers, most prominently of natural rubber. The tear fatigue resistance is considerably elevated, giving natural rubber a significant edge over synthetic rubbers. Thanks to the WAXD analysis, new structure-property relationships, linking the kinetics of strain-induced crystallization to the tear fatigue behaviour, could be established. Furthermore, scattering methods were applied to study cavitations and filler orientation.

Chapter 3 describes the development of new micro-mechanical concepts of fracture mechanics of filled rubbers and demonstrates their validation by experimental data. These concepts are based on the well established Dynamic Flocculation Model for rubber reinforcement which itself combines the well founded tube model of rubber elasticity with specific and detailed features of the morphology and micro-mechanics of the used filler systems. In particular, it is demonstrated that the characteristics of the filler network are closely related to the fracture mechanical properties of filled elastomers. Micro-mechanical material parameters for various elastomeric materials are evaluated by fitting quasi-static stress-strain cycles to the Dynamic Flocculation Model. This allows for a quantitative description of stress softening effects and filler induced hysteresis in the area close to the crack tip by applying numerical simulations with the Finite Element Method (FEM). Thereby, the influences of the cluster structure and the strength of filler-filler bonds on the deformation behaviour close to the crack tip and the fracture mechanical properties are demonstrated.

Linking macroscopic and mesoscopic aspects of crack propagation is topic of Chapter 4. Combining experiments and numerical simulations based on a global energy balance, the influences of particular dissipative processes such as viscoelasticity and deformation-induced changes of filler structures on the tearing energy are investigated. The process of material degradation in the immediate vicinity of the crack front is driven by the rupture of polymer chain segments and by the interaction of resulting defects with the crack front. Characteristic length scales of the fracture process can be estimated from a statistical analysis of fracture surface topography by means of determining characteristic self-affine roughness exponents. Also nonlocal modeling of the material degradation due to rupture of polymer chain segments within the fracture process zone is presented in Chapter 4 taking into account the overloading of chains both in the fully stretched state and due to fast loading.

Characteristic properties like finite deformations, nonlinear elasticity, and local damage as well as rate-dependent and rate-independent dissipative properties of filled elastomers are considered for a realistic representation of the material response within numerical finite element simulations in Chapter 5. Material models formulated on the macroscale are developed which represent finite viscoelasticity and endochronic elastoplasticity of the undamaged continuum. Failure of rubber material is addressed by a micro-continuum-mechanical model. Crack sensitivity of the material is investigated by the so-called material force method where material forces are employed for the computation of fracture mechanical parameters of dissipative rubber material. Arbitrary crack growth is modelled on the structural scale by the use of cohesive elements in combination with an adaptive implementation strategy. Several examples show the numerically simulated crack propagation in rubber specimens under consideration of different fracture criteria as well as the durability analysis of different tire structures.

Chapter 6 contains experimental results of various mechanical and fracture mechanical investigations of different elastomeric materials. Furthermore, results of structural investigations and of micro-fractographical investigations are reported. The fracture mechanics investigations contribute to a further development of the experimental methods in the range of quasi-static and impact-like loading. Especially the performance of impact tests with variation of the test temperature and a subsequent fracture mechanical analysis is a new field in polymer diagnostics. The results show the strong influence of the test temperature on the fracture behavior of filler-reinforced elastomeric materials. The further methodical development of the quasi-static test contributes to the understanding of stable crack initiation and propagation behavior. The experimental results allow for a structure-related discussion of the mechanical and fracture mechanical properties, which is an important factor in material development.

Chapter 7 describes a new method for the analysis of dynamic crack propagation in filled rubber by simultaneous tensile and pure shear mode testing. This method is based upon the mechanical redevelopment of the testing machine TFA and its subsequent equipment with new hardware and, additionally, a camera system for continuously recording of the crack length. It is shown how tearing energy and crack growth rate depend on the test specimen's geometries and crack length. Moreover, the importance of a defined and reproducible notching of elastomeric specimens is also demonstrated in this chapter.

All presented contributions and results in this book are the outcome of the research unit FOR 597 funded by the German Research Foundation (DFG). This support is gratefully acknowledged.

Contents

Rupture Dynamics of Macromolecules

Jaroslaw Paturej[1], Andrey Milchev[2], Vakhtang G. Rostiashvili[3], and Thomas A. Vilgis[3]

[1] Institute of Physics, Univeristy of Szczecin, Wielkopolska 15, 70-451 Szczecin, Poland
jpaturej@univ.szczecin.pl
[2] Institute for Physical Chemistry, Bulgarian Academy of Sciences, Akad. G. Bonchev 11, 1113 Sofia, Bulgaria
[3] Max Planck Institute for Polymer Research, Ackermannweg 10, 55128 Mainz, Germany

Abstract. In a series of studies we consider the breakage of a polymer chain of segments, coupled by anharmonic bonds under applied tensional force or subject to rise in temperature. The chain dynamics at the onset of fracture is studied by means of Molecular Dynamics simulation and also using analytic considerations. A deeper insight into the changes in polymer relaxation dynamics when bonding anharmonicity is taken into account, that is, beyond the limits of the conventional Rouse model description, is gained by comparing analytic results from the Gaussian Self-Consistent approach to data derived from Monte Carlo and Molecular Dynamics simulations. Simulation results on polymer chain rupture are confronted with the predictions of Kramers-Langer theory. Two cases are investigated: a) thermally induced fracture of unstrained chain, and b) rupture of a chain under tensile stress. Cases of both under- and overdamped dynamics are explored. The recently experimentally observed and intensively studied case of covalent bond scission in "bottle-brush" macromolecules adsorbed on a hard surface is modeled and comprehensively investigated with regard to tension accumulation and breakdown kinetics. Eventually we report on our latest studies of the force-induced rupture and thermal degradation of 2D (graphene-like) networks, focusing on the creation and proliferation of cracks during failure.

1 Introduction

Materials failure has been since a long time the standard method in materials testing. Well known from solid state crystalline materials it has been realized how macroscopic materials properties related to failure such as crack propagation, breaking strains, yield stresses are related to crystalline structures, concentrations alloys and other local lattice properties. This knowledge was then used to design metallic alloys or different steels by variation of the carbon content and different quenching temperatures.

In soft materials, especially in polymers similar methods are of wide empiric use. The relation between macroscopic failure and molecular properties is clearly

W. Grellmann et al. (Eds.): *Fracture Mechanics & Statistical Mech.*, LNACM 70, pp. 1–42.
DOI: 10.1007/978-3-642-37910-9_1

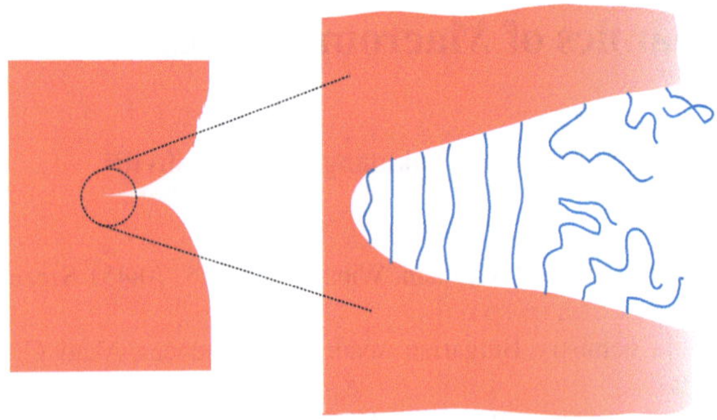

Fig. 1 A simple view of a crack (left) in a polymeric material at different scales. On a local level (right) highly stretched chains break.

visible. Materials with glassy properties show a completely different crack propagation as soft materials in melt states. Glasses are very brittle, whereas rubbers show a viscoelastic behavior and depending of the loading conditions the materials break very differently.

In recent years fracture mechanics has become a significant part of theoretical research, and the prediction of the failure behavior is still a challenge to molecular models. Of course such theories turn out quickly to become more complicated as most of them in classical polymer physics. Close to failure chains become highly stretched and the usual theories such as Gaussian statistic and the corresponding Rouse and reptation theories are no longer valid to discuss materials failure.

Polymer materials break under large stresses and high (local) deformation. This means that chains are locally highly stretched Fig. 1. Gaussian chain statistics and the resulting materials theories, show linear stress strain behavior on local scales and only neo-Hookian stress-strain behavior on molecular scales [1]. However close to the materials break downs finite extensibilities of the chains as well as the non-linear, non Rouse chain motion matter [2].

After a short introduction in Section 1, in Section 2.1 we briefly report on a comprehensive investigation of the relaxation kinetics of an anharmonic polymer chain whereby the characteristic differences with regard to the conventional Rouse dynamics, valid for Gaussian (harmonic) chains are exposed. In the rest of Section 2 we consider the thermal degradation of an unstrained chain as well as the failure of a polymer under constant tension. In Section 3 we examine the scission kinetics of bottle-brush molecules in solution on an adhesive substrate by means of Molecular Dynamics simulation against the background of intensive laboratory experiments. Our investigation is focused on several key questions that determine the bond breakage mechanism such as the distribution of self-induced tension and frequency of scission events along the backbone of a bottle-brush. Eventually, in Section 4 we summarize our investigation concerning detailed Molecular Dynamics simulations of the failure of two-dimensional networks, focusing on graphene-like two-dimensional honeycomb membranes with breakable bonds. We elucidate the frequency of bond scission and the propagation of cracks on various conditions as temperature, applied stress, membrane geometry, etc.

2 Single Polymer

The intuitive examples mentioned above suggest, that it is useful to study the non-linear dynamics and the rupture of individual chains first. This already rises a number of open questions, such as the extension of the linear Rouse dynamics [3] to the realistic cases of anharmonic forces. Evidently, at low degrees of stretching of the chains, their motion will stay linear. At stronger stretching non linearity effects begin to matter whereas beyond some critical extension anharmonic bonds begin to break and material failure sets on. Therefore, the deformation-induced changes in polymer behavior from initially purely elastic (at small elongations) all the way to strongly anharmonic and even rupture at strong stretching have been systematically studied and modeled both analytically and by means of computer experiments (see below).

2.1 Non-linear Chain Dynamics

As a first step in getting insight in the processes of failure of polymer materials we carried out a comprehensive investigation of the relaxation dynamics at different degrees of stretching of a single macromolecule [4]. The study has been focused on the dynamics of a self-avoiding coarse-grained polymer chain at different degrees of stretching. The anharmonicity of covalent bonds was taken into account by replacing the harmonic potential that binds subsequent repeating units (monomers) along the backbone of the macromolecule

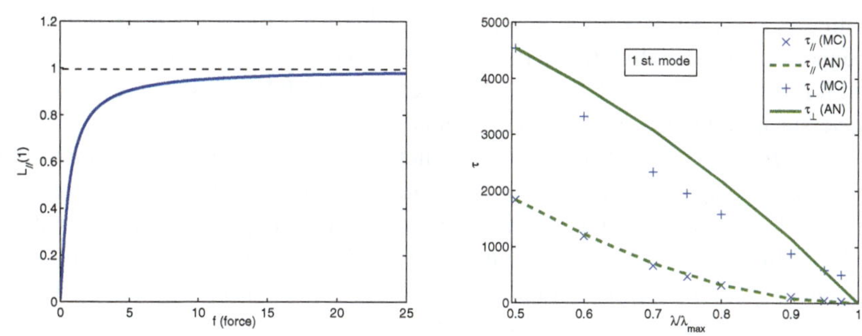

Fig. 2 (left) Variation of the bond length between adjacent beads in a chain with applied tensile force f. Eventually, for large forces the bond length approaches asymptotically its maximum value b_0. (right) Monte-Carlo simulations of the first **mode** relaxation time for the perpendicular $\tau_\perp$ and parallel direction $\tau_\parallel$ as a function of relative degree of stretching $0.5 < \lambda/\lambda_{max} < 1$. Comparison with analytical (AN) and numerical results (MC).

$$U_{\rm H}(r) = \frac{kT}{b^2}(r_i - r_{i-1})^2. \tag{1}$$

(here $r = r_i - r_{i-1}$) by the so called FENE (finitely extensible non-linear elastic) potential which behaves at low degree of stretching as a harmonic potential,

$$U_{\rm F}(r) = kT \ln\{1 - (r_i - r_{i-1})^2/b^2\}. \tag{2}$$

whereas at strong deformations smoothly tends to a maximal length b of the bonds - see Fig. 2 (left panel). The strong anharmonicity of the bonded interaction, Eq. (2), leads to polymer chain dynamics which is qualitatively different from the Rouse dynamics of phantom (Gaussian) chains and is described by highly non-linear equations of motion. Note that many new features that characterize this non-Rouse dynamics typical for anharmonic forces can be conveniently examined even though the bonds, Eq. (2), undergo no scission [4]. If one has to allow for chain rupture, the FENE potential has to be replaced, e.g., by the Morse potential, considered below.

Our analytical treatment as well as extensive computer simulations [4] have provided a rather consistent picture of the qualitative changes that effect the polymer dynamics upon gradual increase of the degree of stretching, λ/λ_{max}, where λ_{max} denotes the maximal extension of the chain, $\lambda_{max} = Nb$. In general, one observes two regimes of chain dynamics, depending on the degree of chain extension. In the first one, that of the friction dominated *overdamped* motion of the monomers, where numerical modeling by means of the Monte Carlo (MC) method is appropriate, both analytic predictions as well as MC results suggest a consistent picture of relaxation time *decrease* with growing stretching of the chain down to a threshold value $\lambda/\lambda_{max} \approx 0.64$. Thereby, the relaxation time parallel to stretching, $\tau_{||}$ is always considerably smaller (about one half) of that in direction perpendicular to stretching, $\tau_{\perp}$, cf. Fig. 2 (right panel). For $\tau_{||}$ the agreement between analytic results from the Gaussian Self-Consistent approximation [4] and MC data is perfect on a quantitative level whereas for $\tau_{\perp}$ it is at most qualitative. The spectrum of oscillation frequencies, corresponding to the independent Rouse modes [4], has been also analyzed for different values of λ/λ_{max}. As expected, the MC results for the relaxation time versus mode index p relationship yields $\tau_p \propto p^{-2}$.

A *transition* regime at $\lambda/\lambda_{max} \approx 0.64$ where friction and inertia terms are of the same order of magnitude separates the *overdamped* regime from an inertial regime at higher degrees of stretching. This latter regime is strongly nonlinear in nature and a faithful description of the polymer dynamics can be produced by means of a Molecular Dynamics (MD) simulation with Dissipative Particle Dynamics (DPD) thermostat as well as by an approximate linearized analytical model.

Among the most salient features of polymer dynamics in this second regime, we find that normal modes are *coupled* to each other and can interchange energy, as expected in a system with strongly anharmonic interactions. The Fourier analysis reveals the existence of principal characteristic frequencies of oscillatory motion for each normal mode. Along with this feature, a broad-band spectrum suggest the existence of a continuum of frequencies around the single peaks, which apparently

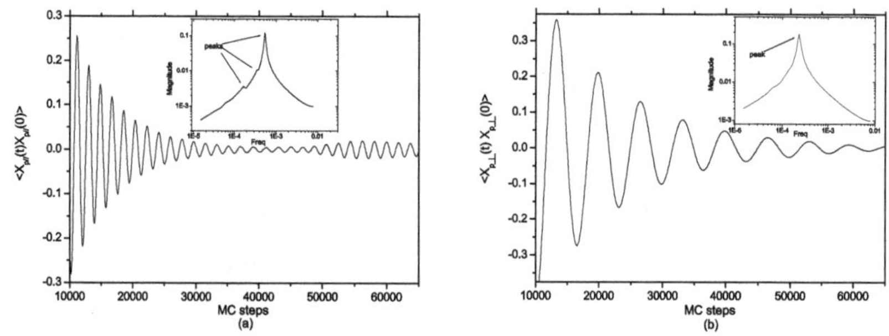

Fig. 3 (left)$\langle X_{p\parallel}(t)X_{p\perp}(0)\rangle$ versus time (in MD time units) for $p = 3$. In the inset the Fourier spectrum of the same function shows a sharp maximum, corresponding to the high basic frequency. Two smaller satellite peaks correspond to lower frequencies, also present in the mode dynamics. (right) The same as in the left panel but in perpendicular direction. The Fourier spectrum reveals the presence of a single well-defined frequency in contrast of other two, less visible, smaller peaks.

make a non-vanishing contribution to the total motion - Fig. 3. These principal characteristic frequencies have been shown by an approximate linear analytical model to scale *linearly* with mode index p in perfect agreement with the simulations. The frequencies also grow upon extension, as expected - Fig. 4. A slight broadening of the half-width of the main frequency peak with increasing extension as well as with increasing mode number at constant extension, is revealed for the parallel direction. This means that as soon as we increase stretching, more and more frequencies appear to contribute in the polymer dynamics, reflected by the single mode correlation functions. This broadening is not evidenced, at least up to $\lambda/\lambda_{max} = 0.95$, for the perpendicular direction.

The relaxation times of the modes depend significantly on the considered direction. While in the parallel direction the relaxation times increase strongly in the transition region but then the possibility to define a single relaxation time for too large extensions $\lambda/\lambda_{max} > 0.8$ becomes questionable, in the perpendicular direction the relaxation times remain more or less constants up to $\lambda/\lambda_{max} \approx 0.95$! It is remarkable that these results are also confirmed qualitatively by our linearized analytic approach. Of course, one must keep in mind that such large extensions of a polymer chain might be of academic interest since bond rupture then may take place as soon as one approaches the tensile strength of the macromolecule. Therefore, in what follows we focus on the rupture of polymer bonds.

2.2 *Thermal Degradation of Unstrained Chain*

The study of degradation and stabilization of polymers is important both from practical and theoretical viewpoints [5]. Disposal of plastic wastes has grown rapidly to a world problem so that increasing environmental concerns have prompted

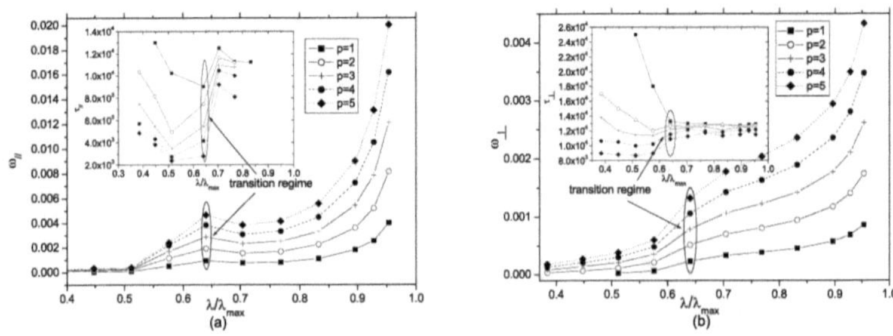

Fig. 4 (left) Characteristic frequency $\omega_{||}$ in the Fourier spectrum of parallel mode correlation functions, $\langle X_{p||}(t)X_{q||}(0)\rangle$ against chain extension. In the inset one shows the variation of relaxation times τ_p for the first five modes with chain elongation λ/λ_{max}. (right) The same as in the left panel for the perpendicular direction.

researchers to investigate plastics recycling by degradation as an alternative [6]. On the other hand, degradation of polymers in different environment is a major limiting factor in their application. Thermal degradation (or, thermolysis) plays a decisive role in the design of flame-resistant polyethylene and other plastic materials [7]. It can also be used in conjunction with chromatography to characterize polymeric structure [8]. Recently, with the advent of exploiting biopolymers as functional materials [9, 10] the stability of such materials has become an issue of primary concern.

Most theoretical investigations of polymer degradation have focused on determining the rate of change of average molecular weight [11, 12, 13, 14, 15, 16, 17, 18, 19, 20]. The main assumptions of the theory are that each link in a long chain molecule has equal strength and equal accessibility, that they are broken at random, and that the probability of rupture is proportional to the number of links present. Experimental study of polystyrene, however, have revealed discrepancies [12] with the theory [11] so, for example, the thermal degradation stops completely or slows down markedly when a certain chain length is reached. Thus, all of the afore-mentioned studies investigate exclusively the way in which the distribution of bond rupture probability along the polymer backbone affects the fragmentation kinetics and the distribution of fragment sizes as time elapses. Only few theoretic studies [21, 22] have recently explored how does the single polymer chain's dynamics itself affect the resulting bond rupture probability. In both studies, however, one has worked with a phantom Gaussian chain (that is, one has used harmonic bond potentials in the simulations) in order to linearize the problem and make it tractable by some analytic approach like the multidimensional Kramers theory, used by Lee [21], and the Wilemski-Fixman approach, employed by Fugmann and Sokolov [22]. In addition, these investigations have been carried out in the heavily damped regime of polymer dynamics where acceleration and inertial effects are neglected.

In this subsection we present the result of simulations concerning the thermal degradation of a linear, unstrained polymer chain where monomers are connected by more realistic non-linear (anharmonic) forces (Morse and Lennard-Jones) using

Langevin molecular dynamics in three dimensions [23]. Main questions to be answered are: How long it will take for this system to break? Where a chain is going to break most frequently?

In order to answer above questions we investigate the average time of bond breakdown $\langle\tau\rangle$, referred to frequently as the Mean First Breakage Time (MFBT) in the literature, regarding its dependence on temperature T, on the number of bonds N in the polymer chain. By changing the friction coefficient of the particles, γ, we examine the scission dynamics of the bonds in both the under- and over-damped cases and find significant qualitative differences. Finally we examine the fragmentation kinetics during thermolysis process of a chain by investigating the mean length of fragments at subsequent times t. While many properties of the thermal degradation process are in agreement with the notion of randomly breaking bonds, the obtained rate histograms of bond rupture indicate unambiguously that the interplay of noise and non-linear interactions are responsible for a certain kind of self-induced multiple-scale-length inhomogeneity regarding the position of the breaking bonds along the backbone of the chain.

2.2.1 Model and Simulation Protocol

We consider a $3d$ coarse-grained model of a polymer chain which consists of N repeatable units (monomers) connected by bonds, whereby each bond of length b is described by a Morse potential,

$$U_M(r) = D\{\exp[-2\alpha(r-b)] - 2\exp[-\alpha(r-b)]\} \tag{3}$$

with a parameter $\alpha \equiv 1$. The dissociation energy of such bonds is D, measured in units of k_BT, where k_B denotes the Boltzmann constant and T is the temperature. The maximum restoring force of the Morse potential, $f^{max} = -dU_M/dr = \alpha D/2$, is reached at the inflection point, $r = b + \alpha^{-1}\ln(2)$. This force f^{max} determines the maximal tensile strength of the chain. Since $U_M(0) = \exp(2\alpha b) - 2\exp(\alpha b) \approx 1.95$, the Morse potential, Eq. (3), is only weakly repulsive and beads could penetrate one another. Therefore, in order to allow properly for the *excluded volume* interactions between bonded monomers, we take the bond potential as a sum of $U_M(r)$ and the so called Weeks-Chandler-Anderson (WCA) potential, $U_{WCA}(r)$, (i.e., the shifted and truncated repulsive branch of the Lennard-Jones potential);

$$U_{WCA}(r) = 4\varepsilon\left[\left(\frac{\sigma}{r}\right)^{12} - \left(\frac{\sigma}{r}\right)^{6} + \frac{1}{4}\right]\Theta(2^{1/6}\sigma - r)) \tag{4}$$

with $\Theta(x) = 0$ or 1 for $x < 0$ or $x \geq 0$, and $\varepsilon = 1$. The non-bonded interactions between monomers are also taken into account by means of the WCA potential, Eq. (4). Thus the interactions in our model correspond to good solvent conditions. Thus, the length scale is set by the parameter $\sigma = 1$ whereby the monomer diameter $b \approx 2^{1/6}\sigma$.

In our MD simulation we use a Langevin equation, which describes the Brownian motion of a set of interacting particles whereby the action of the solvent is split into slowly evolving viscous force and a rapidly fluctuating stochastic force:

$$m\dot{\mathbf{v}}_i(t) = -\gamma\mathbf{v}_i + \mathbf{F}^i_M(t) + \mathbf{F}^i_{WCA}(t) + \mathbf{R}^\mathbf{i}(t). \tag{5}$$

The random force which represents the incessant collision of the monomers with the solvent molecules satisfy the fluctuation-dissipation theorem $\langle R^i_\alpha(t)R^j_\beta(t')\rangle = 2\gamma k_B T\delta_{ij}\delta_{\alpha\beta}\delta(t-t')$. The friction coefficient γ of the Langevin thermostat, used for equilibration, has been set at 0.25. The integration step is 0.002 time units (t.u.) and time is measured in units of $\sqrt{m\sigma^2/D}$ where m denotes the mass of the beads.

We start the simulation with a well equilibrated conformation of the chain as a random coil and examine the thermal scission of the bonds for a free chain. We measure the elapsed time τ until a bond rupture occurs, and average these times over more than 2×10^4 events so as to determine the mean $\langle\tau\rangle$ which we also refer to as Mean First Breakage Time (MFBT). In the course of the simulation we also sample the probability distribution of breaking bonds regarding their position in the chain (a rupture probability histogram) and local bonds strain.

Since in the problem of thermal degradation there is no external force acting on the chain ends, a well defined activation barrier for a bond scission is actually missing, in contrast to the case of applied tensile force which will be studied in details in Sec. 2.3. Therefore, a definition of an unambiguous criterion for bond breakage is not self-evident. Moreover, depending on the degree of stretching, bonds may break and then recombine again. Therefore, in our numeric experiments we use a sufficiently large expansion of the bond, $r_h = 5b$, as a threshold to a broken state of the bond. This convention is based on our checks that the probability for recombination of bonds, stretched beyond r_h, is vanishingly small, i.e., below 1%.

2.2.2 Dependence of τ on Chain Length N and Temperature T

Our consideration of the $\langle\tau\rangle$ vs. N dependence is based on the assumption that bonds in the chain break entirely at random and the scission events happen independent from one another [22]. Consider the survival probability of the i-the bond in the chain, $S_i(t)$, (i.e., the probability that after elapsed time t the bond i is still intact). $S_i(t)$ may be written as $S_i(t) = \exp(-\omega_i t)$, where ω_i is the corresponding scission rate of bond i. Then, for presumably random and independent scission events, the survival probability of the total chain reads

$$S(N,t) = \prod_{i=1}^{N} S_i(t) = \exp(-N\bar{\omega}t) \tag{6}$$

where the average bond scission rate is given by

$$\bar{\omega} = \frac{1}{N}\sum_{i=1}^{N}\omega_i \tag{7}$$

Thus, the MFBT $\langle\tau\rangle$ of the whole chain can be represented as

$$\langle\tau\rangle = -\int_0^\infty t\frac{\partial}{\partial t}S(N,t) = \frac{1}{N\bar{\omega}} \tag{8}$$

where we have used the general relationship between the survival probability and the mean first passage time (see, e.g., Sec. 5.2.7 in ref. [24]). It is worth noting that the product of probabilities in Eq. (6) corresponds to the well known *mean field approximation* in the theory of phase transitions where one neglects the correlations [25].

Our MD simulation results concerning the dependence of MFBT $\langle\tau\rangle$ on chain length N are shown in Fig. 5. Evidently, one observes a power-law decrease, $\langle\tau\rangle \propto N^{-\beta}$, with $\beta \approx 1.0 \pm 0.15$ regardless of temperature. This finding confirms the basic assumption that bonds break entirely at random and the scission events happen independent from one another.

The dependence of the absolute value of the MFBT $\langle\tau\rangle$ on inverse temperature, $1/T$, shown in inset of Fig. 5, appears also in agreement with the general notion of polymer degradation as a thermally activated process. The inset of Fig. 5 demonstrates an expected variation of $\langle\tau\rangle$ with (inverse) temperature, namely an Arrhenian-dependence $\tau \propto \exp(\Delta E_b/k_BT)$ and allows to estimate activation barrier of thermally-induced scission event to the value of around $0.8D$ where D is depth of the potential well, Eq. (3).

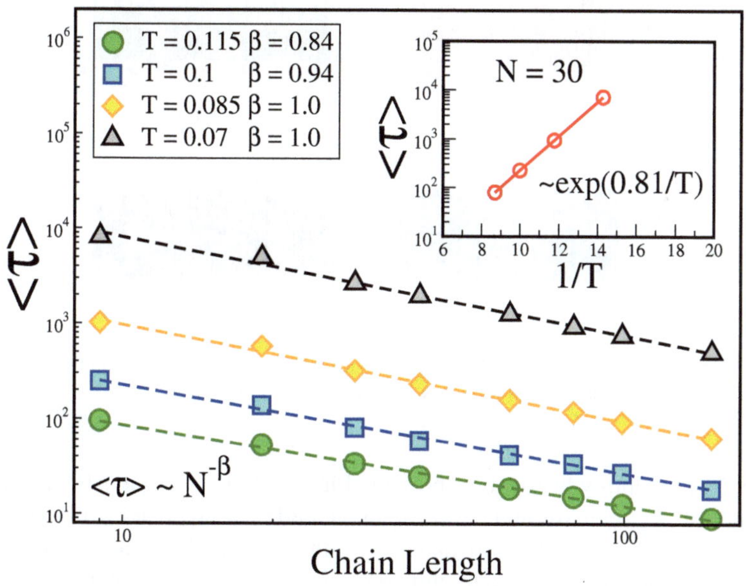

Fig. 5 Mean first passage time $\langle\tau\rangle$ vs chain length N for different temperatures of the heat bath ($\gamma = 0.25$). Dashed lines with a slope $\beta \approx 1.0$ represent fitting curves. The inset displays variation of $\langle\tau\rangle$ with inverse temperature $1/T$ for a chain with $N = 30$ beads. Fitting lines corresponds to an Arrhenian relationship, $\tau \propto \exp(\Delta E_b/k_BT)$.

2.2.3 Preferential Scission of Bonds along the Backbone

In this subsection we present the distribution of scission events along the polymer backbone. In Fig. 6 we display the (normalized) probability that a certain bond n along the polymer backbone will break within a time interval.Two salient features of the probability histograms, shown in Fig. 6, appear characteristic:

i) terminal bonds seldom happen to break.
ii) a well expressed modulation of the rupture frequency along the consecutive bond number, which is best visible in the case of a chain with $N = 30$. There one observes a modulation within an interval that seems to encompass roughly 15 bonds. In longer chains such modulation still persist albeit the periodic pattern gets distorted.

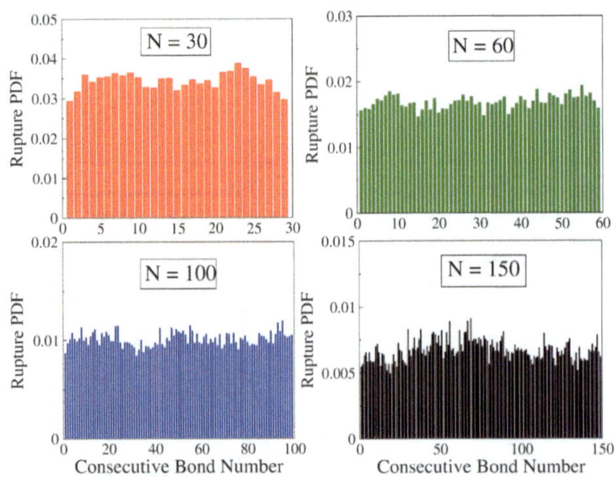

Fig. 6 Overview of rupture probability vs consecutive bond number for free and grafted chains composed of $N = 30, 40, 60$ and 80 beads ($T = 0.1$, $\gamma = 0.25$).

While the probability histograms, presented in Fig. 6, unambiguously indicate the existence of persistent differences in the likelihood of bond breaking in regard with the consecutive number of a particular bond along the chain backbone, the origin and the physical background of such inhomogeneity is not self-evident. On the ground of the observed modulation of the scission probability one may speculate that this self-induced inhomogeneity results from the interplay of thermal noise and the non-linearity of the bond-potential. It is furthermore conceivable that both control parameters such as the temperature T and the nonlinearity of the interactions, as well as other factors (e.g., the friction γ, i.e., under / over-damped dynamics) would affect the multi-scale-length inhomogeneity. Presumably this finding presents an example of more general phenomena where different spatio-temporal order induced by the noise in nonlinear systems [26].

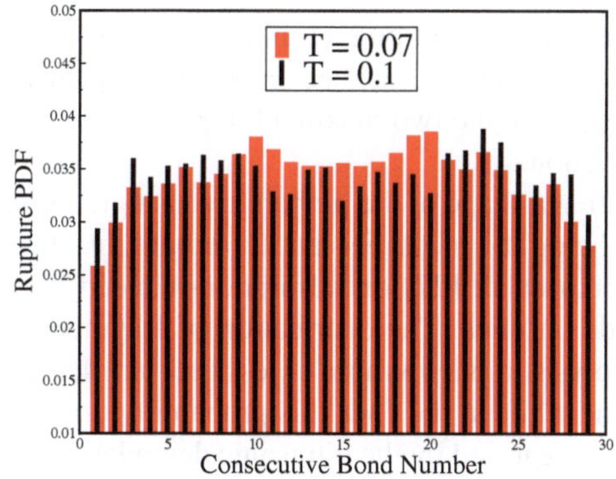

Fig. 7 Comparison of rupture histograms for chains degraded at different temperature of the heat bath. Here $N = 30$, $\gamma = 0.25$.

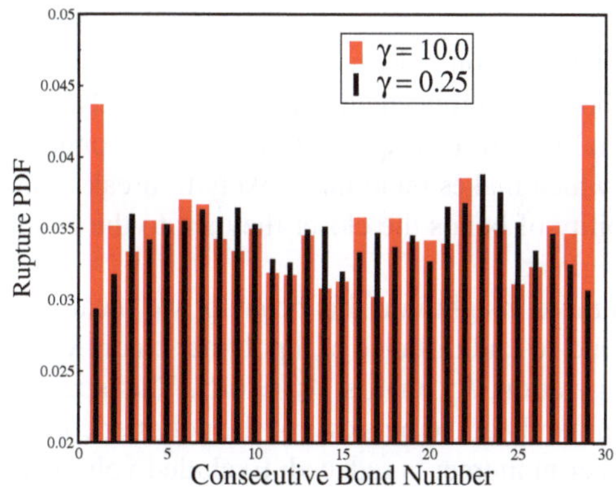

Fig. 8 Rupture histograms for free chains composed of $N = 30$ particles bonded by Morse potential in the under-damped, $\gamma = 0.25$, and overdamped, $\gamma = 10$, regimes ($T = 0.1$).

Since temperature is a major factor in thermal degradation, we present in Fig. 7 the probability histogram for two different temperatures, $T = 0.07$ and $T = 0.1$. One can see that the rupture histogram visibly changes shape due to temperature. Evidently, a decrease of temperature results in changing the positions of the local maxima which shift closer to each other while the modulation of rupture PDF grows. Also, at lower temperature the histogram becomes less flat and the non-uniformity in the rupture probability increases.

Another interesting point concerns the different impact of under- (over)-damped dynamics on the distribution of breaking events along the polymer backbone. The impact of friction γ is displayed in Fig. 8. Indeed, it is evident from Fig. 8 that a change of the system dynamics from under- to over-damped one by a 40−fold increase in γ leads to qualitative changes in bond breakage. The most striking effect of dynamics is the change in the frequency of scissions at both ends of a free chain. When inertial effects are strong, terminal bonds are the least likely to break while in the overdamped regime these become the most vulnerable ones. The well-pronounced dip in between the two maxima for $\gamma = 0.25$ also appears to vanish in the case of strong friction.

One should note that a shape of the probability histograms, as for $\gamma = 0.25$, with the terminal bonds being the most resistant to rupture, has been inferred from the experiments of Sheiko et al. [27] on adsorption-induced thermal degradation of carbon-carbon bonds on mica. The distribution of fragment lengths with time in the course of thermolysis implies a more or less constant scission probability along most of the backbone while at the ends this probability drops significantly [27]. A similar conclusion is suggested by the ultrasonic degradation experiments of Glynn et al. [28] with polystyrene who found that the bonds in the middle of the chain break preferentially to those at the ends. Probably, in the case of under-damped dynamics over-stretched terminal bonds can quickly restore their equilibrium length when friction is low and the restoring force needs only to pull back few segments at the chain end.

In contrast, if chain motion is heavily damped, one might argue that the over-stretched terminal bonds of a free chain need comparatively significantly more time to attain their normal length. During this time they are longer exposed to destructive thermal kicks which makes them more likely to break. This would explain the increased vulnerability of bonds the closer they are to the chains ends in the over-damped dynamic regime.

In order to gain more insight into this behavior we recall that in a good solvent the mean squared end-to-end distance of a polymer chain R_e^2 is significantly larger than the radius of gyration, R_g^2, that is, $R_e^2 \approx 6R_g^2$. This suggests that the terminal beads 'live' predominantly at the outskirts of the polymer coil where the chance for collision with another monomer is reduced. Excluded volume interactions are thus weaker on the average, and, correspondingly, terminal bonds are on the average less stretched, i.e., they are less likely to break.

As far as our Langevin MD simulation deals essentially with anharmonic (Morse and Lennard-Jones) interactions between chain monomers, it appears useful to make a comparison with a reference system, a purely harmonic chain Fig. 9, that has been studied recently [21, 22]. As in [22], the threshold r_h for rupture of such a Gaussian chain is set arbitrary to some extension of the harmonic bond - bonds longer than r_h are then considered broken. It is seen from Fig. 9, that shape of the rupture histogram in the overdamped case, $\gamma = 10$, is similar to that of our Morse chain in the same dynamic regime. Our simulation result reproduces very well the recent observations of Fugmann and Sokolov [22], cf. Fig. 2 in [22], who modeled the thermally induced breakdown of a Gaussian chain. In the under-damped regime, however, the Gaussian chain histogram is strongly leveled off as in [21], retaining only a very weak (symmetric) increase in the rupture probability of the individual bonds as one moves away from the center of the chain and approaches the free ends. Any trace of self-induced inhomogeneity as in the case of non-linear interactions, Fig. 6, is absent. This supports again our assumption that the observed inhomogeneity in the

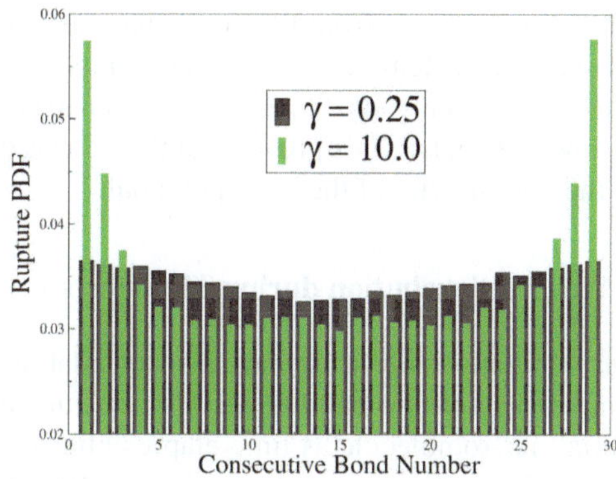

Fig. 9 Rupture histograms of a 30-particles free harmonic chain for under- ($\gamma = 0.25$) and over-damped cases ($\gamma = 10$). Here $T = 0.15$.

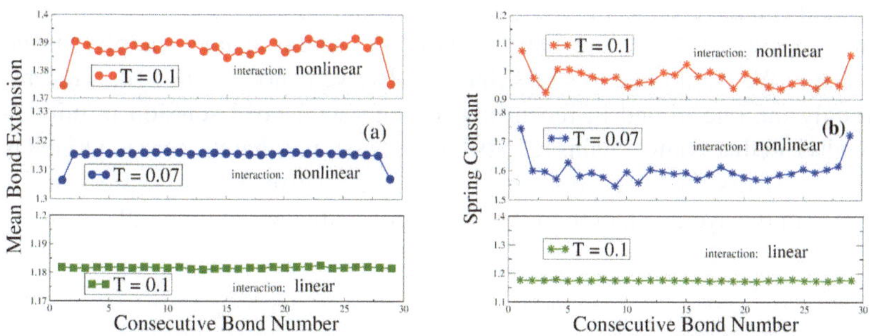

Fig. 10 Mean extension of bonds $\langle b_n \rangle$ (left panel) and mean effective spring constant $\lambda_n = k_B T / \langle \Delta b_n^2 \rangle$ of the individual bonds (right panel) vs. consecutive bond number n for a free chain with $N = 30$-particles. Here $T = 0.1$, $\gamma = 0.25$.

rupture probability distribution among individual bonds occurs as a result from interplay between the thermal noise and the nonlinearity.

A test of this conclusion is suggested by the study of another property - the average strain of the bonds $\langle b \rangle$ with respect to the consecutive bond number, shown in Fig. 10. One can see that this quantity resembles the behavior of rupture probability regarding n. The terminal bonds are less stretched than the other ones and therefore break seldom. Moreover, the *effective* spring constant λ_n of the individual bonds which is given by the variance of the strain, $\langle \Delta b_n^2 \rangle$, shown in the right panel in Fig. 10, behaves similarly. Indeed, the Hamiltonian $H = \frac{1}{2} \sum_{n=1}^{N-1} \lambda_n (\Delta b_n)^2$ of the chain defines a bond length probability distribution function $\mathscr{P} \propto \exp\{-H/k_B T\}$. The distribution of individual strain is Gaussian, $\mathscr{P}_n(\Delta b_n) \propto \exp[-(\Delta b_n)^2 / k_B T \langle \Delta b_n^2 \rangle]$ with $\langle \Delta b_n^2 \rangle = k_B T / \lambda_n$. Thus one can see that in an uniform chain there appear regions of effectively "stiffer" bonds (at the ends of the chain), and of "softer" bonds (away from the ends) that are less or more likely to stretch and break, respectively. Remarkably, in a harmonic Gaussian chain, where excluded volume effects are absent, both $\langle b \rangle$ and λ_n are seen to be entirely uniformly distributed, Fig. 10 (lowest panels). This proves that the observed inhomogeneity is indeed attributed to the nonlinearity of the bond potential.

2.2.4 Molecular Weight Distribution during Thermolysis

In the course of our simulations we examined the degradation kinetics which is manifested by a time-dependent probability distribution function, $P(n,t)$, of the fragments of the initial macromolecule as time elapses after the onset of the process - Fig. 11. The initial length of our polymer is $N = 100$ and the temperature $T = 0.1$. Averages of $P(n,t)$, obtained over 10^4 cycles of scission - Fig. 11 - are shown to evolve in time from a $\delta-$like distribution at $t = 0$ to a rather flat distribution with a rapidly growing second maximum around sizes of $n \approx 2 \div 5$. After a short time of ≈ 250 t.u., the initial chain has already disintegrated into small clusters whose length is most probably $n \approx 1$.

At late times, $t \approx 10^4$ $t.u.$, the distribution $P(n,t)$ attains again a sharply-peaked $\delta-$like shape. We would like to stress that the observed variation of $P(n,t)$ resembles strongly the one, found in recent experiments with bond scission in poly(2-hydroxyethyl methacrylate) chains [27, 29]. If one assumes that the scission kinetics is described by a first-order reaction, then one may derive an analytic expression for the decrease in the average length of the fragments with elapsed time [29] as

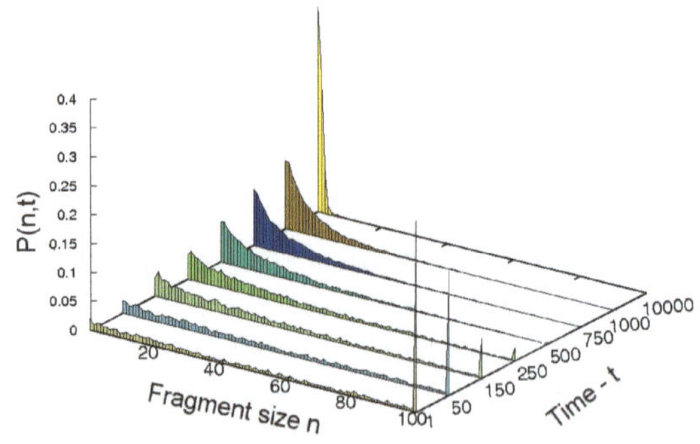

Fig. 11 Probability distribution of fragment sizes $P(n,t)$ at different times t (in MD time units) after beginning of the thermal degradation process for a chain of length $N(t=0) = N_0 = 100$ with $T = 0.1$ and $\gamma = 0.25$. At late times $t \approx 10^4$ $t.u.$ the chain disintegrates into single segments, $N_\infty = 1$

$$\left(\frac{1}{N(t)} - \frac{1}{N_\infty}\right) = \left(\frac{1}{N_0} - \frac{1}{N_\infty}\right) e^{-kt}, \tag{9}$$

where N_0 is the initial contour length at $t = 0$ and N_∞ is the mean contour length of polymer chains at infinite time. k is a first-order rate constant. This result, Eq. (9), is compared with our simulation data in Fig. 12. The final fragment size is $N_\infty = 1$. Upon a closer look, however, the exponential decay $\propto \exp(-kt)$ is found to deviate slightly at late times $t > 2 \times 10^4$ t.u. We interpret the observed discrepancy as an indication that fragment recombination may occasionally take place at late times when the fragments become sufficiently small and mobile. Recombination comprises a second-order (binary) reaction which adds to the dominant first-order reaction of decay and, therefore, contributes to the observed deviations. Nonetheless, it appears that our simulation model faithfully accounts for the degradation kinetics.

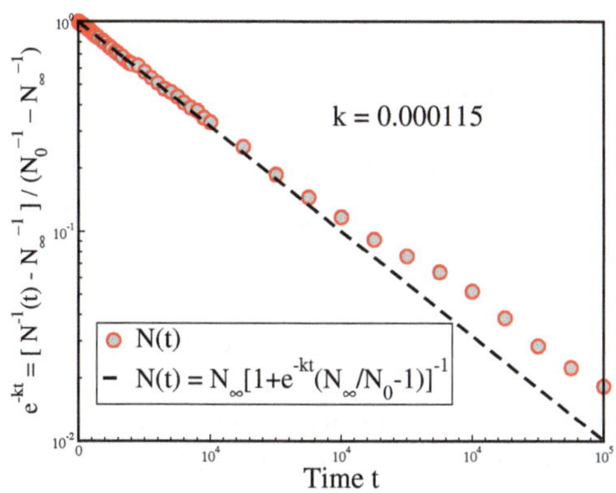

Fig. 12 Variation of the mean fragments length $N(t)$ after the onset of thermal degradation. The dashed line denotes the theoretical result, Eq. (9) with respect to $\exp(-kt)$ and proves that the process evolves predominantly as a first order chemical reaction.

2.3 Polymer Chain at Constant Tension

A great variety of problems both in material and basic science rely on the fundamental understanding of the intramolecular dynamics and kinetics of fragmentation (bond rupture) of linear macromolecular subject to a tensile force. Typical examples comprise material failure under stress [30, 31], polymer rupture [32, 33, 34, 35, 36], adhesion [37], friction [38], mechanochemistry [39, 40], and biological applications of dynamical force microscopy [41, 42].

In particular, the problem of polymer fragmentation has got a longstanding history in scientific literature. The treatment of bond rupture as a kinetic process dates back to the publications of Bueche [43] and Zhurkov *et al.* [44]. In the recent years these seminal papers have been complemented by a variety of computer experiments. Molecular Dynamics (MD) simulations of chain rupture at constant stretching *strain* has been carried out, whereby harmonic [21, 45], Morse [46, 47, 48] or Lennard-Jones [49, 50, 51, 52] interactions have been employed. A theoretical interpretation of MD results, based on an effectively one-particle model (Kramers rate theory) has been suggested [51, 52]. On the other hand, an analytical treatment of a polymer fragmentation under *constant stress* has been proposed in terms of many-particle version of transition state theory (TST) [48]. Nevertheless, despite the multidimensional nature of TST, it does not take into account properly the collective unstable mode development, which leads, in our opinion, to the essential overestimation of the breaking rate.

The collectivity effect has been recently treated [53] for constant strain and periodic boundary conditions (a ring polymer) on the basis of the multidimensional Kramers approach [54, 55]. Within this approach the development of a collective unstable mode and the effect of dissipation can be described consistently. It has been shown that in this case the effective break frequency is of the same order of magnitude as the one observed in the simulation.

In this subsection the results [56, 57] of further development of this approach will be presented for the case of a tethered Morse chain, consisting of N segments and subjected to a constant tensile force f applied at its free end. For comparison with theoretical prediction, we also perform extensive MD simulations in both one- $1D$, and three dimensions, $3D$, and witness significant differences in the fragmentation behavior of the chain.

2.3.1 Kramers-Langer Theory and Scission of a Polymer under Tension

Let's consider a tethered one-dimensional string of N beads which experiences a tensile force f at the free end as depicted in Fig. 13. Successive beads are joined

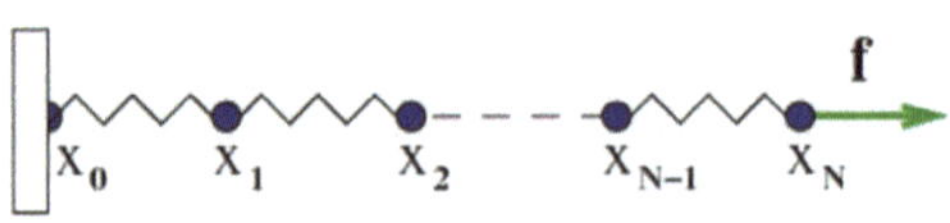

Fig. 13 Schematic representation of a tethered string of beads, subject to a pulling force f. The corresponding coordinates are marked as $x_1, x_2, \ldots x_N$.

by bonds, governed by the Morse potential, $U_M(y) = D(1 - \mathrm{e}^{-\alpha y})^2$, where parameters D and α has been already defined in Eq. 3. The total potential energy is $V(\{x_i\}) = \sum_{i=1}^{N} U_M(x_i - x_{i-1}) - f x_N$ where we set $x_0 = 0$ (see Fig. 13). Upon change of variables, $y_i = x_i - x_{i-1}$, one gets $V(\{x_i\}) = \sum_{n=1}^{N} [U_M(y_n) - f y_n] = \sum_{n=1}^{N} U(y_n)$ so the combined one-bond potential then reads $U(y) = D(1 - \mathrm{e}^{-\alpha y})^2 - f y$. Both potentials, with and without external pulling force have been plotted in Fig. 14. From this figure one can see that the pulling force gives rise to metastable minimum which is separated from another state by the energy barrier ΔE_b. It can be shown that ΔE_b [56] declines with the growing force f. In equilibrium all the bonds with the extension corresponding to metastable minimum lay at the bottom of the well.

Our main aim is to study chain dynamics on the onset of fracture with the help of Kramers-Langer theory developed originally to study dissociation of molecules. Within this approach a single bond rupture is seen as a thermally activated escape from the bottom of a potential well. The life time τ before a bond scission takes place, is determined by diffusive crossing of an energy barrier E_b that is reduced under the applied external force f. The adopted theoretical treatment [56] assumes a single collective unstable modes as being mainly responsible for chain breakage. Such unstable mode peaks around an "endangered" bond of negative spring constant and decays exponentially towards both chains end.

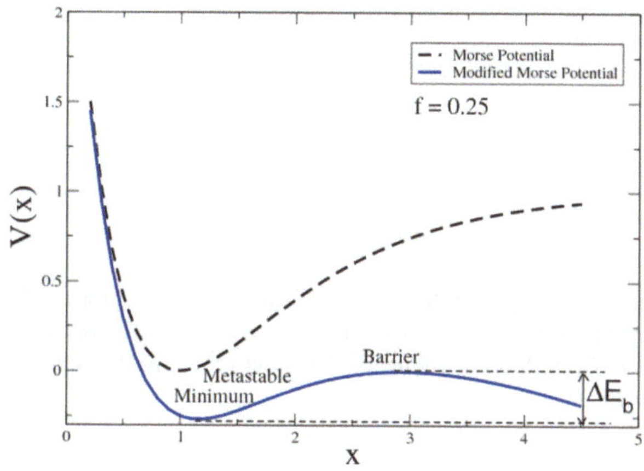

Fig. 14 Comparison of bonded Morse potential with (dashed line) and without pulling force (solid line). External force gives rise to a metastable minimum and a barrier ΔE_b.

2.3.2 Simulation Results

We compare here the simulation results with our theoretical prediction for the rupture probability of the n-th bond in a chain with N bonds. The simulation model is essentially the same as the one used for the case of thermal degradation of a chain presented in the Sec. 2.2.1. The only difference is to add external constant stress in Eq. 5 which acts on the last bead in the chain. We start the simulation with all beads placed at distance b from each other, and then we let the chain equilibrate in the heat bath at temperature low enough that the chain stays intact. Due to the presence of the external pulling force, the equilibrium configuration of the chain is more or less stretched and deviates markedly from coil shape (see Fig. 15a). Once equilibration is achieved, the temperature is raised to the working one, time is set to zero and one measures the elapsed time τ before any of the bonds exceeds certain extension r_h (Fig. 15b), which sets the criterion for considering such bond broken. We use a large value for the critical bond extension, $r_h = 5b$, which is defined as a threshold to a broken state. This convention is based on our checks that the probability for

Fig. 15 Snapshots of a chain with 30 beads fixed at the left end while the right end is pulled by a constant force: (a) an equilibrated initial conformation, (b) a broken chain with the beads at the scission site shown in white.

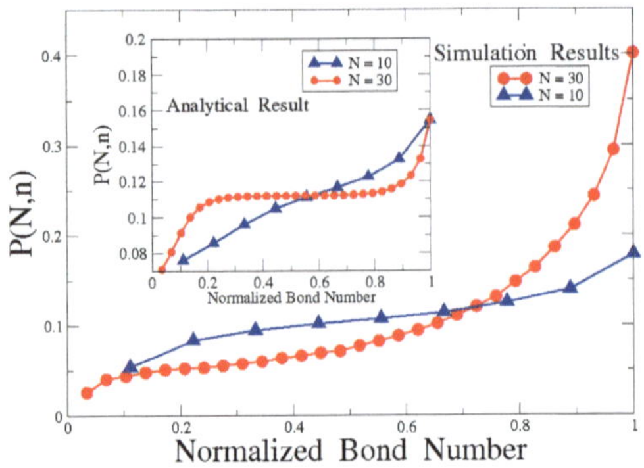

Fig. 16 Normalized rupture probability vs consecutive bond number for $1D$ chains with length N, subject to tensile force $f = 0.25$, and friction $\gamma = 0.25$. The consecutive number of the bonds is normalized as n/N for convenience. The inset shows the theoretical prediction. Here $T = 0.05$, $\gamma = 0.25$.

recombination of bonds, stretched beyond r_h, is vanishingly small. We repeat this procedure for a large number of events 5×10^4 so as to determine the mean rupture time $\langle \tau \rangle$.

In Fig. 16 the normalized rupture probability for $1D$ chains (with $N = 10$ and $N = 30$) is shown with respect to the consecutive number of the individual bonds. The theoretical prediction of Kramers-Langer approach is given in the inset. Both the theory- and MD-results indicate that the pulled end of the chain and the bonds in its vicinity break more frequently due to more freedom than those around the fixed end. Generally, the probability of rupture decreases steadily from the pulled end to the fixed end. For the longer chain, the end effects are not felt by the middle part of the chain and the probability of rupture $P(N,n)$ is nearly uniform forming a plateau-like region all over the length of the chain except at the ends. This feature is more pronounced in the theoretical rather than in the MD results.

In Fig. 17 we present the probability for bond scission of individual bonds both in $1D$ and $3D$ for several strengths of the pulling force. From the inspection of Fig. 17 one sees that the preferential scission of the bonds with particular consecutive bond number essentially depends on the value of force and dimensionality of the chain. For strong pulling $f = 0.3$ one finds that the terminal bond which is subjected to pulling as well as the bonds in its neighborhood break more frequently than whose around fixed end. A similar scission scenario is visible also for the $1D$ chain as

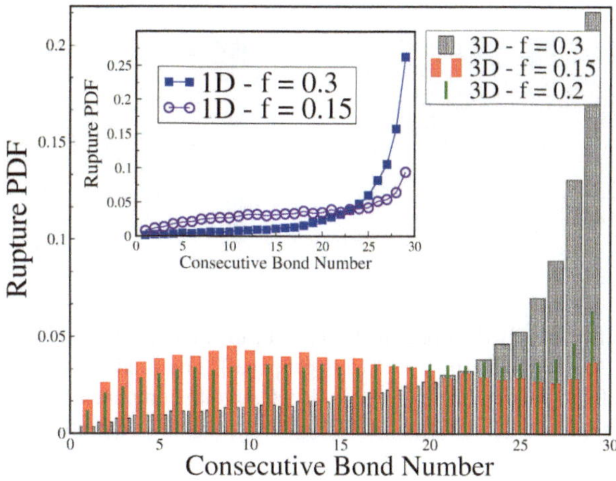

Fig. 17 Rupture probability histograms for $1D$ and $3D$ chains composed of $N = 30$ for different pulling forces as indicated. Here $T = 0.05$, $\gamma = 0.25$.

shown in the inset of Fig. 17. In contrast, as the stretching force is decreased, the corresponding rupture histogram for a $3D$ chain becomes flatter. For $f = 0.2$ the distribution of scission rates becomes uniform except for the bonds in the vicinity of both ends. A further decrease of the pulling force results in a qualitative change in the distribution. Evidently, for $f = 0.15$ the bonds in the middle of the chain, which are also somewhat closer to the fixed chain end, become more vulnerable as compared to those at the chain ends. Note that for the smallest pulling force ($f \lesssim 0.15$) the rupture histogram already resembles the respective histogram in the case of thermal degradation of a polymer (see Fig. 6 for comparison) which takes place in the absence of externally induced tension.

How can such an inhomogeneity in the probability of bond rupture be understood? A possible explanation of the change in the location of preferential breakdown sites along the chain may be gained by Fig. 18. In this figure we present maps of the density distribution $P(x,r)$ of bead positions where x is measured in direction of the pulling force f whereas $r = \sqrt{y^2 + z^2}$ denotes the radial component. Fig. 18 [left panel] indicates that at high stretching ($f = 0.3$) the most probable position of the beads is along the direction of the tensile force. For $f = 0.3$ (strong stretching) the chain conformation corresponds to a quasi-$1D$ structure, and the transversal fluctuations are reduced. In contrast, when the pulling force is weak ($f = 0.15$) one finds from Fig. 18 [right panel] that the individual beads are free to make big excursions in space – $P(x,r)$ is roughly two times broader in the middle. Thus, Fig. 18 suggests that the density maps comply with the rupture histograms given in Fig. 17. For $f = 0.3$, due to larger freedom around the pulled end, the end bonds break more easily in the terminal part of the chain. When the force is weak, $f = 0.15$, the beads become more mobile around the center of the polymer which in turn leads to increased bonds scission rate there.

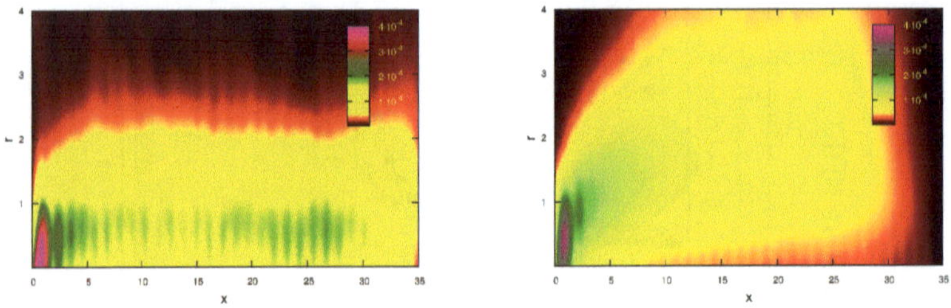

Fig. 18 Probability density distribution $P(x,r)$ of beads in a $3D$ chain with $N = 30$ particles at force: $f = 0.3$ [left panel], and $f = 0.15$ [right panel]. The x-axis coincides with the direction of pulling force whereas $r = \sqrt{y^2 + z^2}$ denotes radial component of the bead position. Different colors indicate the value of the PDF as indicated in the legend.

The dependence of the MFBT $\langle \tau \rangle$ on external force f for $1D$ and $3D$ chains composed of $N = 30$ beads is shown in Fig. 19. Evidently for sufficiently strong stretching forces $f \gtrsim 0.175$ an exponential decay $\langle \tau \rangle \propto e^{(E_0 - \alpha f)/k_B T}$ is observed. The main reason for this is the following: As the pulling force grows, the energy barrier, which separates intact bonds from the broken ones, declines. As a consequence, $\langle \tau \rangle$ decreases. One should note that the parameters α and E_0 change only slightly with the coupling parameter γ of the thermostat. Fig. 19 indicates also a considerable difference in the values of α between $1D$ and $3D$. In the inset of Fig. 19 we present $\langle \tau \rangle$ as a function of the ratio E_b/T of the barrier height to temperature. This finding is in agreement with the understanding of the polymer rupture as a thermally activated process [43, 44] and is manifested by an Arrhenian relationship – $\langle \tau \rangle \propto e^{E_b/T}$, where $E_b = E_0 - \alpha f$.

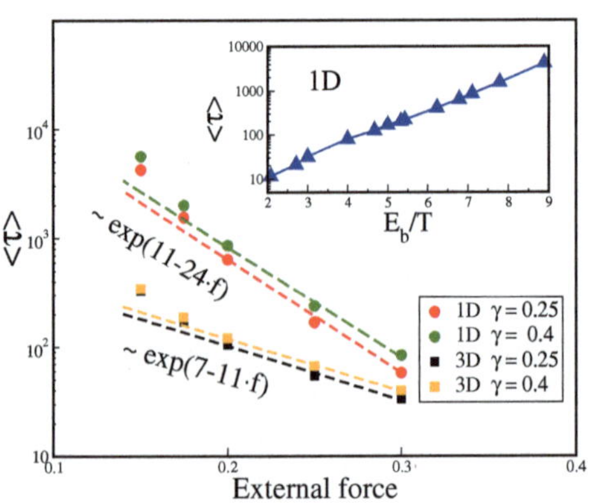

Fig. 19 Force-dependent mean first breakage time for a $1D$ and $3D$ chains with $N = 30$. The inset shows of $\langle \tau \rangle$ vs. E_b/T for $1D$ system.

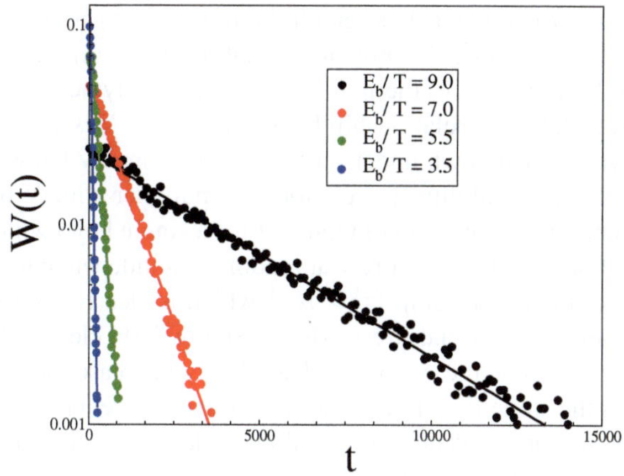

Fig. 20 Life-time probability distributions $W(t)$ for different height of the energy barrier E_b/T in $1D$. Here the chain length is $N = 30$, the pulling force $f = 0.15$, and $\gamma = 0.25$.

In Fig. 20 we display the probability distribution function $W(t)$ of the observed scission times t for several ratios E_b/T of the barrier height to temperature in the case of $1D$ chain composed of $N = 30$ beads. It appears that $W(t)$ goes asymptotically as $W(t) \propto e^{-t/\tau(E_b/T)}$ in accordance with theoretical prediction (solid lines)

In Fig. 21a) and the inset of Fig. 21a) we present numerical results for $\langle\tau\rangle$ as a function of the number of beads N for chains that are stretched in the interval $0.1 \geq f \geq 0.3$ of pulling forces both in $1D$ and $3D$. Regardless of dimensionality of the examined systems, for a given value of f one observes a power-law decrease, $\langle\tau\rangle \propto N^{-\beta}$. This relationship is found for sufficiently long chains (asymptotic limit) – $N \gtrsim 80$, where finite-size effects do not play a role. Furthermore, Fig. 21b)

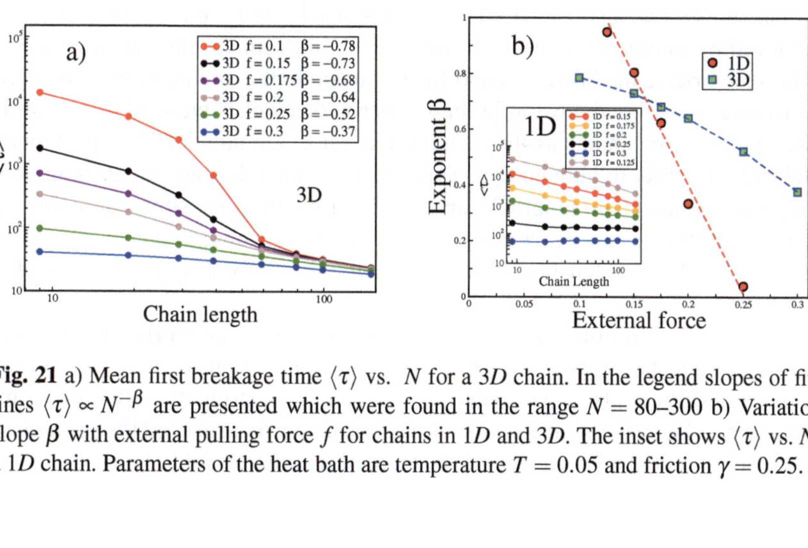

Fig. 21 a) Mean first breakage time $\langle\tau\rangle$ vs. N for a $3D$ chain. In the legend slopes of fitting lines $\langle\tau\rangle \propto N^{-\beta}$ are presented which were found in the range $N = 80$–300 b) Variation of slope β with external pulling force f for chains in $1D$ and $3D$. The inset shows $\langle\tau\rangle$ vs. N for a $1D$ chain. Parameters of the heat bath are temperature $T = 0.05$ and friction $\gamma = 0.25$.

indicates that with growing tensile strength the life time $\langle \tau \rangle$ becomes nearly independent of N which is among the most important results of this study. It should be noted that in the limiting case of thermal degradation of polymers ($f = 0$) discussed previously in Sec. 2.2.2 the relationship between $\langle \tau \rangle$ and N is very different from the case of polymer breakdown under tension. As we already know in the case of thermolysis the total probability for scission of a polymer with N bonds within a certain time interval is N times larger than that for a single bond which is what one would expect if bonds do break entirely at random and independent of one another. The latter leads to the relationship $\langle \tau \rangle \propto N^{-1}$ which is clearly not satisfied for the chain under influence of external stress. Moreover, Fig. 21b) clearly shows that with increasing pulling force f the exponent β gradually decreases within the interval $0 < \beta < 1$. Thus the slope β can be treated as a quantitative measure of the degree of cooperativity in rupture events . As the slope β decreases, the nature of scission events become more and more collective.

3 Bottle-Brush

One of the most outstanding challenges in modern material sciences is the design and synthesis of "smart" macromolecules with stress-activated functions [58, 59]. During the last decade one observes thus a rapidly escalating interest in the field of novel polymer mechanochemistry which, in contrast to the traditional (non-selective) one, allows to control bond tension on molecular length scales [58, 60, 61, 62]. Examples related to these advances enable, for instance, rupture of specific chemical bonds [29, 63].

In a series of recent experiments, a strong enhancement of the tension in the (typically, polymethacrylate with degree of polymerization $L = 3600$) backbone of bottle-brush polymers with side chains of poly($n-$butyl acrylate) of length $N = 140$, self-induced upon adsorption on a solid surface (mica), was reported [27, 29, 64]. An experimental method for control and manipulation of the bond-cleavage in bottle-brush backbones was also proposed [64]. Thus, a selectivity of bond breakage can be achieved by tuning the molecular size of such macromolecules which makes it possible to fabricate the brush so as to focus tension in the middle of the molecule. The increase of the bond tension in these macromolecules is induced by the steric repulsion of the side chains as they tend to maximize the number of contacts with the substrate in order to gain energy. This tension, which depends on grafting density σ_g, on the side chain length N, and on the strength of substrate attraction ε_s, effectively lowers the energy barrier for bond scission. As observed in experiments, self-induced build up of tension proves sufficient to instantly sever covalent bond in the backbone.

The effect of adsorption-induced bond scission might have important implication for surface chemistry, in general, and for specific applications of new macro- and supramolecular materials, in particular, for example, by steering the course of chemical reactions. One may use adsorption as a convenient way to exceed the strength of covalent bonds and invoke irreversible fracture of macromolecules, holding the key to making molecular (DNA) architectures that undergo well-defined fragmentation upon adsorption.

The possibility for breaking strong covalent bonds is also an interesting problem from the standpoint of fundamental physics. Amplification of tension in branched polymers has been considered theoretically by Panyukov and collaborators in several recent works [65, 66, 67] by means of scaling theory and Self Consistent Field techniques. Numerous possible regimes of brush-molecule behavior in terms of N, σ_g and ε_s have been outlined [65]. It was argued that polymers with branched morphology, physically adsorbed on an attractive plane, allow focusing of the side-chain tension on the backbone whereby at given temperature T the tension in the backbone becomes proportional to the length of the side chain, $f \approx f_s N$. Here f_s denotes the maximum tension in the side chains, $f_s \approx k_B T/b$, with k_B being the Boltzmann constant, and b - the Kuhn length (or, the monomer diameter for absolutely flexible chains). However, a comprehensive understanding of covalent bond breaking in adsorbed branched polymers still has to be reached. Many of the detailed theoretical predictions can hardly be measured directly experimentally.

In this section we report on our studies of chain fragmentation in desorbed and adsorbed bottle-brush macromolecules by means of a coarse-grained bead-spring model and Langevin dynamics [68, 69, 70].

3.1 Computational Model

We consider a three-dimensional coarse-grained model of a bottle-brush macromolecule which consists of L monomers in the backbone connected by bonds - Fig. 22 (left panel). Moreover, two side chains of length N are grafted to every σ_g^{-1}-th repeatable unit of the backbone (except for the terminal beads of the backbone where there are three side chains anchored). In this way a grafting density σ_g, which gives the number of side chains *pairs* per unit length, is defined. Thus, the total number of monomers in the brush molecule is $M = L + 2N[(L-1)\sigma_g + 2]$. The bonded U_M and nonbonded interaction U_{WCA} for the monomers in the bottle-brush backbone are the same as previously used for the case of single chain (see Eq. 3 and 4 in Sec. 2.2.1). For the bonded interaction in the side chains we take the frequently used Kremer-Grest potential, $U_{KG}(r) = U_{WCA}(r) + U_{FENE}(r)$, with the so-called 'finitely-extensible non-linear elastic' (FENE) potential,

$$U_{FENE}(r) = -\frac{1}{2} k r_0^2 \ln\left[1 - \left(\frac{r}{r_0}\right)^2\right]. \tag{10}$$

In Eq. (10) $k = 30$, $r_0 = 1.5$, so that the total potential $U_{KG}(r)$ has a minimum at bond length $r_{bond} \approx 0.96$. Thus, the bonded interaction, $U_{KG}(r)$, makes the bonds of the side chains in our model unbreakable whereas those of the backbone may and do undergo scission.

The substrate in the present study is considered simply as a structureless adsorbing plane, with a Lennard-Jones potential acting with strength ε_s in the perpendicular $z-$direction, $U_{LJ}(z) = 4\varepsilon_s\left[\left(\frac{\sigma}{z}\right)^{12} - \left(\frac{\sigma}{z}\right)^{6}\right]$. In our simulations we consider as a rule the case of *strong* adsorption, $\varepsilon_s/k_B T = 5.0$.

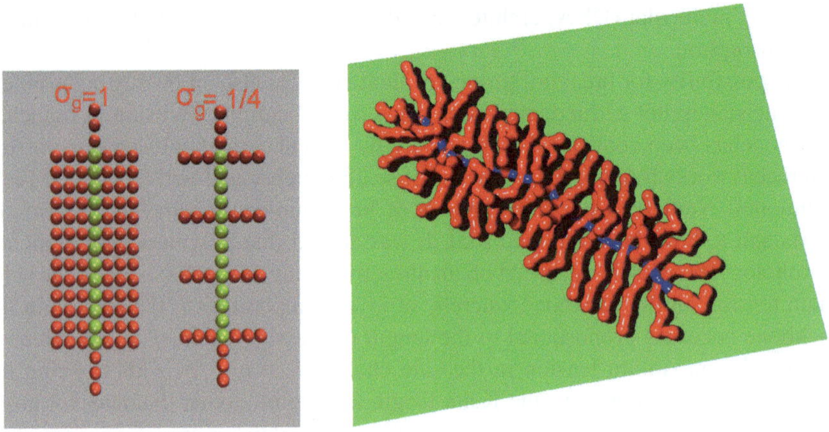

Fig. 22 (left) Staring configuration of a bottle-brush molecule (a "centipede") with $L = 13$ (backbone) and $N = 3$ (side chain), so that for grafting density $\sigma_g = 1$ the total number of segments $M = 97$, and for $\sigma_g = 1/4$ one has $M = L + 2N[(L-1)\sigma_g + 2] = 43$. (right) A snapshot of a thermalized "centipede" with $L = 20$ backbone monomers (blue) and 42 side chains (red) of length $N = 4$. The total number of beads is $M = 188$. Here $k_B T = 1$ and the strength of adsorption $\varepsilon_s = 9.5$. Side chains which are too strongly squeezed by the neighbors when the backbone bends are seen occasionally to get off the substrate in order to minimize free energy.

The initially created configurations, Fig. 22 (left panel), are equilibrated by integration of equations Eq. 5 for a period of time so that the mean square displacement of the polymer center-of-mass moves a distance several $(3 \div 5)$ times larger than the polymer size (i.e., larger than the radius of gyration R_g). During this period no scission of backbone bonds may take place. We then start the simulation with a well equilibrated conformation of the chain and allow thermal scission of the bonds. We measure the mean life time τ until the first bond rupture occurs, and average these times over more than 2×10^4 events so as to determine the mean $\langle \tau \rangle$. As in the case of the single chain in the course of the simulation we also sample the probability distribution of bond breaking regarding their position in the chain (a rupture probability histogram).

3.2 Simulation Results

First we examine the distribution of scission probability (the probability of bond rupture) along the polymer backbone for the case of a strong adsorption, $T = 0.125$, $\varepsilon_s = 0.5$ in Fig. 23. One can readily verify from Fig. 23a that for a given contour length L the shape of the probability histogram changes qualitatively as the length of side chains N and the grafting density σ_g is varied. While for $N = 1$, $\sigma_g = 1.0$ the scission probability is uniformly distributed along the backbone (being significantly diminished only in the vicinity of both terminal bonds), for

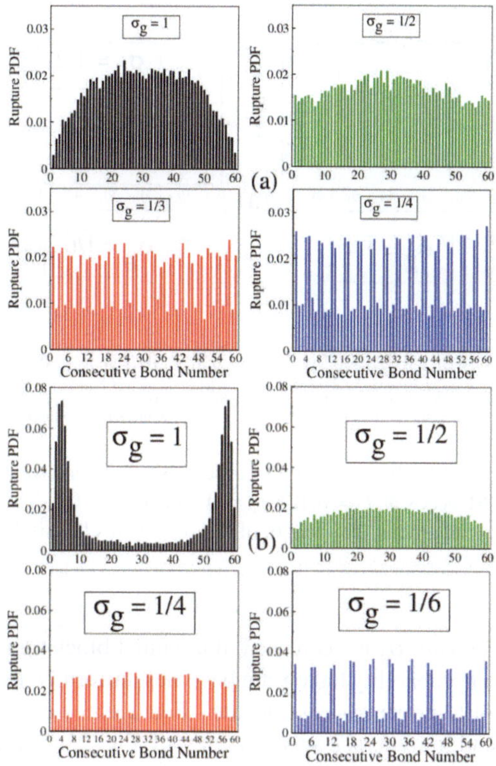

Fig. 23 (a) Scission probability histogram for a polymer backbone with $L = 61$, length of the side chains $N = 1$, and different grafting density $0.25 \leq \sigma_g \leq 1.0$. (b) Variation of the scission probability histogram with grafting density σ_g for brush molecules with fixed side chain length $N = 4$.

$N = 4$, $\sigma_g = 1.0$, in contrast, one observes a well expressed minimum in the probability in the middle of the chain in between the two pronounced maxima ("horns") close to the chain ends. Evidently, at the highest grafting density the side chains for $N > 1$ become mutually strongly squeezed whereby their mobility is suppressed and no additional tension in the respective bonds of the backbone is induced. Such mutual blocking of side chains is absent for $N = 1$, of course, since they are too short to block one another. Thus, it appears that there should exist some necessary free volume around the side chains which would enable their motion and, therefore, permit the generation of increased tension that would ultimately lead to bond rupture. In very long bottle-brush molecules such areas of enhanced mobility would exists in the vicinity of the macromolecule ends as well as around bends and kinks in the conformation.

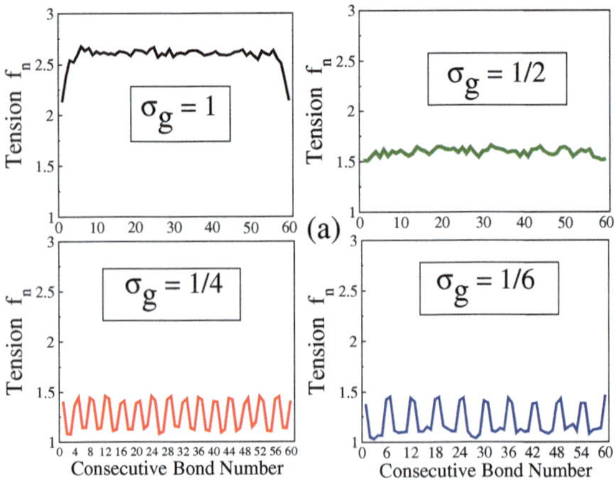

Fig. 24 Mean tension f_n in the bonds with consecutive number n of an adsorbed macromolecules with $L = 61$ beads at $T = 0.125$, $\gamma = 0.25$ and $\varepsilon_s = 0.5$. The respective grafting density is indicated in the graphs. The length of the side chains here is $N = 4$.

As the grafting density σ_g is decreased, the mutual blocking is relieved and the shape of the scission probability histogram becomes uniformly distributed along the backbone of the bottle-brush macromolecule - Fig. 23. For $\sigma_g < 0.5$ one observes alternatively high and low average scission rates, cf. Fig. 23, whereby the high rates appear always in pairs because the induced large tension is transmitted to the bonds immediately connected to each grafting site.

In the course of our MD simulation one has also the possibility to measure directly the tension f induced by the steric repulsion of side chains on the covalent bonds that comprise the macromolecule backbone. It is interesting to see how this tension is distributed along the backbone of the macromolecule and whether it correlates with the distribution of scission rates, Fig. 23. In Fig. 24 we show the distribution of the mean tension f_n along the bonds with consecutive number n along the backbone of adsorbed bottle-brush macromolecule. Evidently, away from both terminal bonds the tension is uniformly distributed along the inner bonds for $\sigma_g \geq 0.5$. In fact, such a distribution is assumed in the interpretation of all experimental observations [29]. For smaller grafting density the tension is seen to alternate in compliance with the bond scission distribution, shown in Fig. 23 so that one can prove the existence of direct relationship between f_n and the ensuing probability of bond rupture.

In Fig. 25a we show the dependence of the mean time $\langle \tau \rangle$ elapsed before any of the backbone bonds breaks on the contour length L and on the total number of segments in the bottle-brush molecule $M = L + 2N[(L-1)\sigma_g + 2]$. The mean life time $\langle \tau \rangle$ of the macromolecule was obtained as a first moment of the probability distribution of life times, $W(\tau)$, (not shown) which strongly resembles a Gaussian distribution with a slight asymmetry (a somewhat longer tail at the large times).

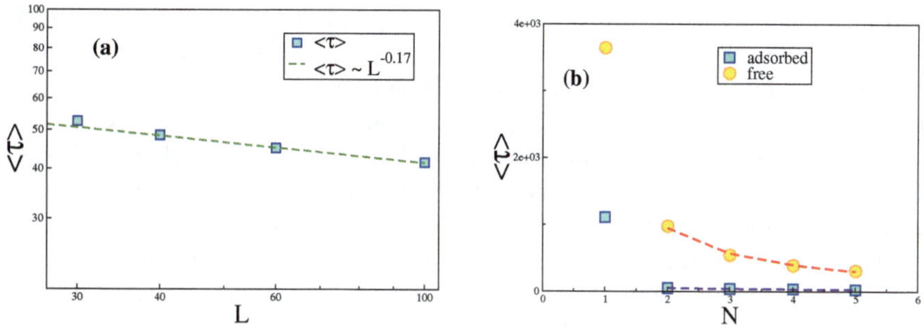

Fig. 25 (a) Variation of the MFBT $\langle\tau\rangle$ with contour length L and with total number of monomers M of the brush molecule (inset) for length of the side chains $N = 2$. Here $k_BT = 0.10$ and $\varepsilon_s = 0.50$. (b) Mean life time $\langle\tau\rangle$ vs N for a desorbed (free) and adsorbed brush molecule with $L = 30$.

Evidently, in Fig. 25a one observes a well expressed power law, $\langle\tau\rangle \propto L^{-\beta}$ with exponent $\beta \approx 0.17$. Since for large L one has $M \propto L$, the variation of $\langle\tau\rangle$ with the total number of segments M is the same.

This finding is important because it indicates that $\langle\tau\rangle$ depends rather weakly on the total number of bonds that might break, in clear contrast to thermal degradation of polymers without side chains discussed in Sec. 2.2 where $\beta = 1$. Indeed, when bonds break entirely at random, the probability that *any* of the L bonds may undergo scission within a certain time interval should be proportional to the total number of bonds, and therefore $\langle\tau\rangle \propto 1/L$. We already know (see Sec. 2.3 for details) that in cases of chain scission when a constant external force pulls at the ends of the polymer, however, one finds typically $\beta < 1$ whereby the value of β steadily decreases as the force strength grows. This suggests a gradual crossover from a predominantly individual to a more concerted mechanism of bond scission. In adsorbed bottle-brush molecules it is the side chains that induce tension in the polymer backbone and thus lead to rupture behavior similar to that with external force.

In Fig. 25b we compare the dependence of $\langle\tau\rangle$ on length N of the side chains for the case of non-adsorbed (free) and adsorbed brush molecules of length $L = 30$. Generally, adsorption alone is found to diminish the mean rupture time by more than an order of magnitude, at least for $N > 1$. As mentioned before, the case $N = 1$ where neighboring side chains almost do not overlap is qualitatively different so, upon adsorption, the MFBT shortens by a factor of three only.

4 Polymerized Membrane

4.1 Thermal Degradation

Understanding the interplay between elastic and fracture properties is even more challenging and important in the case of $2D$ polymerized networks (elastic-brittle sheets). A prominent example of biological microstructure is *spectrin*, the red blood cell membrane skeleton, which reinforces the cytoplasmic face of the membrane. In erythrocytes, the membrane skeleton enables it to undergo large extensional deformations while maintaining the structural integrity of the membrane. A number of studies, based on continuum- [71], percolation- [72, 73, 74], or molecular level [75, 76] considerations of the mechanical breakdown of this network, modeled as a triangular lattice of spectrin tetramers, have been reported so far. Another example concerns the thermal stability of isolated graphene nanoflakes. It has been investigated recently by Barnard and Snook [77] using *ab initio* quantum mechanical techniques whereby it was noted that the problems "has been overlooked by most computational and theoretical studies". Many of these studies can be viewed in a broader context as part of the problem of thermal decomposition of gels [78], epoxy resins [79, 80] and other 3D networks, studied both experimentally [78, 79, 80], and by means of simulations [81] as in the case of Poly-dimethylsiloxane (PDMS). In most of these cases, however, mainly a stability analysis is carried out whereas still little is known regarding the collective mechanism of degradation, the dependence of rupture time on system size, as well as the decomposition kinetics, especially as far as (2D) polymer network sheets are concerned. It is also interesting from the standpoint of basic physics to compare the degradation process to the one taking place in linear polymers which has been already presented in Sec. 2.2. Therefore, in this section we extend our investigations to the case of (2D) polymer network sheets, embedded in 3D-space, and study as a generic example the thermal degradation of a suspended membrane with honeycomb orientation [82].

4.1.1 Computational Model

We study a coarse-grained model of honeycomb membrane embedded in three-dimensional (3D) space. In this investigation we consider generally *symmetric* hexagonal membranes (*flakes*) (Fig. 26). In a very few cases we also discuss fracture of a *ribbon* shape membranes. The membrane flake consists of N spherical particles (beads, monomers) of diameter σ connected in a honeycomb lattice structure whereby each monomer is bonded with three nearest-neighbors except of the monomers on the membrane edges which have only two bonds (see Fig. 26a). The total number of monomers N in such a membrane is $N = 6L^2$ where by L we denote the number of monomers (or hexagon cells) on the edge of the membrane (i.e., L characterizes the linear size of the membrane). There are altogether $N_{bonds} = (3N - 6L)/2$ bonds in the membrane.

We find it appropriate to divide the two-dimensional membrane network so that all the beads and bonds are distributed into different subgroups presented by concentric "circles" with consecutive numbers (see Fig. 26b) proportional to the radial distance from the membrane center. To *odd* circle numbers thus belong beads and bonds that are nearly tangential to the circle. *Even* circles contain no beads and only radially oriented bonds (shown to cross the circle in Fig. 26). The total number of circles C in the membrane of linear size L is found to be $C = (2L - 1)$. We use this example of dividing the beads and the bonds composing the membrane in order

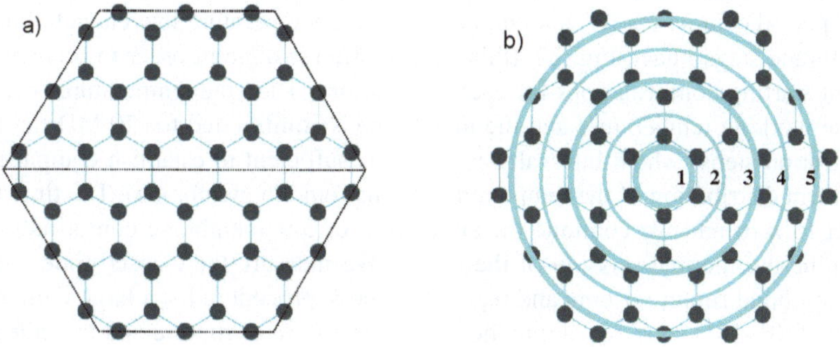

Fig. 26 a) A model of a membrane with honeycomb structure that contains a total of $N = 54$ beads and has linear size $L = 3$ (L is the number of beads or hexagonal cells on the edge of the membrane). b) An example of subdivision of beads and bonds, composing a membrane with $L = 3$, into subgroups ("circles"). The total number of circles C in the membrane of linear size L is $C = 2L - 1$.

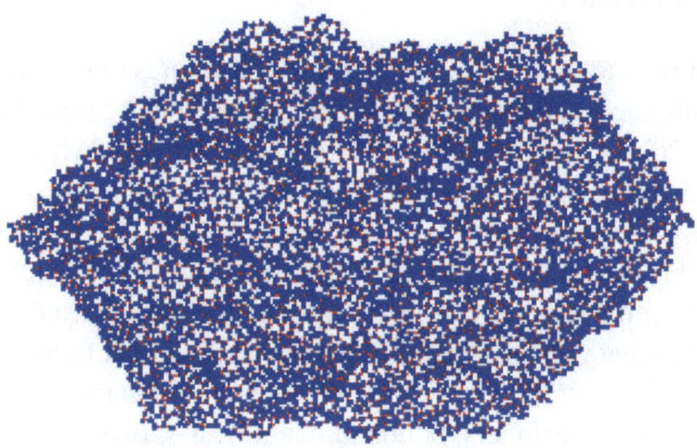

Fig. 27 A snapshot of a typical conformation of an intact membrane with $L = 30$ containing 5400 monomers after equilibration. Characteristic ripples are seen to cross the surface.

to represent our simulation results in appropriate way which relates them to their relative proximity to membrane's periphery.

As in the case 1D polymer chain and bottle-brush the nearest-neighbors monomers in the membrane are connected to each other by "breakable bonds" described by a Morse potential given by Eq. 3. Nonbonded interaction between the monomers are taken by means of WCA potential, Eq. 4. Velocity-Verlet algorithm is used in order to integrate Langevin equations of motion, Eq. 5. Simulations are carried out in the following order. First, we prepare an equilibrated membrane conformation, starting with a fully flat configuration shown schematically in Fig. 26, where each bead in the network is separated by a distance $r_{min} = 1$ equal to the equilibrium separation of the bond potential. Then we start the simulation with this prepared conformation and let the membrane equilibrate in the heat bath for a very long period of time ($\approx 10^7$ integration steps) at a temperature low enough that the membrane stays intact, Fig. 27, (this equilibration is done in order to prepare different starting conformations for each simulation). Then the temperature is raised to the working temperature and the membrane is equilibrated for 20 MD t.u. (10^4 integration steps) which interval was found as sufficient to establish equipartition (uniform distribution of the temperature throughout the membrane). The time then is set to zero and we continue the simulation of this membrane conformation to examine the thermal scission of the bonds. We measure the elapsed time τ until the first bond rupture occurs and repeat the above procedure for a large number of events (10^3–10^4) so as to sample the stochastic nature of rupture and to determine the mean $\langle\tau\rangle$. In the course of simulation we also calculate properties such as the probability distribution of breaking bonds regarding their position in the membrane (a rupture probability histogram), the mean extension of the bonds with respect to the consecutive circle number in the membrane, as well as other quantities of interest.

4.1.2 Simulation Results

In Fig. 28a we show the distribution of bond scission rates among all bonds of the honeycomb membrane for flakes (with zig-zag pattern on the periphery) of several sizes $L = 10$, 15, 20. Somewhat surprisingly, one finds the overwhelming fraction of bond breaking occurs at the outer-most rim of the membrane where monomers are bound by only two bonds to the rest of the sheet. We have also sampled the rupture histograms for ribbon-like square membranes, Fig. 28b. Interestingly, we observe no difference between the rupture rates of zig-zag and armchair edges whereas the bonds in the four corners of such membrane expectedly break more frequently. The difference in the relative stability of the bonds becomes clearly evident in Fig. 28c where the frequency of periphery bonds appears nearly two orders of magnitude larger when compared to bonds in the 'bulk' of the membrane where each monomer (node) is connected by three bonds to its neighbors. One may therefore conclude that a moderate increase in the coordination number of the nodes (by only 33% regarding the maximum coordination of a node) leads to a major stabilization of the supporting bonds and much stronger resistance to fracture. Our additional simulation in the strongly damped regime for $\gamma = 10$ indicates no qualitative changes compared to $\gamma = 0.25$ except an absolute overall increase of the rupture times which is natural for a more viscous environment.

Note that the question of where and which bonds predominantly break is by no means trivial. For example, in the case of linear polymer chain thermal decomposition the rate of bond rupture is least at both chain ends although the end monomers, in contrast to those inside the chain, are bound by a single bond only as already discussed in Sec. 2.2.3. This interesting feature holds also for the honeycomb membrane flake, provided the rim is clamped and left immobile during the simulation (not shown). In this case the highest frequency of bond scissions grows towards the membrane center.

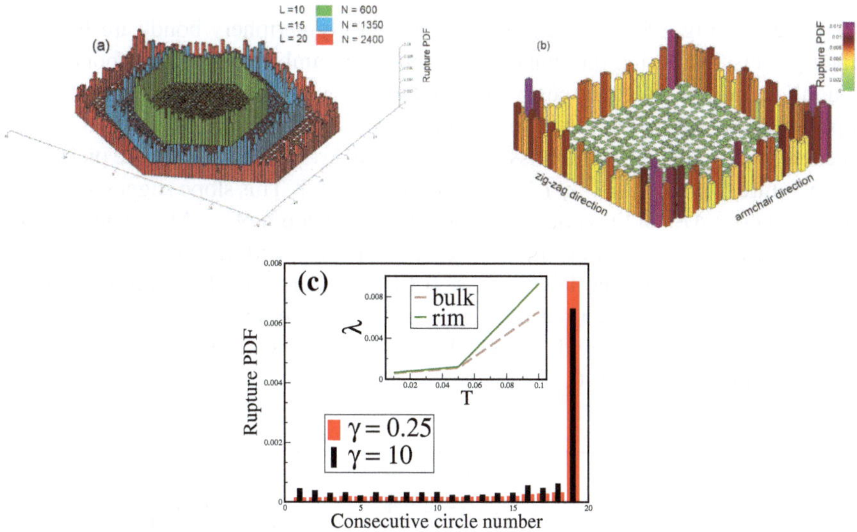

Fig. 28 (a) Rupture probability histograms for thermally induced scission events in flexible honeycomb flake of different linear size L as shown in the legend. (b) Rupture histogram for a ribbon-like square honeycomb membrane with 496 nodes. (c) Probability distribution of breakage events as a function of consecutive circle number for a membrane flake with $N = 600$ and two different friction coefficients $\gamma = 0.25$ and 10.0. Here $T = 0.1$. The inset shows estimated values of Lyapunov exponents λ vs T for beads located in the rim/bulk of membrane as indicate. Here $N = 5400$, $\gamma = 0.25$.

In order to provide deeper insight into the mechanism of temperature-induced bond breaking, in the inset to Fig. 28c we present the temperature variation of Lyapunov's exponent λ for membrane nodes located in the bulk and in the rim of the sheet. Evidently, beyond a cross-over temperature $T \approx 0.05$ one observes a significant growth of λ_{rim} as compared to λ_{bulk}. This indicates that the trajectories of nodes at the membrane periphery attain much faster chaotic features at higher temperature than those of the bulk nodes. Moreover, we should note that beads in the vortices have values of λ which exceed those in the rim by about 5%. Therefore this analysis of trajectory stability at characteristic locations in the membrane clearly demonstrate that bond rupture is induced by intermittent motion of the respective nodes.

The variation of the MFBT $\langle \tau \rangle$ of a bond with membrane size N during thermolysis for both hexagonal and square shapes of the 2D sheet is displayed in Fig. 29. Evidently, one observes for $\langle \tau \rangle$ a well pronounced power law behavior, $\langle \tau \rangle \propto N^{-\beta}$ with an exponent $\beta \approx 0.50 \pm 0.03$. It turns out that the scaling exponent β remains insensitive to changes in the geometric shape of the membrane sheet. This value of β might appear somewhat surprisingly to deviate from the expected exponent of unity, given that in the absence of external force all bonds are supposed to break completely at random so that the total probability for a bond scission (i.e., the chance that *any* bond might break within a time interval) is additive and should be, therefore, proportional to the total number of available bonds, $N_{bonds} = (3N - 6L)/2$. As suggested by Fig. 28, however, predominantly only periphery bonds are found to undergo scission during thermal degradation. The number of periphery bonds goes roughly as $\propto \sqrt{N}$ which agrees well with the observed value $\beta \approx 0.5$ and provides a plausible interpretation of the simulation result, Fig. 29. From the inset in Fig. 29 one may verify that the bond scission displays an Arrhenian dependence on inverse temperature, $\tau \propto \exp(\Delta E_b/k_B T)$, with a slope $\Delta E_b \approx 1$. This slope suggests a dissociation energy ΔE_b of the order of the potential well depth of the Morse interaction, Eq. 3 where $\varepsilon_M = 1.0$. In our model we deal typically with $E_b/(k_B T) \approx 10$ which at 300 K and typical bond length $r_{\min} \approx 0.144$ nm, corresponds to ultimate tensile stress ~ 0.6 GPa. This is a reasonable value for our membrane which is considerably softer than graphene with ~ 100 GPa [83] and is ranged between typical values for rubber materials 0.01–0.1 GPa.

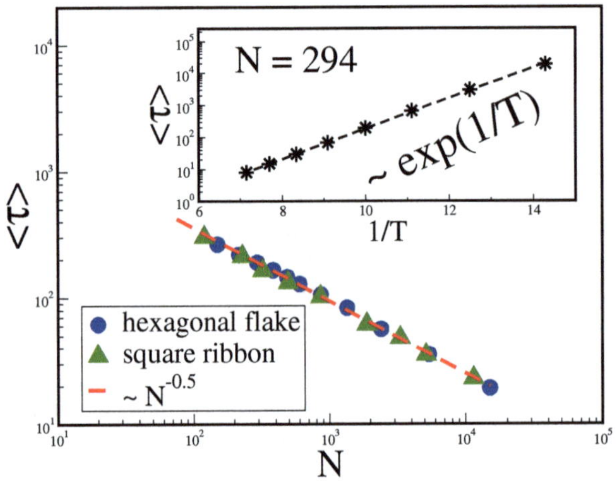

Fig. 29 Mean first breakage time $\langle \tau \rangle$ vs. number of beads N for two different membrane shapes: a hexagonal flake and a square ribbon. Solid line represents a fit by power law with an exponent $\beta = 0.5$ in both cases. The inset shows the variation of $\langle \tau \rangle$ with inverse temperature $1/T$ for a flake membrane with $N = 294$ particles. The fitting line yields an Arrhenian relationship, $\langle \tau \rangle \propto \exp(\Delta E_b/k_B T)$ with dissociation energy barrier $\Delta E_b \approx 0.95$.

4.2 Constant Force

Fracture in engineering materials is a long-standing topic of research due to problems that arise with technological applications and the ensuing economical implications. Thus, for decades a lot of attention has been focused on understanding the macroscopic and microscopic factors which trigger failure. Recently, the interest and the need for better understanding of the interplay between elastic and fracture properties of brittle materials has been revived due to the rapidly developing design of advanced structural materials. Promising aspects for applications include reversible polymer networks [84, 85], and also graphene, that shows unusual thermomechanical properties [86, 83]. Among other things, graphene can be used as anti-corrosion gas barrier protective coating [87], in chemical and bio-sensors [88], or as efficient membrane for gas separation [89]. In all possible applications the temperature and stress-dependent fracture strength of this 2D-network is of crucial importance. Besides numerous analytical and laboratory investigations, computer simulations [90, 92, 91] have provided meanwhile a lot of insight in aspects that are difficult for direct observations or theoretical treatment - for a review of previous works see Alava et al. [93]. Most of these studies focus on the propagation of (pre-existing) cracks, relating observations to the well known Griffith's model [94] of crack formation. A number of important aspects of material failure have found thereby little attention. Thus only a few simulations examine the rate of crack nucleation which involves long time scales necessary for thermal activation - see, however, [95, 96, 97, 98]. Effects of system size on the characteristic time for bond rupture have not been examined except in a recent MD study by Grant et al. [53]. In view of the possible applications as anti-corrosion and gas barrier coating, in this section we consider a radially-spanned sheet of regular hexagonal flake shape so as to minimize effects of corners and unequal edge lengths that are typical for ribbon-like sheets. Tensile constant force is applied on the rim of the flake, perpendicular to each edge. By varying system size, tensile force and temperature, we collect a number of results which characterize the initiation and the course of fragmentation in stretched 2D honeycomb networks [99].

4.2.1 Computational Model and Simulation Results

In our study of membrane fragmentation subjected to external stress we use eventually the same model as the one used in thermolysis of honey-comb sheet (see Sec. 4.1 for details). The only difference with respect to the case analyzed in Sec. 4.1 is addition of external stretching force f which is applied to the monomers at the membrane rim perpendicular to respective edge, see Fig. 30.

We examine the scission of bonds between neighboring nodes in the network sheet with honeycomb topology, assuming thermal activation as a driving mechanism in agreement with early experimental work by Brenner [100] and Zhurkov [101]. In Fig. 31 we show a series of representative snapshots of a membrane of size $L = 10$ with $N = 600$ monomers taken at different time moments during the process of decomposition. Typically, the first bonds that break are observed to belong to the last (even) most remote circle as, for example, at $t \approx 171 t.u.$ in Fig. 31. As mentioned above, these are the radially oriented bonds which belong to concentric circles of even number. Gradually a line of edge beads is then severed from the rest of the membrane and a crack is formed which propagates into the bulk until eventually a piece of the network sheet is ripped off, as in Fig. 31 at $t \approx 370 t.u.$ As we shall see below, this mechanism of membrane failure, whereby an initial crack is formed parallel to the edge monomers, yet perpendicular to the tensile force, dominates largely the process of disintegration under constant tensile force. The process is, therefore, mainly described by two characteristic times, $\langle \tau \rangle$ and $\langle \tau_r \rangle$, which mark

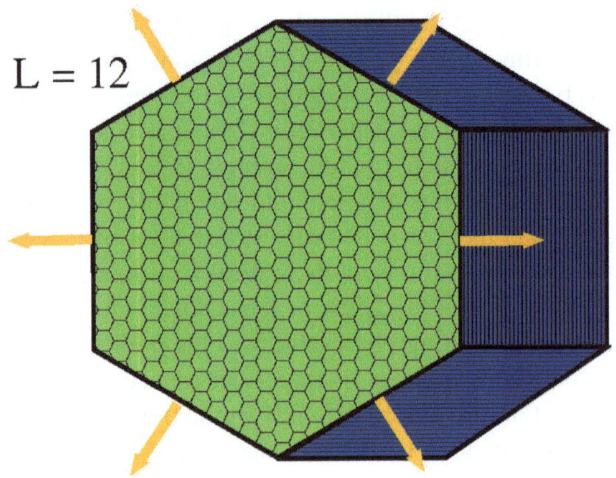

Fig. 30 A protective honeycomb network is spanned at the orifice of a prism whose size may vary due to thermal expansion. Tensile forces acting on the membrane periphery are indicated by arrows.

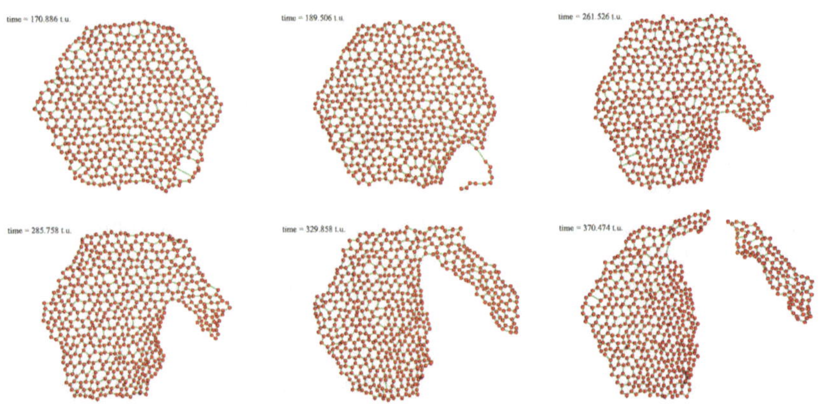

Fig. 31 Snapshots illustrate the process of bond breakage (crack generation) in different time moments for a membrane with $N = 600$ particles subject to external tensional force $f = 0.15$ at $T = 0.05$ and $\gamma = 0.25$. The force is applied to periphery monomers only and stretches the network perpendicular to its original edges.

the occurrence of the first scission of a bond (MFBT) and that of the eventual breakdown of the flake into two distinct parts.

In Fig. 32a we show the probability distribution of a first rupture events for *all* bonds in the honeycomb membrane flake as a 3D plot. It is seen that the scission rate is localized in the outer-most circle of radial bonds whereas bonds in the inner part of the membrane practically hardly break. Note that this is not a trivial effect since tension is distributed uniformly over all bonds in the equilibrated membrane so there is no additional propagation of the tension front from the rim towards the center.

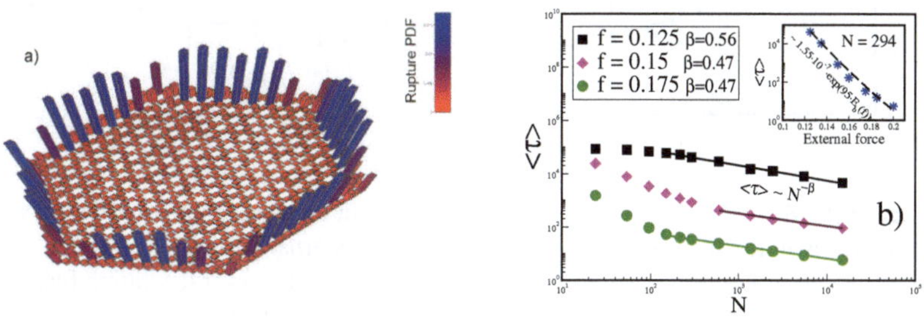

Fig. 32 (a) Rupture probability histogram of flexible hexagonal membrane subjected to external tensile stress $f = 0.15$. Here $N = 600$, $T = 0.05$ and $\gamma = 0.25$. (b) Mean first breakage time $\langle \tau \rangle$ vs. number of particles N in the membrane pulled with different tensile stress f as indicated. Symbols represent simulation data whereas solid lines stand for fitting functions $\langle \tau \rangle \sim N^{-\beta}$. The inset shows force-dependent $\langle \tau \rangle$ for a membrane composed of $N = 294$ beads.

The variation of the MFBT $\langle \tau \rangle$ with system size N (i.e., with the number of monomers in the membrane $N = 6L^2$ where L denotes the linear size of the flake) is shown in Fig. 32b. For sufficiently large membranes one observes a power law decline of the MFBT, $\tau \propto N^{-\beta}$ with an exponent $\beta \approx 0.5 \pm 0.03$ for the tensile forces studied. If thermally activated bonds break independently from one another and entirely at random, then $\langle \tau \rangle$ measures the interval before *any* of the available intact bonds undergoes scission, that is, either the first bond breaks, or the second one, and so on which, at constant rate of scission, would reduce the MFBT $\langle \tau \rangle \propto 1/N$ as observed for instance in the case of thermal degradation of a linear polymer chains, Sec. 2.2. This simple result can be derived by means of the classical theory of Weibull. In the present system of a honeycomb membrane the bonds that undergo rupture are nearly all located at the rim of the flake and their number is proportional to L so that with $\beta \approx 0.5$, cf. Fig. 32b and $N \propto L^2$, one obtains eventually the important result $\langle \tau \rangle \propto 1/L$. This observation is in agreement with recent results of Grant et al. [53] who studied the nucleation of cracks in a brittle $2D$-sheet. One can also see from the inset in Fig. 32b that $\langle \tau \rangle$ decreases rapidly with growing stress f, that is, the energy barrier for rupture declines with f in agreement Zhurkov's experiments [101].

In the course of our simulations we were also able to measure time needed to disintegration of the membrane into two separate parts which we refer as mean failure time $\langle \tau_r \rangle$. The variation of τ_r, the mean failure time of the membrane with system size N, shown in Fig. 33a, displays also a power-law dependence on system size N, $\langle \tau_r \rangle \propto N^{-\phi}$, whereby ϕ undergoes a cross-over to a lower value beyond roughly

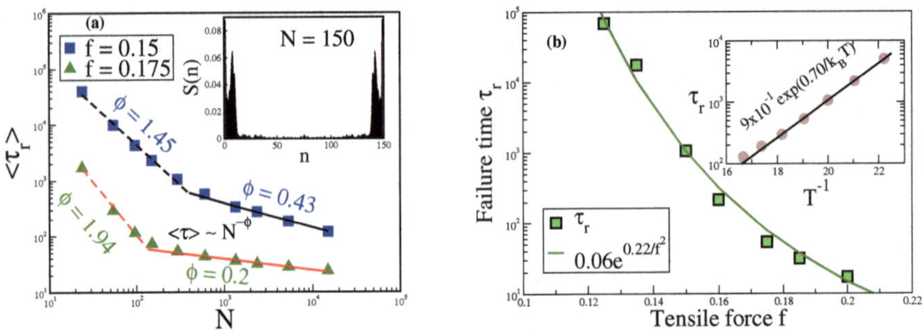

Fig. 33 (a) Mean failure time $\langle \tau_r \rangle$ (time needed to split membrane into two pieces) vs. number of particles in the membrane for two values of the external pulling force f at $T = 0.05$ and $\gamma = 0.25$. Symbols denote simulation results and represents power law fitting function $\langle \tau_r \rangle \sim N^{-\phi}$. The inset shows PDF of number of particles in the moment of splitting for a membrane composed of $N = 150$ beads. (b) Failure time $\langle \tau_r \rangle$ vs f in the case of $N = 294$. The inset shows variation of $\langle \tau_r \rangle$ with inverse temperature (Arrhenian plot).

$N > 300$. However, $\langle \tau_r \rangle$ has different physical meaning. Following Pomeau [102], the failure time can be approximately identified with the nucleation of a crack of critical size l_c given by Griffith's critical condition [94, 103] assuming that crack propagation is much faster than the nucleation time. For a $2D$-geometry consisting of a flat brittle sheet with a crack perpendicular to the direction of stress, the potential energy per unit thickness of the sheet reads $U = -\frac{\pi l^2 f^2}{4Y} + 2\varepsilon l + U_0$ where Y is the Young modulus, ε is the surface energy needed to form a crack of length l, and U_0 is the elastic energy in the absence of stress ($f = 0$). This energy reaches a maximum for a critical crack length $l_c = \frac{4\varepsilon Y}{\pi f^2}$ beyond which no stable state exists except the separation of the sheet into two broken pieces. Thus, with a crack nucleation barrier $\Delta U = \frac{4\varepsilon^2 Y}{\pi f^2}$ (in $3D$ $\Delta U \propto f^{-4}$), the failure (rip-off) time $\tau_r = \tau_0 \exp(\Delta U_0 / k_B T)$ as found in experiments with bidimensional micro crystals by Pauchard and Meunier [104] and in gels by Bonn et al. [105]. In Fig. 33b we present the variation of τ_r for membrane failure with stress f in good agreement with the expected relationship $\Delta U \propto f^{-2}$. In addition, we show the variation of τ_r with temperature, see inset in Fig. 33b, which is found to follow a well expressed Arhenian relationship with inverse temperature, in agreement with earlier studies [53, 103].

The end of the sheet rupturing process is marked as a rule by disintegration into two pieces of different size so it is interesting to asses the size distribution of such fragments upon failure. In the inset in Fig. 33a we show a probability distribution $S(n)$ of the sizes of of both fragments upon membrane rip-off. In a membrane composed of N beads one observes a sharp bimodal distribution with narrow peaks at sizes $N_1 \approx 10$ and $N_2 \approx 140$. Evidently, for the adopted nearly radial direction - cf. Fig. 30 - of the applied tensile force one always finds a pair of one small and another very large fragment.

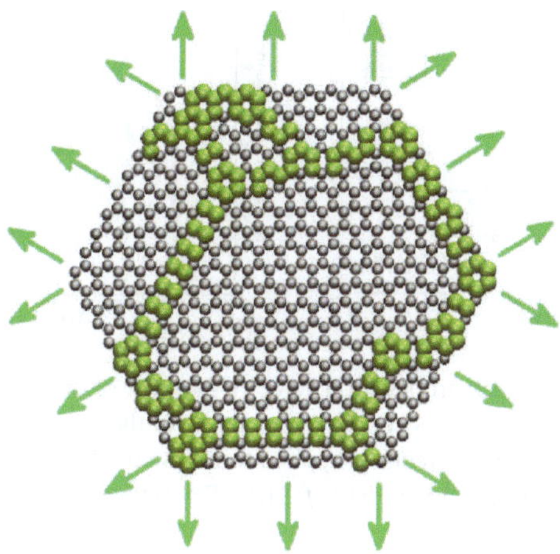

Fig. 34 Typical pattern of cracks observed in a honeycomb membrane composed of $N =$ 600 particles. Cracks are marked in color on the geometrically undistorted arrangement of network nodes for better visibility. Parameters of a heat bath are $T = 0.05$ and $\gamma = 0.25$.

One can readily verify from the typical topology of the observed cracks in the membrane, presented in Fig. 34, that (i) cracks emerge as a rule perpendicular to the direction of applied stress, and (ii) it is almost always the first row of nodes to which the tensile force is immediately applied that gets ripped off upon failure. Cracks that break the network sheet in the middle occur very seldom, in compliance with the sampled distribution of fragment sizes, $S(n)$ in the inset of Fig. 33a. One would, therefore, predict a breakup of a protective cover spanned on the orifice of tube like the one shown in Fig. 34 to proceed immediately at the fixed orbicular boundary where the tensile force applies to the network. It is interesting to note that the geometry of cracks in the membrane shown in Fig. 34 appears very similar to the one observed in drying induced cracking of thin layers of materials subject to structural disorder [109].

The emerging cracks are expected to propagate with speed that increases as the strength of the external force is increased as the inset in Fig. 35a indicates. In fact, in Fig. 35a one observes typical curves comprising a series of short intervals with steep growth of the number of broken bonds per unit time and longer horizontal 'terraces' preceding the nucleation of a new crack. Even though the data, presented in Fig. 35a, is not averaged over many realizations, and, as Fig. 35b suggests, individual realizations of propagating cracks may strongly differ even at the same stress f, a general increase of the propagation velocity with growing external force f - see inset - can be unambiguously detected, in agreement with earlier observations [90].

For our model membrane with computed Young modulus $Y \approx 0.02$ we get for the Rayleigh wave speed $c_R \approx 0.14$. Thus for most of the applied tensile stress values we observe crack propagation at speed both below and above c_R - inset in Fig. 35b.

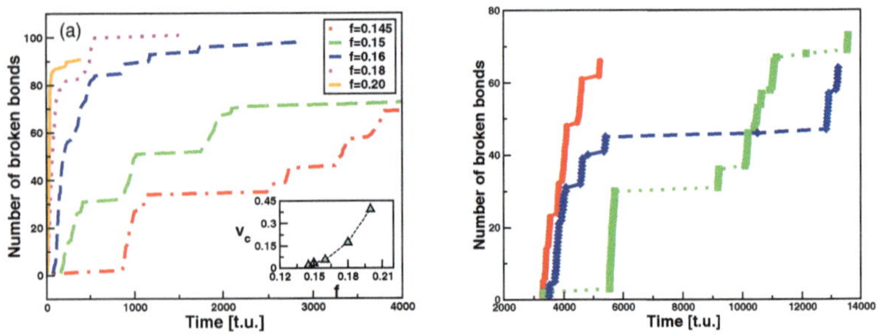

Fig. 35 (Color online) (a) Crack propagation velocity (number of broken bonds per unit time) for a membrane with $N = 600$ beads at different strength of the external force f as indicated. (b) Three different realizations of cracks at applied force $f = 0.14$. The inset shows a variation of the mean crack propagation velocity with f. Here $T = 0.05$ and $\gamma = 0.25$.

As argued by [106] propagation speed cannot exceed c_R because crack splits off into multiple cracks before reaching c_R. In contrast, Abraham and Gao in Ref. [107] have reported on cracks that can travel faster than the Rayleigh speed. Thus, our rough estimates (inset in Fig. 35) agree well with data from literature. Converting our results to proper metric units, with bond length $\sigma \approx 0.144$ nm and energy $\approx 20k_BT$ which yields 1 MD t.u. $\approx 10^{-12}$ s, we estimate the typical crack propagation speed $v_c \approx 50$ m/s. Note that mean crack speed for natural latex rubber was given as 56 m/s [108].

5 Summary

In the present contribution we have summarized our studies concerning kinetics of bond rupture for linear chains [4, 23, 56, 57], bottle-brushes [68, 69, 70], and membranes [82, 99]. The most important conclusions that can be drawn from our results can be summarized as follows:

- The bond relaxation dynamics changes qualitatively as compared to the polymer Rouse dynamics, valid for idealized Gaussian polymer chains, in the case of bonding by anharmonic forces. Among the most salient features of anharmonic bonds we observe mode coupling and energy transfer between the Rouse modes as well as strong anisotropy in relaxation times along and perpendicular to the direction of elongation. Relaxation times decline with growing stretching of the bonds differently in both directions.
- The basic notion of the thermal degradation of a single polymer chain is a result of random and independent scission of bonds. This observation is supported by the dependence of the mean lifetime of a bond $\langle\tau\rangle$ on the chain length N as $\langle\tau\rangle \propto 1/N$. In addition, the variation of the life time, $\langle\tau\rangle$, with (inverse) temperature turns out to be an Arrhenian-law, $\langle\tau\rangle \propto \exp(\Delta E_b/k_BT)$ whereby the activation energy is rather close to the potential well depth of the Morse interaction. The distribution of the bond rupture probability reveals the existence of

a multiple length-scale inhomogeneity which is self-induced presumably as a result of the interplay between thermal noise and the nonlinearity of the bond potential. This inhomogeneity does not exist in the Gaussian chain model where the forces depend linearly on distance between monomers which supports the notion of force anharmonicity (i.e., non-linearity) as an origin of the observed inhomogeneity.

- The mean life time of the polymer chain at constant tensile force depends on chain length like $\langle\tau\rangle \propto N^{-\beta}$ whereby the power law exponent β varies in the interval $0 < \beta < 1$. Generally, it appears that the exponent β systematically declines as the external pulling force f grows. This behaviour indicates a growing degree of cooperativity during the chain breakage as the pulling force f is increased. The rates of bond rupture are distributed differently along the polymer backbone in the $1D$ and $3D$ chain models. In a $1D$ chain the rupture rate steadily grows as one approaches the free chain end where the external pulling force is applied whereas in a $3D$ chain bonds break predominantly in the middle of the chain. Bond rupture histograms correlate with the degree of spreading in the monomer density distribution, indicating that scissions occur most frequently in those parts of the macromolecule which undergo large fluctuations in position.
- In the process of thermal degradation of strongly adsorbed bottle-brush molecules the mean life time of a bond $\langle\tau\rangle$ decreases by more than an order of magnitude upon adsorption of a free bottle-brush molecules on an adhesive surface. The probability distribution for rupture depends on both grafting density σ_g and length of the side chains N. It is sensitive to the degree of steric repulsion of the side chains - the shape of the scission probability distribution resembles the experimentally established one only for weaker repulsion when the side chains do not mutually block one another.
- In the case of thermal degradation of 2D polymerized sheet the probability of bond scission is highest at the periphery of the membrane sheet where nodes are connected by two bonds only. The mean life time $\langle\tau\rangle$ until a bond undergoes scission event declines with the number of nodes N (with membrane size) by a power law as $\langle\tau\rangle \propto N^{-0.5}$ independently of the geometric shape of membrane sheet.
- In the case of polymerized subjected to external pulling f the failure time $\langle\tau_r\rangle$ until a brittle sheet disintegrates into pieces follows a power law, $\tau \propto N^{-\phi(f)}$, and an exponential decay $\langle\tau_r\rangle \propto \exp(\mathrm{const}/f^2)$ upon increasing strength of the pulling force, in agreement with Griffith's criterion for failure. Cracks emerge in the vicinity of membrane edges and typically propagate parallel to the edges, splitting the sheet in two pieces of size ratio of $\approx 7\%$. Crack propagation speed is observed to increase rapidly with tensile force.

References

1. Vilgis, T.A., Klueppel, M., Heinrich, G.: Reinforcement of Polymer Nano Composites, Theory, Experiments, Applications. Cambridge University Press, Cambridge (2009)
2. Vilgis, T.A., Singh, N.: Kautschuk Gummi und Kunststoffe 60, 444 (2007)

3. Doi, M., Edwards, S.F.: The Theory of Polymer Dynamics. Oxford University Press, N. Y. (1986)
4. Febbo, M., Milchev, A., Rostiashvili, V., Dimitrov, D., Vilgis, T.A.: J. Chem. Phys. 129, 154908 (2008)
5. Allen, N.S., Edge, M.: Fundamentals of Polymer Degradation and Stabilization. Elsevier Applied Science, New York (1966)
6. Madras, G., Smith, J.M., McCoy, B.J.: Ind. Eng. Chem. Res. 35, 1795 (1996)
7. Nyden, M.R., Forney, G.P., Brown, G.E.: Macromolecules 25, 1658 (1992)
8. Flynn, J.H., Florin, R.E.: In: Liebman, S.A., Levy, E.J. (eds.) Pyrolysis and GC in Polymer Analysis, p. 149. Marcel Dekker Inc., New York (1985)
9. Sarikaya, M., Tamerler, C., Jen, A.K., Schulten, K., Banyex, F.: Nature Mater. 2, 577 (2003)
10. Han, T.H., Kim, J., Park, J.S., Park, C.B., Ihee, H., Kim, S.O.: Adv. Mater. 19, 3924 (2007)
11. Simha, R.: J. Appl. Phys. 12, 569 (1941)
12. Jellinek, H.H.G.: Trans. Faraday Soc. 1944, 266 (1944)
13. Ballauff, M., Wolf, B.A.: Macromolecules 14, 654 (1981)
14. Ziff, R.M., McGrady, E.D.: Macromolecules 19, 2513 (1986); McGrady, E.D., Ziff, R.M.: Phys. Rev. Lett. 58, 892 (1987)
15. Cheng, Z., Redner, S.: Phys. Rev. Lett. 60, 2450 (1988)
16. Blaisten-Barojas, E., Nyden, M.R.: Chem. Phys. Lett. 171, 499 (1990); Nyden, M.R., Noid, D.W.: J. Phys. Chem. 95, 940 (1991)
17. Doerr, T.P., Taylor, P.L.: J. Chem. Phys. 101, 10107 (1994)
18. Wang, M., Smith, J.M., McCoy, B.J.: AIChE Journal 41, 1521 (1995)
19. Hathorn, B.C., Sumpter, B.G., Noid, D.W.: Macromol. Theory Simul. 10, 587 (2001)
20. Doruker, P., Wang, Y., Mattice, W.L.: Comp. Theor. Polym. Sci. 11, 155 (2001)
21. Lee, C.F.: Phys. Rev. E 80, 031134 (2009)
22. Fugmann, S., Sokolov, I.M.: Phys. Rev. E 81, 031804 (2010)
23. Paturej, J., Milchev, A., Rostiashvili, V.G., Vilgis, T.A.: J. Chem. Phys. 134, 224901 (2011)
24. Gardiner, C.W.: Handbook of Stochastic Methods. Springer, Heidelberg (2004)
25. Goldenfeld, N.: Lectures on Phase Transitions and the Renormalization Group. Addison-Wesley Publishing Company, New York (1992)
26. Sagués, F., Sancho, J.M., Garcia-Ojalvo, J.: Rev. Mod. Phys. 79, 829 (2007)
27. Sheiko, S.S., Sun, F.C., Randall, A., Shirvanyants, D., Rubinstein, M., Lee, H.-I., Matyjaszewski, K.: Nature Lett. 440, 191 (2006)
28. Glynn, P.A.R., van der Hoff, B.M.E., Reilly, P.M.: J. Macromol. Sci. - Chem. A6, 1653 (1972)
29. Lebedeva, N.V., Sun, F.C., Lee, H.-I., Matyjaszewski, K., Sheiko, S.S.: J. Am. Chem. Soc. 130, 4228 (2007); Park, I., Sheiko, S.S., Nese, A., Matyjaszewski, K.: Macromolecules 42, 1805 (2009)
30. Kausch, H.-H.: Polymer Fracture. Springer, New York (1987)
31. Crist, B.: Ann. Rev. Mater. Sci. 25, 295 (1995)
32. Garnier, L., Gauthier-Manuel, B., van der Vegte, E.W., Snijders, J., Hadziioannou, G.: J. Chem. Phys. 113, 2497 (2000)
33. Grandbois, M., Beyer, M., Rief, M., Clausen-Schaumann, H., Gaub, H.E.: Science 283, 1728 (1999)
34. Saitta, A.M., Klein, M.L.: J. Phys. Chem. A 105, 6495 (2001)
35. Maroja, A.M., Oliveira, F.A., Ciesla, M., Longa, L.: Phys. Rev. E 63, 061801 (2001)
36. Rohrig, U.F., Frank, I.: J. Chem. Phys. 115, 8670 (2001)

37. Gersappe, D., Robins, M.O.: Europhys. Lett. 48, 150 (1999)
38. Filipov, A.E., Klafter, J., Urbakh, M.: Phys. Rev. Lett. 92, 135503 (2004)
39. Aktah, D., Frank, I.: J. Amer. Chem. Soc. 124, 3402 (2002)
40. Beyer, M.K., Clausen-Schaumann, H.: Chem. Rev 105, 2921 (2005)
41. Harris, S.A.: Contemp. Phys 45, 11 (2004)
42. Perverzev, Y.V., Prezhdo, O.V.: Phys. Rev. E 73, 050902 (2006)
43. Bueche, F.: J. Appl. Phys. 29, 1231 (1958)
44. Zhurkov, S.N., Korsukov, V.E.: J. Polym. Sci. (Pol. Phys. Ed.) 12, 385 (1974)
45. Doerr, T.P., Taylor, P.L.: J. Chem. Phys. 101, 10107 (1994)
46. Stember, J.N., Ezra, G.S.: Chem. Phys. 337, 11 (2007)
47. Bolton, K., Nordholm, S., Schranz, H.W.: J. Phys. Chem. 99, 2477 (1995)
48. Puthur, R., Sebastian, K.L.: Phys. Rev. B 66, 024304 (2002)
49. Oliveira, F.A., Taylor, P.L.: J. Chem. Phys. 101, 10118 (1994)
50. Oliveira, F.A.: Phys. Rev. B 52, 1009 (1995)
51. Oliveira, F.A., Gonzalez, J.A.: Phys. Rev. B 54, 3954 (1996)
52. Oliveira, F.A.: Phys. Rev. B 57, 10576 (1998)
53. Sain, A., Dias, C.L., Grant, M.: Phys. Rev. E 74, 046111 (2006)
54. Langer, J.S.: Ann. Phys (N.Y.) 54, 258 (1969)
55. Hänngi, P., Talkner, P., Borkovec, M.: Rev. Mod. Phys. 62, 251 (1990)
56. Ghosh, A., Dimitrov, D.I., Rostiashvili, V.G., Milchev, A., Vilgis, T.A.: J. Chem. Phys. 132, 204902 (2010)
57. Paturej, J., Milchev, A., Rostiashvili, V.G., Vilgis, T.A.: EPL 94, 48003 (2011)
58. Caruso, M.M., Davis, D.A., Shen, Q., Odom, S.A., Sottos, N.R., White, S.R., Moore, J.: Chem. Rev. 109, 5755 (2009)
59. Black, A.L., Lenhardt, J.M., Craig, S.L.: J. Mat. Chem. 21, 1655 (2011)
60. Beyer, M.K., Clausen-Schaumann, H.: Chem. Rev. 105, 2921 (2005)
61. Liang, J., Fernández, J.M.: ACS Nano 3, 1628 (2009)
62. Huang, Z., Boulatov, R.: Chem. Soc. Rev. 40, 2359 (2011)
63. Yang, Q.-Z., Huang, Z., Kucharski, T.J., Khvostichenko, D., Chen, J., Boulatov, R.: Nat. Nanotechnol. 4, 302 (2009)
64. Park, I., Nese, A., Pietrasik, J., Matyjaszewski, K., Sheiko, S.S.: J. Mat. Chem. 21, 8448 (2011)
65. Panyukov, S., Zhulina, E.B., Sheiko, S.S., Randall, G.C., Brock, J., Rubinstein, M.: J. Chem. Phys. B 113, 3750 (2009)
66. Panyukov, S., Sheiko, S., Rubinstein, M.: Phys. Rev. Lett. 102, 148301 (2009)
67. Sheiko, S., Panyukov, S., Rubinstein, M.: Macromolecules 44, 4520 (2011)
68. Milchev, A., Paturej, J., Rostiashvili, V.G., Vilgis, T.A.: Macromolecules 44, 3981 (2011)
69. Paturej, J., Kuban, L., Milchev, A., Vilgis, T.A.: EPL 97, 4371 (2012)
70. Paturej, J., Kuban, L., Milchev, A., Rostiashvili, V.G., Vilgis, T.A.: Macromol. Symp., vol. 316, p. 112 (2012)
71. Hansen, J.C., Skalak, R., Chien, S., Hoger, A.: Biophys. J. 70, 146 (1996)
72. Beale, P.D., Srolovitz, D.J.: Phys. Rev. E 37, 5500 (1988)
73. Saxton, M.J.: Biophys. J. 57, 1167 (1990)
74. Boal, D.H., Seifert, U., Zilker, A.: Phys. Rev. Lett. 69, 3405 (1992)
75. Monette, L., Anderson, M.P.: Modelling Simul. Mater. Sci. 2, 53 (1994)
76. Dao, M., Li, J., Suresh, S.: Mater. Sci. Eng. C 26, 1232 (2006)
77. Barnard, A.S., Snook, I.K.: J. Chem. Phys. 128, 094707 (2008)
78. Argyropoulos, D.S., Bolker, H.I.: Macromolecules 20, 2915 (1987); Macromol. Chem. 189, 607 (1988)

79. Barral, L., Diez, F.J., Garcia-Garabal, S., Lopez, J., Montero, B., Montes, R., Ramirez, C., Rico, M.: Europ. Polym. J. 41, 1662 (2005)
80. Zhang, Z., Liang, G., Ren, P., Wang, J.: Polym. Composites 28, 755 (2007)
81. Chenoweth, K., Cheung, S., van Duin, A.C.T., Goddard III, W.A., Kober, E.M.: J. Am. Chem. Soc. 127, 7192 (2005)
82. Paturej, J., Popova, H., Milchev, A., Vilgis, T.A.: J. Chem. Phys. 137, 054901 (2012)
83. Zhao, H., Min, K., Aluru, N.R.: Nano Letters 9(8), 3012 (2009); Zhao, H., Aluru, N.R.: Jour. Appl. Phys. 108, 064321 (2010); Min, K., Aluru, N.R.: Appl. Phys. Lett. 98, 013113 (2011)
84. Sijbesma, R.P., Beijer, F.H., Brunsveld, L., Folmer, B.J.B., Hirschberg, J.H.K.K., Lange, R.F., Lowe, J.K.L., Meijer, E.W.: Science 278, 1601 (1997)
85. Sijbesma, R.P., Meijer, E.W.: Chem. Commun. 1, 5 (2003)
86. Neek-Amal, M., Peeters, F.M.: Phys. Rev. B 81, 235437 (2010); Phys. Rev. B 82, 085432 (2010); Appl. Phys. Lett. 97, 153118 (2010)
87. Chen, S., Brown, L., Levendorf, M., Cai, W., Ju, S.-Y., Edgeworth, J., Li, X., Magnuson, C.W., Velamakanni, A., Piner, R.D., Kang, J., Park, J., Ruoff, R.S.: ACS Nano 5, 1321 (2011)
88. Pumera, M.: Materials Today 14, 308 (2011)
89. Jiang, D.-E., Cooper, V.R., Dai, S.: Nano Lett. 9, 4019 (2009)
90. Holland, D., Marder, M.: Phys. Rev. Lett. 80, 746 (1998)
91. Mattoni, A., Colombo, L., Cleri, F.: Phys. Rev. Lett. 95, 115501 (2005)
92. Buehler, M.J., Tang, H., van Duin, A.C.T., Goddard III, W.A.: Phys. Rev. Lett. 99, 165502 (2007)
93. Alava, M.J., Nukala, P.K., Zapperi, S.: Adv. Phys. 55, 349 (2006)
94. Griffith, A.A.: Philos. R. Soc. London A 221, 163 (1920)
95. Santucci, S., Vanel, L., Guarino, A., Scorretti, R., Ciliberto, S.: Europhys. Lett. 62, 320 (2003)
96. Wang, Z.-G., Landman, U., Blumberg Selinger, R.L., Gelbart, W.M.: Phys. Rev. B 44, 378 (1991)
97. Blumberg Selinger, R.L., Wang, Z.-G., Gelbart, W.M., Ben-Shaul, A.: Phys. Rev. A 43, 4396 (1991)
98. Gagnon, G., Patton, J., Lacks, D.J.: Phys. Rev. E 64, 051508 (2001)
99. Paturej, J., Popova, H., Milchev, A., Vilgis, T.A.: Phys. Rev. E 85, 021805 (2012)
100. Brenner, S.S.: J. Appl. Phys. 33, 33 (1962)
101. Zhurkov, S.N.: Int. J. Fract. Mech. 1, 311 (1965)
102. Pomeau, Y.: C. R. Acad. Sci. (Paris) 314, 553 (1992)
103. Rabinovich, A., Friedman, M., Bahat, D.: Europhys. Lett. 67, 969 (2004)
104. Pauchard, L., Meunier, J.: Phys. Rev. Lett. 70, 3565 (1993)
105. Bonn, D., Kellay, H., Prochnow, M., Ben-Djemiaa, K., Meunier, J.: Science 280, 265 (1998)
106. Fineberg, J., Gross, S.P., Marder, M., Swinney, H.L.: Phys. Rev. Lett. 67, 457 (1991)
107. Abraham, F.F., Gao, H.: Phys. Rev. Lett. 84, 3113 (2000)
108. Petersan, P.J., Deegan, R.D., Marder, M., Swinney, H.L.: Phys. Rev. Lett. 93, 015504 (2004)
109. Villalobos, G., Kun, F., Munoz, J.D.: Phys. Rev. E 84, 041114 (2011)

In-Situ Structural Characterization of Rubber during Deformation and Fracture

Karsten Brüning[1,2], Konrad Schneider[1], and Gert Heinrich[1,2]

[1] Leibniz-Institut für Polymerforschung Dresden e.V., Hohe Str. 6, 01069 Dresden, Germany
[2] Technische Universität Dresden, Institut für Werkstoffwissenschaft, Helmholtzstr. 7, 01069 Dresden, Germany

Abstract. To obtain an in-depth understanding of the mechanical and fracture behaviors of elastomers, a thorough material characterization is inevitable to establish structure-property relationships. Synchrotron X-ray diffraction and scattering were used to characterize filled and unfilled natural rubber under deformation in situ, using tailor-made miniature tensile testing machines for quasistatic, dynamic cyclic and impact loads. The degree and kinetics of strain-induced crystallization, which accounts for the excellent mechanical and tear properties of natural rubber, were determined under various loading cases with unraveled resolution in time and space. Employing a strain-step method, a new model was established to describe the kinetics of strain-induced crystallization. On-line characterization during cyclic loading at 1 Hz gave insight into structural processes on fast time scales under realistic loading conditions, enabling a better understanding of static and dynamic crack growth. Information about filler orientation and cavitation was obtained by ultra small-angle X-ray scattering (USAXS). A direct connection between the magnitude of uniaxial strain and the orientation of anisometric model filler particles was found and implemented into a model, enabling a quantitative analysis of the scattering patterns from first principles. The reversibility of filler orientation under fatigue loading was shown. Cavitational processes in filled rubbers are thought to be crucial for the dissipation of strain energy, but also are potential precursors of macroscopic cracks. Cavities significantly increase the scattering intensity due to the formation of new phase boundaries. Cavitation processes were identified in the bulk and in the vicinity of crack tips of stretched carbon black-filled rubbers by scanning USAXS, employing a three-phase model to quantify the extent of cavitation. These new insights into the structural changes in elastomers under strain open new possibilities for the development of physically sound constitutive mechanical models.

W. Grellmann et al. (Eds.): *Fracture Mechanics & Statistical Mech.*, LNACM 70, pp. 43–80.
DOI: 10.1007/978-3-642-37910-9_2

1 Introduction

As pointed out in the previous chapters, fracture phenomena in rubbers can be studied with various techniques, each accessing the issue from a different direction and thus contributing to the overall goal, which is to obtain a complete understanding of the fracture behavior, all the way from the molecular basis to the implications for the rubber and tire industry. To gain this understanding, it is desirable to obtain in-situ quantitative and representative information about the processes occurring in the bulk and in the vicinity of the crack tip with good spatial and time resolution. Since classical techniques like electron microscopy cannot provide real time information, the methods of choice for these investigations are X-ray scattering and diffraction.

Especially since the advent of powerful synchrotron sources over the last two decades, numerous researchers have taken advantage of small-angle X-ray scattering (SAXS) and wide-angle X-ray diffraction (WAXD) techniques to study structural changes in polymers under deformation. By the combination of SAXS and WAXD, structural changes of various length scales can be observed, ranging from a few Å to almost a µm. Thus processes like the orientation of the network under deformation, the ordering of chain segments into crystallites, the alignment of anisometric filler particles along the tensile direction, cavitation and void formation are directly accessible. The major drawback of scattering techniques is that all information is provided in reciprocal space, i. e. in order to be accessible to common human perception, either a transformation to real space has to be done or certain quantitative features have to be extracted directly from reciprocal space.

The goal of this work was to study the effect of strain field and time on the structure of the rubber. In the context of strain-induced crystallization (SIC), WAXD was used to study the crystallization under uniaxial and biaxial load, the local crystallinity around a crack tip and the kinetics of the crystallization process. Unprecedented time resolution in the ms range was achieved by using high-flux third generation synchrotron sources. To make the step to the filler structure, the degree of crystallinity was used as an indicator for the local strain in the matrix in filled rubbers and thus a strain amplification factor was derived. The orientation of fillers was followed by SAXS and USAXS (ultra small-angle X-ray scattering). Due to the formation of new phase boundaries, USAXS and SAXS also are ideally suited to quantitatively study the formation of cavities, which are hypothesized to serve as nuclei for a possible macroscopic failure. Scanning electron microscopy during in-situ tensile tests was used to complement the data with real space information.

1.1 Introduction to Strain-Induced Crystallization in Natural Rubber

The earliest studies of structural changes in rubbers under strain date back to the 1920s, when Katz published his report on strain-induced crystallization (SIC)[1] in natural rubber (NR) observed by WAXD. [47] Even though the fundaments of macromolecular chemistry were still lacking at that time, he concluded that highly stretched natural rubber consists of two phases and that the phase transition is reversible when the strain is removed. Since the early 2000s, the number of papers dealing with SIC has been increasing owing to the availability of more powerful synchrotron sources. SIC attracts so much attention in the research community because it makes a big contribution to the outstanding properties of NR. Crystallites can be thought of as additional crosslinking points or as rigid reinforcing particles, strengthening the otherwise amorphous rubber. [76, 108] Thanks to SIC, the strength at break and the crack propagation behavior of NR are superior to those of non-crystallizing synthetic rubbers. Nowadays, we know the precise structure of the cis-1,4 polyisoprene chain segments in the crystalline lattice. [15, 78, 80, 81, 107]

While SIC can be analyzed by a range of methods, e. g. DSC, [90, 92] NMR, [65, 89] IR spectroscopy [65] and dilatometry, [90] the most widely used method is WAXD. The relations between strain, temperature and crystallinity have been studied extensively. [113] With increasing strain, the melting temperature increases. [89] Undeformed natural rubber undergoes significant thermal crystallization (TIC) at temperatures below 0 °C with a maximum crystallization rate around −28 °C. [115, 116] The thermodynamic melting point lies considerably above 0 °C. [20, 52] At room temperature, crystallization sets in at a strain of roughly 300 % in unfilled natural rubber. [42] It should be noted that, despite having the same crystalline structure, the morphologies of thermally induced crystallization (TIC) and SIC are completely different. [5, 7] Quiescent crystallization forms spherulites, [48, 64] whereas SIC promotes the formation of highly oriented fibrillar structures. [7, 67] In the following, only SIC is considered.

At first sight, the fact that in unfilled rubber the SIC sets in at strains as large as 300 %, suggests that SIC might be of little relevance in real-life loading conditions of most rubber products. However, in reality the local strain often considerably exceeds the nominal strain. In filled rubbers, due to the rigidity of the filler phase, the matrix phase has to bear the complete deformation, such that the matrix strain is considerably larger than the external strain, especially when the rubber is highly filled. This effect is referred to as strain amplification. Thus, in filled rubbers, the SIC onset strain can be reduced to around 150 % strain. Second, certain geometries give rise to local strain concentration, e. g. around a crack tip. Theoretically, when approaching a crack tip, one encounters a strain singularity. In reality, the local strain

[1] Katz called it *fibering* what is now termed *strain-induced crystallization.*

concentration leads to a crystalline zone around the crack tip which locally reinforces the material and thus slows down the crack growth.

1.2 Kinetics of Strain-Induced Crystallization

Besides the effects of strain and temperature, the time-dependency of SIC should not be overlooked. Since most rubber products, especially tires, are subjected to dynamic loads, the structure under realistic loading conditions can considerably deviate from what is suggested by quasistatic experiments. This is due to the finite crystallization kinetics and was shown for the first time in 1932 by Acken, who utilized a stroboscopic technique to accumulate WAXD diffractograms over several deformation cycles to reach sufficient exposure. [1] He found that the crystallinity in the dynamically stretched rubber was considerably suppressed as compared to quasistatic experiments. This type of setup was later optimized by Kawai et al. [41, 49] Using a variable phase shift between the stroboscope and the mechanical dynamical stretching device, the complete crystallinity vs. strain curve could be reconstructed. Recently, Candau et al. extracted a crystallinity vs. time curve from stroboscopic experiments. [16] Several of these studies postulated an incubation time of the order of 100 ms.

The alternative approach to cyclic loading is to apply a steplike loading and then to follow the crystallization over time. The pioneering work in this respect was done by Dunning in 1967, who took advantage of a continuous steplike strain when a rubber band is passed over two rolls rotating at different speeds. [24] He found similarly long incubation times. Mitchell utilized the enthalpy of crystallization and used thermal techniques to measure the crystallinity after a steplike loading. [77] Only in the last few years, owing to the availability of more powerful synchrotrons, did the direct measurement of crystallinity after a step strain become possible. Tosaka et al. performed WAXD experiments with a pattern acquisition rate of approximately 12 Hz. [109–111] Despite relatively slow steplike loading, they reported that the crystallization process only begins after the loading step is finished. After roughly 10 s the crystallinity was found to approach a steady state. In the framework of this work, we could show that even after less than 10 ms after completion of a very fast strain step, the crystallization process has progressed considerably. We propose that the nucleation step of the crystallization process occurs instantaneously during stretching, which is followed by a growth step.

1.3 Strain Amplification

As outlined above, the concept of strain amplification is frequently used to describe the matrix strain in a filled system. Since the filler particles are much stiffer than the matrix, they are usually taken to be rigid. From a theoretical point of view, assuming constant volume, the ratio of the matrix strain over the external strain then follows as $(1-\phi)^{-1}$ with ϕ being the filler volume fraction. [22] The stiffening of the composite simply due to the reduction of the effectively deformable volume fraction is termed as hydrodynamic reinforcement and represents the simplest approach to describe filler reinforcement. From an experimental point of view, direct evaluation of the matrix strain is a challenge. The most widely used methods include mechanical tests, WAXD and NMR. The underlying assumption is that a unique relation between the measured quantity ψ (e. g. stress, crystallinity, orientation) and strain ϵ exists, which applies to the unfilled material as well as to the matrix of the filled material: [38, 114]

$$A_{\psi,\text{filled}}(\epsilon) = \left.\frac{\epsilon_{\text{unfilled}}}{\epsilon_{\text{filled}}}\right|_{\psi} = \left.\frac{\epsilon_{\text{matrix}}}{\epsilon_{\text{filled}}}\right|_{\psi}, \tag{1}$$

where $A_{\psi,\text{filled}}$ is the strain amplification factor of the filled material for the quantity ψ. For instance, let the strain in a filled rubber be half of the strain in an unfilled reference at the same stress, then the mechanical strain amplification factor is two. The concept of the mechanical strain amplification has proven to be useful as long as the reinforcing capability of the filler is high, such as in carbon black filled systems, [38] and the strain is larger than the Payne regime but below the SIC onset. [25] Given that hydrodynamic effects are not the only source of reinforcement, and also numerous other mechanisms like filler-filler networks and adsorption of polymer at the filler surface (immobilized layer of bound rubber, filler particles as additional network crosslinks) have to be considered, quantitative deviations from a constant strain amplification are to be expected and have indeed been observed. [25, 89]

In general, the strain amplification factors obtained by different methods do not agree. This can be ascribed to the fact that they are sensitive to different local phenomena. When the SIC onset serves as input variable for the calculation of the strain amplification, the amplification is larger than suggested by the mechanics, since due to the heterogeneity of the network a large local strain in some highly confined regions is sufficient to initiate SIC. On the other hand, NMR averages over the orientation of all chains, and thus gives a lower strain amplification. [25]

We used the degree of crystallinity Φ as an indicator of the strain in the matrix of a filled NR. Using different fillers, we could show that the hydrodynamic strain amplification cannot be regarded as the only source of reinforcement. Instead, the reinforcement heavily depends on the nature of the filler.

1.4 Filler Structure under Deformation

It is well known that the structure of the filler strongly influences the mechanical properties of the composite. In the case of carbon black, numerous grades for specific purposes are available, classified by their surface area and structure. The former is measured by nitrogen adsorption, the latter by dibutylphthalate absorption. As shown in fig. 1, carbon black is organized in a hierarchical structure, with the spherical primary particles of 20 to 60 nm in diameter representing the smallest length scale. During the production process, groups of 8 to 60 primary particles cluster together to form aggregates. [43, 51, 75, 97, 99] Due to the strong interaction between the primary particles, the aggregates cannot be broken up when mixed into the rubber. Typically during the processing of the rubber composite, aggregates cluster together to form agglomerates, which are more loosely bound. On an even larger length scale, provided the filler content is sufficiently high, a network superstructure can be established.

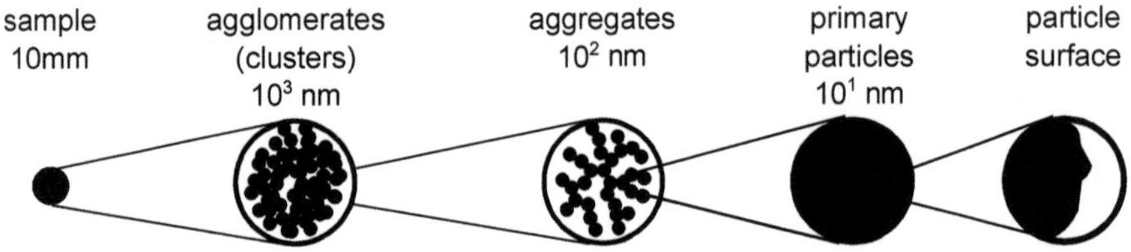

Fig. 1 Schematic representation of the hierarchical structure in a carbon black filled rubber (adopted from Schneider [95])

Due to the polydispersity of the aggregates and wide distribution of characteristic sizes from nanometers to microns, no distinct correlation peaks can be observed in the SAXS and USAXS regime. Instead, a continuous decay of scattering intensity over scattering angle is detected. The most popular approach to interpret the slope of this decay is the fractal model. According to this model, the surface of primary particles is made up of a surface fractal structure, and the organization into aggregates and agglomerates can be represented by mass fractals. The fractal dimension is then directly related to the slope of a logarithmic plot of scattering intensity over scattering vector magnitude. [9, 12, 94]

Assuming a cluster-cluster aggregation process for the formation of aggregates in the furnace reactor, carbon black aggregates should theoretically have a mass fractal dimension $D_{\mathrm{m,ar}}$ of 1.78. [12, 50] Since the agglomerates are clusters of the aggregates, their mass fractal dimension $D_{\mathrm{m,al}}$ must accordingly be smaller than $D_{\mathrm{m,ar}}$. The fractal interpretation is frequently subject to discussion, especially since the power law region in scattering data often extends over less than an order of magnitude. [8, 40] This is not surprising since typically an aggregate is made up of no more than roughly 60 primary particles. It was also shown that the scattering curves are very much affected by polydispersity and the apparent surface fractal region can alternatively simply be explained by an ensemble of polydisperse spheres with a certain size distribution without making any assumptions of a fractal order. [74] Even though the fractal property of the surface is seemingly in line with theoretical models of random deposition surface growth, [50] other interpretations appear in the literature. [91]

However, the widely used model of a self-similar fractal cannot describe anisotropies, which are commonly observed in carbon black filled rubbers under strain . [26, 82, 83, 100] A self-affine fractal can be transformed into a self-similar fractal by an affine transformation along at at least one of the spatial dimensions. However, the definition of the fractal dimension of a non self-similar fractal is not unique. [27, 28] Recently, Schneider developed a model to describe the anisotropy observed in the USAXS patterns of silica filled elastomers under deformation following the concept of self-affine fractals. [95–98]

Due to the complications involved in the interpretation of scattering data from carbon black filled rubbers, we resorted to some geometrically simple model fillers, specifically nanosized aluminum oxide to represent spherical particles and aragonite to represent rodlike particles.

1.5 Cavitation

It is commonly accepted that the presence of cavities in deformed rubber interferes with the mechanical behavior of the material. On the one hand, the energy dissipation involved in the formation of cavities toughens the material. On the other hand, cavities can impair the mechanical properties and the growth of cavities is considered to play a major role in the early stages of crack propagation. [2, 23, 29, 31, 57] Due to its long history and due to the importance of the subject, it has been reviewed several times. [29, 30, 56]

The most widely used method to detect cavitation is dilatometry during deformation. [19, 85, 90, 102, 103, 116] It was pointed out that due to the diffusion properties of rubber, gas dilatometers cannot be used to determine cavitation. [90] Also the superposition of volume changes by cavitation and strain-induced crystallization has to be accounted for. SIC leads to a decrease in volume by up to 2.7 %. [116] Another popular method is scattering, [44, 46, 88, 117] either from X-rays or light. Due to the large difference in electron density between the cavities and the surrounding matrix, cavitation is reflected in a strong increase in total scattering intensity. The presence of elongated cavities is observed as a streak in the scattering pattern. The disadvantage of scattering methods is that they can only detect cavities within a certain narrow size range. Recently, optical methods gained popularity as a method for volume measurement. [55, 58, 59] Furthermore, cavitation was studied by SEM, [57, 101] tomography, [10] acoustic emission analysis [45] and NMR. [2]

Despite the multitude of investigations, no clear picture about the quantity of cavitation could be obtained so far (fig. 2). However, most reports agree that in unfilled rubber, cavitation is negligible. [19, 35, 85, 88, 102] Few

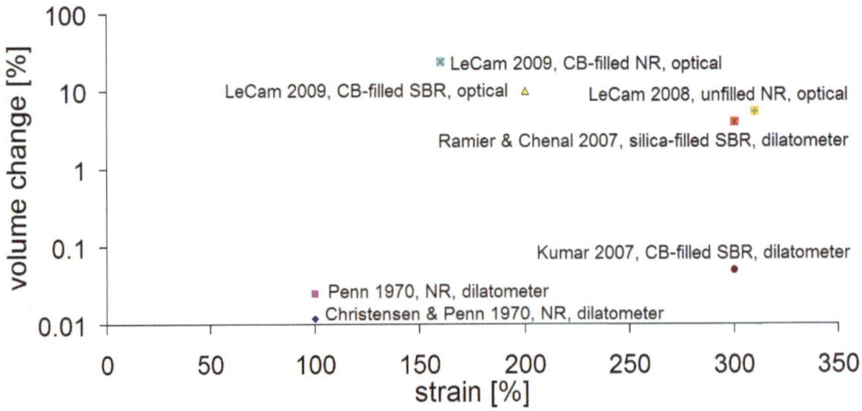

Fig. 2 Literature overview of volume changes in elastomers under strain, measured by various methods and on different materials. Le Cam 2008, [58] Le Cam 2009, [59] Ramier & Chenal 2007, [17, 88] Kumar 2007, [54] Penn 1970, [85] Christensen & Penn 1970 [19]

authors reported a significant volume increase in unfilled rubber, [17, 58, 59] which might be ascribed to the high zinc oxide content in the *unfilled* samples under study. For filled rubbers, most studies reported volume increases between 1 % and 5 %, [17, 19, 88, 90, 102] whereas LeCam found a volume increase by as much as 25 %. The results suggest that cavitation is more prominent when filler loadings are high, the adhesion between filler and matrix is low and the strain is large. [33, 88, 101]

The mechanical criteria for cavitation are still subject to discussion. The most famous approach is due to Gent (1958), who argued that the critical stress is $\frac{5}{6}E$, with E being the tensile modulus. [32] Despite the good agreement with experimental results, he later noted that this simple relation only holds for a certain limited particle size range. [18, 30, 34, 117] A different approach follows the Griffith criterion in the derivation of a cavitation criterion. [29] It supports experimental evidence that a certain minimum initial flaw size is required in order to expand the flaw to a cavity. It is commonly assumed that defects of a size larger than 100 nm exist in any rubber material, such that the cavitation problem can be reduced to the growth of these defects.

In analogy to Zhang et al., [117] we performed USAXS and SAXS experiments to follow cavitational processes in the bulk and around a crack tip.

1.6 Structure-Property Relationships: Crack Propagation

Crack propagation is probably the most instructive case to illustrate the interplay between structure and mechanical properties. It is common knowledge that natural rubber has superior crack resistance over most other rubbers, [36] which is ascribed to strain-induced crystallization. One of the most illustrative demonstrations of the reinforcing ability of SIC in NR was recently given by Zhang et al. [118] They subjected a planar rubber specimen to a uniaxial load and then introduced a cut at one of the strained edges. Contrary to common intuition, the cut growth was slowed down if the sample had been stretched to a sufficiently large extent. This is due to the reinforcing effect of SIC, which sets in above a certain strain threshold at the crack tip.

From an application point of view, crack growth under dynamic fatigue loading is more frequently encountered than static crack growth. The subject of fatigue crack growth has recently been reviewed twice. [73, 86] By definition, mechanical fatigue involves crack nucleation and growth due to fluctuating loads. [73] In the case of rubbers, crack propagation is considered to be more critical than crack nucleation since initial defects are typically present in a material but they can often withstand severe loadings without growth. [93] As pointed out in the previous chapters, crack propagation is typically studied by the application of an oscillating load to a notched rubber sample, oftentimes of pure shear geometry. [70] The most striking impact of SIC is the effect of the R-ratio, which is defined as $R = \frac{P_{\mathrm{min}}}{P_{\mathrm{max}}}$, where P is some mechanical quantity, typically the strain energy density. When increasing R, while maintaining P_{max} constant, the crack growth rate is reduced considerably. [37, 69–73, 93] For instance, Mars pointed out that even $R = 0.06$ can slow down the crack growth rate by two orders of magnitude. [73] In static crack growth, representing the extreme case of $R = 1$, the energy required to drive a crack through NR is 100 times larger than under dynamic conditions. [62] This peculiarity is only observed in strain-crystallizing elastomers. [73] This phenomenon can be explained by two contributions: a) The onset strain and kinetics of SIC play a role. Below a certain strain, NR is amorphous and the crack can propagate more easily. This strain cannot directly be deduced from quasistatic tensile tests or cracktip scans, but the crystallization kinetics also need to be accounted for, since the strain rate at a crack tip under dynamic load is very high. b) SIC indirectly affects the strain field around the crack tip. Due to the large stress-strain hysteresis caused by SIC, the strain field around the crack tip is considerably different under dynamic load compared to quasistatic load. In high-hysteresis materials, the stress concentration around the crack tip is less than commonly thought since relaxing material from behind the crack tip can compensate for the stress build-up at the newly developed crack front. [6]

The consequences of a) were more closely investigated by Saintier et al. [93] They found crack branching, crack deviation and microcrack formation in strain-crystallizing rubber. These processes dissipate a lot of energy and thus reduce the apparent crack growth as measured by the length of the macroscopic crack. Under certain conditions, characteristic striations were observed on the crack surface. [60] These were ascribed to the crack making a detour around the crystalline zone at the crack tip, and hence the authors deduced the size of the crystalline zone from the period of the striations.

A more direct method to study SIC around the crack tip is scanning WAXD. Employing a microsize beam, the sample is scanned and at each position a diffraction pattern is recorded, from which the crystallinity is calculated. Given the large number of diffraction patterns required in a single scan, this technique is only feasible at synchrotron sources. Lee and Donovan were the first to carry out this type of experiment in 1987, however they only performed line scans. [61] They found a crystalline zone of a size of approximately 2 mm. This value depends on the geometry of the sample. Later, Trabelsi et al. [112] found a crystalline zone of roughly the same size. In filled rubbers, due to the lower SIC onset strain, this zone is larger than in gum rubber [61] and thus contributes to a further reinforcement.

Despite the longstanding research, no quantitative relationship between mechanical behavior and microstructure has been established so far. The description of fatigue crack growth in general is purely phenomenological, no physical theory is available. [86] Mars developed an empirical model to describe the influence of SIC on fatigue crack growth. [69] However, it neglects the time-dependency of SIC. Recently, one of the first constitutive models dealing with SIC has been introduced. [53]

In conclusion, one can rephrase the statement by Zhang et al.: [118] *Quantitative models of crystallization and its role in the elastic, viscous and fracture response of rubber remain to be developed.* In order to better understand the role of SIC on crack propagation, we performed quasistatic crack tip scans in a fashion similar to Lee and Donovan and Trabelsi. [61, 112]

2 Experimental Section

2.1 Materials and Sample Preparation

In order to achieve consistency in the materials and to avoid ambiguity in the interpretation of the data obtained by the numerous experimental methods employed by different project members of the FOR597 research unit, experiments were performed on materials from a central material pool whenever possible. In addition, some recipes were specifically designed to serve certain purposes. For instance, the presence of zinc oxide has to be avoided if the material is to be characterized by USAXS, since zinc oxide has a broad particle size distribution and due to its large scattering contrast considerably contributes to the scattering signal in the USAXS region and thus obscures scattering signals from the filler. [79] Because zinc oxide is essential in sulfur curing systems, peroxide was frequently used as a curing agent. Unless the

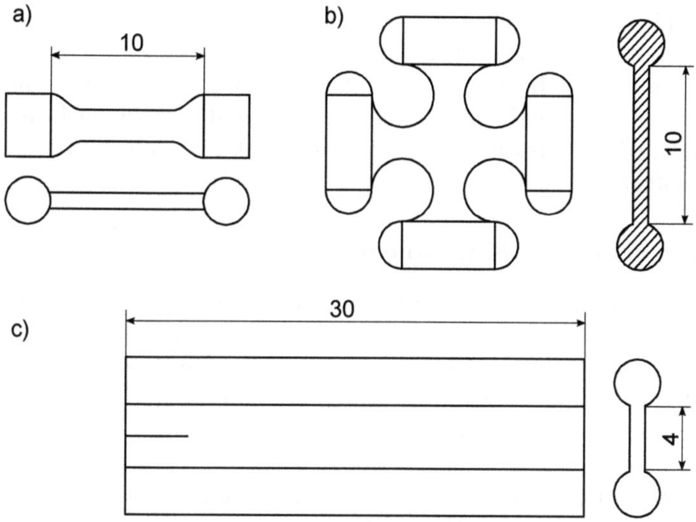

Fig. 3 Schematic representation of the tensile specimens for: a) uniaxial tensile tests, b) biaxial tensile tests, c) pure shear tests and tear fatigue analysis (notched). Dimensions are given in mm. The thickness is 1 mm and the bead diameter is 3 mm.

material was obtained as a ready-to-use compound from the central material pool, it was prepared in a two step process. First, all ingredients except the crosslinker were mixed in an internal mixer (Haake Rheocord PolyLab 300p with HAAKE Rheomix 600p, chamber volume 78 mL, typically 7 min at 70 rpm and 70 °C), afterwards the crosslinker was added on a two-roll mill (Servitec Polymix 110L). Vulcameter tests were performed on a Scarabeus SIS-V50. The vulcanization time was taken as t_{90} and additional time was added depending on the sample thickness and mold geometry. The mold was preheated to the vulcanization temperature before being loaded.

A combined mold was used for pure shear specimens and strips, which were then die-punched to obtain tensile bars. The geometry of the tensile bars is shown in fig. 3 a). The beads in the clamping area minimize sample slippage in the grips. The grips were specifically machined to optimize the form fit between sample and grips. Biaxial tests were performed on a cruciform-shaped specimen, which is shown as obtained from the mold without further cutting or die-punching in fig. 3 b). The pure shear specimen for pure shear tensile tests and tear fatigue experiments is depicted in fig. 3 c).

2.2 Mechanical Experiments

Laboratory Experiments

Mechanical experiments in the laboratory were carried out on an Instron E1000 Electropuls electrodynamic tensile machine. This machine is capable of performing quasistatic and dynamic tests. As in all other experiments, strain data were obtained optically. For uniaxial tests, ink marks were applied on the sample. In order to gain information about the two-dimensional strain field, e. g. around a crack tip, a speckle pattern was airbrushed onto the sample, which was then evaluated using GOM Aramis digital image correlation software.

In-Situ Synchrotron Experiments

All in-situ experiments were carried out on specifically designed tensile machines to meet the requirements imposed by the synchrotron: limited space and weight, versatility without long dead times and full operability by remote control from the control room. A sketch of the beamline setup is shown in fig. 4. The machine for quasistatic testing was developed and built in-house and is depicted in fig. 5. It consists of two electric linear motors, which can be operated independently and allow a maximum displacement of 100 mm and a maximum force of 250 N. During a tensile experiment, the two motors run in opposite directions to hold the sample center fixed with respect to the beam position. The strain rate for quasistatic tests is around $0.005\,s^{-1}$. The machine is equipped with a load cell and can be operated under displacement control as well as under load control. A sliding mechanism allows for fast sample change. The sample can be rotated around two axes (each perpendicular to the X-ray beam) to better explore the reciprocal space.

In order to avoid parallax effects when measuring the strain and aligning the sample with respect to the beam, it is desirable to observe the sample exactly along the beam. To do so, a mirror is placed at a 45° angle between the sample and the guard slits (the final aperture of the X-ray beam). A hole in the mirror lets the X-ray beam pass through. Thus, via the mirror, the sample can be directly observed by a videocamera. The camera as well as the tensile machine are controlled by a software written with the Testpoint software package running on a laptop PC. This PC is remotely controlled from

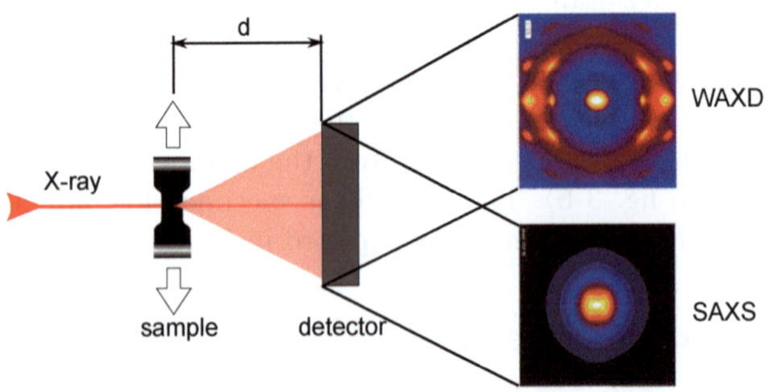

Fig. 4 Setup of in-situ tensile tests at a synchrotron beamline. The detector distance d is variable and is around 150 mm for WAXD and is between 3 m and 14 m for SAXS and USAXS.

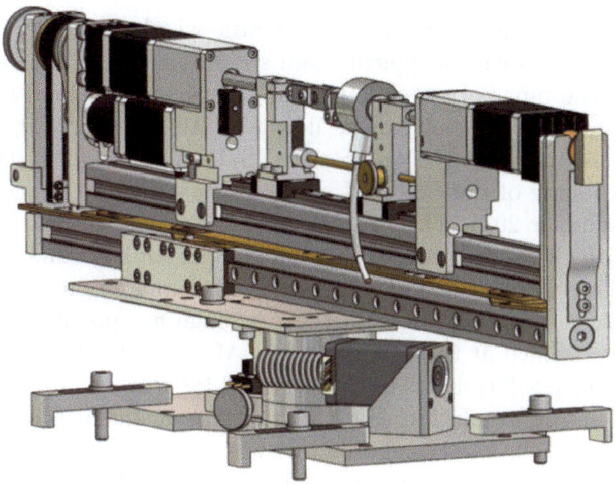

Fig. 5 Miniature tensile testing machine for quasistatic tensile tests at the synchrotron

the control room, via a remote desktop control. The graphic user interface facilitates the synchronous operation of several motors. Furthermore, automatized data processing in the later stages of the data evaluation is facilitated by coupling the tensile machine and the camera to the X-ray detector.

In order to perform biaxial experiments, the quasistatic machine was coupled to a second machine (Kammrath und Weiss, modified) mounted vertically.

Since the maximum displacement rate of the quasistatic tensile machine is 3 mm/s, two additional machines were designed to perform dynamic cyclic tensile tests and impact tensile tests. The dynamic machine is driven by an electric motor connected to an eccentric, which via three levers connects to both grips, stretching the sample in a sinusoidal manner, keeping the sample center fixed with respect to the X-ray beam. The stroke can be adjusted manually before starting the experiment. The sine frequency can be adjusted by the voltage applied to the motor, but was typically kept around 1 Hz.

The tensile impact machine consists of a servomotor acting on a lever which releases a spring that is connected to one of the grips. It stretches the sample to a predetermined displacement within less than 10 ms. The machine can be equipped with a dynamic load cell (Instron dynacell, time resolution 0.5 ms) for experiments in the lab.

2.3 Synchrotron Scattering

Besides a few preparatory experiments on a lab source, all scattering experiments were carried out at Hasylab, DESY (Deutsches Elektronen Synchrotron), Hamburg. Experiments prior to spring 2011 were done at the BW4 beamline at the Doris storage ring, a versatile wiggler beamline for WAXD, SAXS, and USAXS with a typical beam size of 300 µm (FWHM). Exposure times are in the range of seconds (typically between 10 s and 60 s), and thus only quasistatic experiment can be done. Usually, for WAXD, a PSI Pilatus 100K detector was used and for SAXS/USAXS a MarCCD 165. After the commissioning of the Petra III storage ring in late 2010, user operation started in summer 2011. The P03 MiNaXS beamline can be considered the successor of the BW4; however since it is an undulator beamline, the brilliance is increased by three orders of magnitude. In combination with fast detectors, very short exposure times in the ms range are possible. All dynamic tests were performed at the P03 beamline. Also, due to the small beam size (minimum 10 µm × 20 µm), it is ideally suited for scanning experiments with an excellent spatial resolution. For dynamic experiments without the need of a good spatial resolution, the photon rate was further increased by a factor of 10 by increasing the beam size. A PSI Pilatus 300K detector was used with a readout time of 3 ms. For cyclic dynamic tests, an exposure time of 17 ms was used (giving an image aqcuisition rate of $50\,\mathrm{s}^{-1}$). For impact tensile tests, the exposure time was set to 4 ms, resulting in a pattern acquisition rate of approximately $140\,\mathrm{s}^{-1}$.

Data Processing

Data processing is done with the help of self-written code in the digital image processing language PV-wave based on work by Stribeck. [105, 106] Background correction is performed based on reference patterns without a sample. The data streams from the tensile machine (displacement, force, temperature, position), the camera (optical strain), the beamstop photodiode (if available, beam intensity after the sample), the ionization chamber (if available, the beam intensity before the sample), the beamline motors (e.g. scan position) and the X-ray patterns are synchronized and stored in one data package for each pattern containing all relevant information. Patterns are normalized to unit sample volume based on sample transmission. The beamstop and other invalid areas are masked and, if required for further quantitative data evaluation, reconstructed with symmetric data from other quadrants. Fiber symmetry is assumed for tensile samples and has been checked on selected samples.

Selected example diffraction patterns are shown in fig. 6 a) to c). The stretching direction is horizontal along the s_3-axis.[2] The one-dimensional diffraction curve in part d) is obtained from an equatorial slice along the s_{12}-axis. The degree of crystallinity Φ was taken to be proportional to the ratio of the area under the (120) peak to the area under the amorphous halo. This

[2] The scattering vector s is defined as $s = \frac{2\sin\Theta}{\lambda}$, where λ is the wavelength and 2Θ is the scattering angle between the incoming and scattered beam. An alternative definition of the scattering vector, often found in literature, is $q = 2\pi s$. The units of the scattering vector are reciprocal length, usually nm^{-1}.

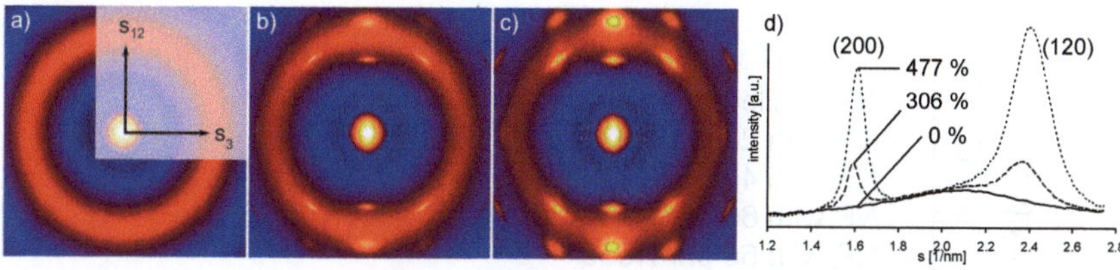

Fig. 6 Two-dimensional diffraction patterns (after processing) of unfilled, sulfur-vulcanized (2 phr) NR at different strains: a) 0 %, b) 306 %, c) 477 %. The stretching direction is horizontal. d) One-dimensional diffraction curves obtained from slices along the equator.

relative number was normalized with literature values from absolute methods (e. g. dilatometry) to obtain an approximate absolute degree of crystallinity.

2.4 Scanning Electron Microscopy

Scanning electron microscopy (SEM) was done on a Zeiss Gemini Ultra Plus. A microtensile machine made in-house was used to perform in-situ tensile tests. Samples with a high carbon black filler content were used to make the sample sufficiently conductive to avoid charge build-up. Coating of the sample is not useful because the conductive layer breaks up when the sample undergoes large deformations during observation.

3 Crystallization under Quasistatic Load

3.1 Characterization of Samples from Central Sample Pool

Filler Content

As is well known from literature, the presence of reinforcing fillers shifts the onset of SIC to lower strains (figs. 7 & 8). This was also observed in the NR samples from the central sample pool (fig. 7). Even though crystallization sets in at lower strains, the crystallinity at a given strain can be lower in filled rubbers as compared to unfilled because the increase in crystallinity with strain is steepest in unfilled rubbers. This behavior can be ascribed to the inhomogeneous strain distribution in filled rubbers. Due to the rigidity of the fillers, the local deformation becomes non-affine and heterogeneous, such that due to the local confinement the strain is locally sufficiently large enough to allow for crystallization, while the external strain is still low.

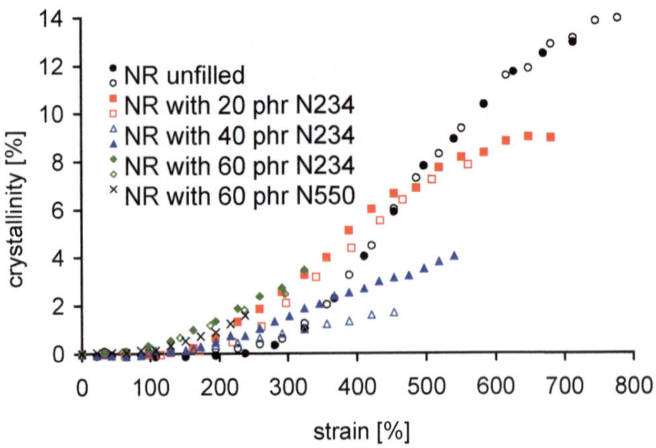

Fig. 7 Crystallinity vs. strain for NR samples with various filler contents. For all samples (except NR with 60 phr N550), two data series from different beamtimes with different detector setups are displayed.

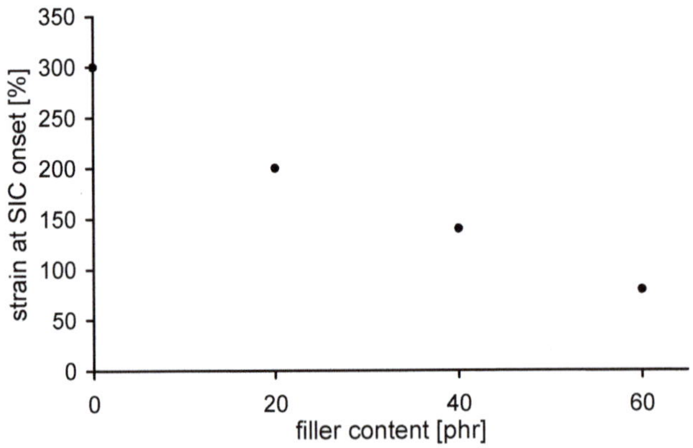

Fig. 8 Onset of strain-induced crystallization vs. filler content for the sample series containing N234 carbon black (unfilled, 20 phr, 40 phr, 60 phr)

Pure Shear vs. Simple Tension Experiments

In agreement with Beurrot et al., [11] we found that the degree of crystallinity under pure shear or simple tensile loads follow the same dependency on strain. The crystallization mechanism appears to be unaffected by the confinement of the lateral contraction to one dimension, which is essentially the difference between simple tensile and pure shear loading. This is exemplified in fig. 9. Usually, the maximum degree of crystallinity in pure shear samples is below

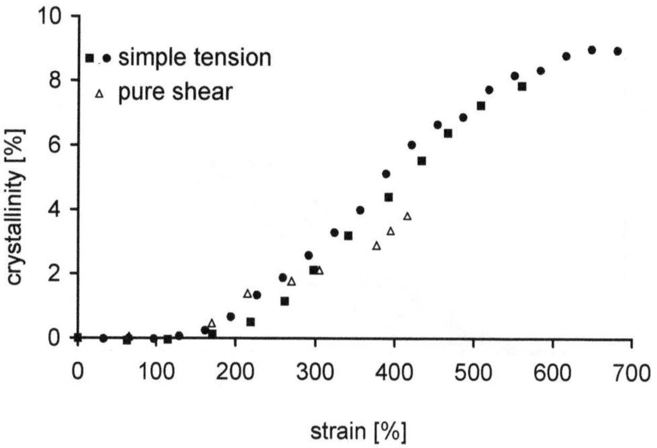

Fig. 9 Comparison of strain-induced crystallization in pure shear and simple tension geometries (fig. 3). NR with 20 phr N234.

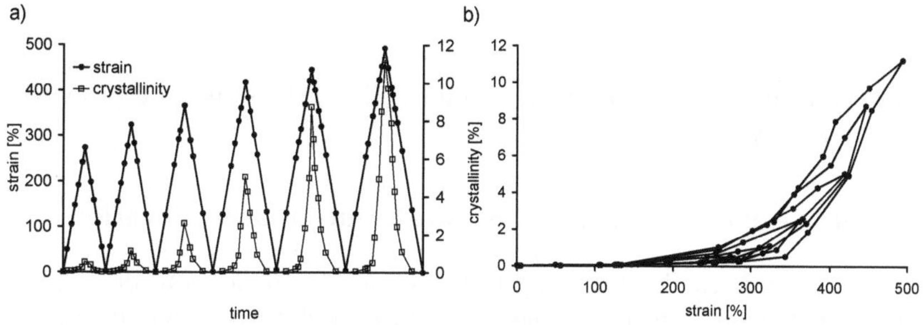

Fig. 10 Strain-induced crystallization during repeated loading in unfilled NR. a) Strain history over time and resulting crystallinity. b) Crystallinity vs. strain.

5 %, because the elongation at break is significantly reduced compared to simple tensile samples.

Crystallization under Repeated Loading

The crystallization process itself is well reproducible and insensitive to previous loading histories. As shown in fig. 10 b), there is some hysteresis between the loading and unloading branch, but the crystallinity curves for any of the loading cycles coincide. When performing the same experiment on carbon black filled rubber, the result clearly reflects the Mullins effect (fig. 11). When loading the sample repeatedly, larger strains are required to obtain the same degree of crystallinity. We hypothesize that local overstrains are reduced from cycle to cycle due to the breakdown of filler clusters. In analogy to

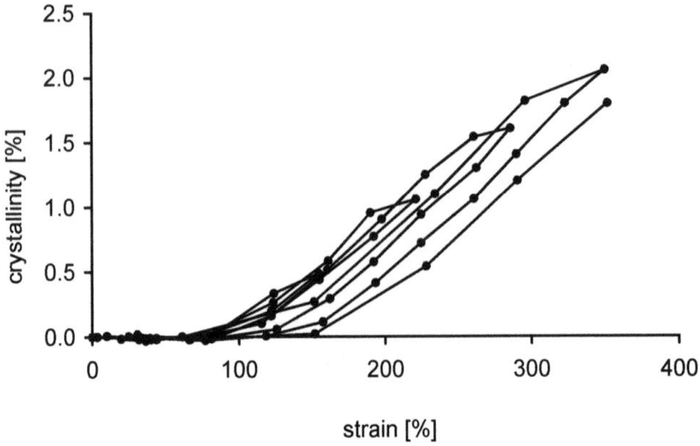

Fig. 11 Strain-induced crystallization during repeated loading in NR filled with 60 phr N234 carbon black. The Mullins effect results in an evolution of crystallinity over the number of loading cycles.

the mechanical consequences of the Mullins effect, the structural properties settle after a few cycles.

Effect of Mechanical Aging on Strain-Induced Crystallization

In order to mimic the loading conditions in real-life applications or in a tear fatigue experiment, tensile samples were subjected to repeated loading at 1 Hz for approximately 1×10^5 cycles. The amplitude was chosen to be as large as possible, such that the sample would not fail before reaching 1×10^5 cycles. In none of the materials was an effect of mechanical aging on the crystallization behavior observed, as exemplified in fig. 12.

Strain in Crystallites

Deformations in the crystalline structure are reflected in a broadening and shifting of the diffraction peaks. Under strain, a shifting of the (120) and (200) peaks towards larger scattering angles 2Θ by approximately 0.15° to 0.3° is observed. At the same time, the (201) peak shifts to smaller angles. In conclusion, the maximum strain in the crystallites along the c-direction is 2 %, while in the transversal a and b-directions, the contraction amounts to roughly 1 %. Thus, the common approximation of an incompressible crystalline phase is well justified. These results are in line with Poompradub et al. [87]

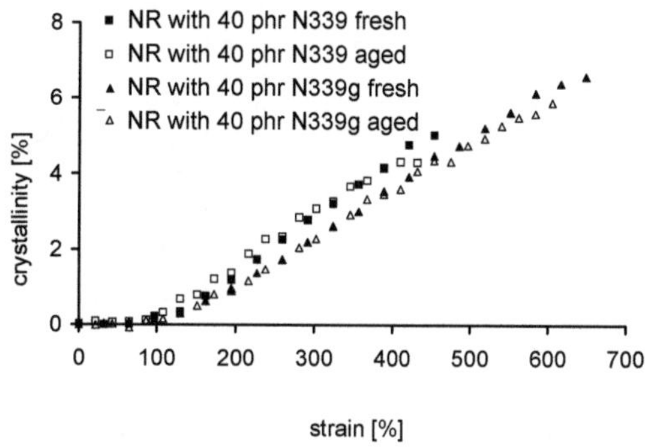

Fig. 12 Comparison of SIC in mechanically aged and fresh samples. *g* stands for graphitized N339 carbon black. (For details regarding the graphitization, see Klüppel. [50])

3.2 Strain Amplification in Filled Rubbers

As outlined in section 1.3, the concept of strain amplification serves to study the micromechanical behavior of the soft phase in a composite system. The degree of crystallinity Φ was employed as a quantity to derive the local strain state in the matrix of a filled elastomer. The underlying assumption is that a unique relation exists between the degree of crystallinity and the strain, and that this relation holds in an unfilled material as well as in the matrix of a filled material. Then one can calculate the strain amplification factor A_Φ from eq. 1. The results are shown in fig. 13. It is clear that a reinforcing filler, such as carbon black, in general has a strain amplificaton factor A_Φ greater than unity, as expected from theory. However, A_Φ is not strain-independent. In general, it decreases with strain. This can be ascribed to the method, taking the degree of crystallinity as relevant quantity. Of course, there is a finite limit of maximum crystallinity, and when this is approached, A_Φ is expected to decrease. What is more interesting is that there is no clear dependency of A_Φ on the filler content. This becomes even clearer when drawing the comparison between carbon black filled materials and aragonite filled ones. Aragonite was chosen as a non-reinforcing filler. According to the simplified hydrodynamic strain amplification model, only the volume content of the filler governs the matrix overstrain, as long as no cavitation occurs. Clearly, in the aragonite filled samples, crystallization is shifted to higher strains by the pure presence of aragonite, as can be seen from the 1 phr-sample, where the low filler loading can be assumed to have a negligible effect on the matrix strain. This shows that the degree of crystallinity should be used with caution to derive the matrix strain.

Thus, we can make the conclusion that hydrodynamic strain amplification might be one of the factors contributing to reinforcement, but it is certainly not the major one.

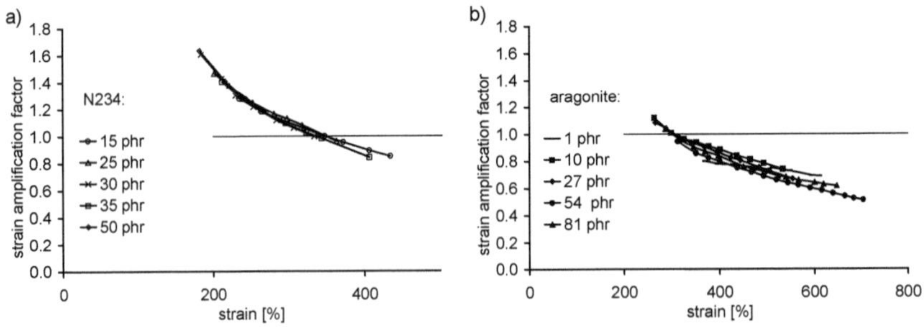

Fig. 13 Strain amplification factor A_Φ derived from strain-induced crystallization. a) carbon black (N234) filled NR, b) aragonite filled NR. The horizontal line at $A_\Phi = 1.0$ is for orientation.

3.3 Local Structure around a Crack Tip

In the context of tear fatigue, the local structure in the vicinity of the crack tip governs the crack propagation behavior. Scanning WAXD with a microbeam is the method of choice to study strain-induced crystallization around the crack tip with a spatial resolution in the micron range. Technically, the resolution is limited by the thickness of the sample, since every diffraction pattern acquired in transmission mode gives an average over the sample thickness at the point of beam incidence. Thus, in order to optimize the spatial resolution, a reduction of the sample size is desirable. On the other hand, this weakens the scattering signal, calling for longer exposure times and hence increasing the probability of beam damage. As a tradeoff a sample of 1 mm thickness was used, combined with a beam size of 25 µm and a scanning stepwidth of 100 µm. The results are presented in fig. 14. The regions subjected to the largest strains also exhibit the highest degree of crystallinity. A further refinement in spatial resolution down to 30 µm revealed a narrow region directly below the crack surface with substantially higher crystallinity.

In notched samples of pure shear geometry (fig. 3), the crystalline region is much smaller and never extends over the complete sample. Instead, catastrophic crack growth occurs. The maximum extension of the crystalline zone in the crack growth direction is around 400 µm. On the other hand, even at external strains as low as 20 %, a small crystalline area at the crack tip is observable.

3.4 Crystallization under Biaxial Load

As compared to classic uniaxial testing, multiaxial strain more closely resembles the loading conditions in real rubber parts. By changing the ratio between the displacements of the two axes, one can gradually move from uniaxial strain to equibiaxial strain. When applying an equibiaxial displacement to the cruciform specimen (fig. 3), the sample undergoes SIC starting in the

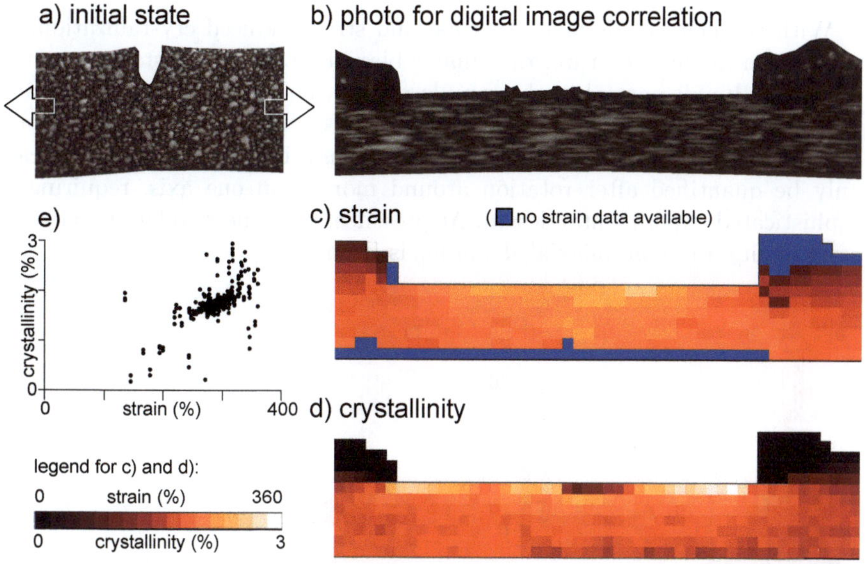

Fig. 14 Scanning WAXD performed on a strained tensile sample with a crack. The sample consists of carbon black reinforced (60 phr N234), sulfur vulcanized (1 phr) natural rubber. a) Photograph of the sample under very moderate strain. The speckle pattern serves for the strain field evaluation by digital image correlation. Arrows indicate the stretching direction. b) Photograph of sample under high strain, the crack has propagated. c) Strain field analysis (strain along the tensile direction). d) Crystallinity obtained by scanning WAXD. Each pixel represents 100 µm × 100 µm. e) Plot of crystallinity over strain extracted from the scan data. The relation is similar to the simple tensile case.

arms and along the edges. In these regions, uniaxial strain prevails and thus fiber symmetry (with a varying orientation of the fiber axis) was assumed. All crystallites are highly oriented along the principal strain directions. One quadrant of the sample stretched by 35 mm along both axes is shown in fig. 15. The figure was obtained by scanning WAXD with a microbeam of 25 µm diameter and a stepwidth of 500 µm in both directions, i. e. each pixel in fig. 15 represents an area of 500 µm × 500 µm in real space. For each of the 999 WAXD patterns, the crystallinity and orientation were computed. Considering that during a scan the material is exposed to the beam for roughly 30 min,[3] larger elongations than 35 mm lead to sample failure during the scan due to beam damage and thus no scans of highly biaxially stretched samples could be obtained. The central part of the cruciform specimen remained amorphous.

[3] The exposure time was set to 0.5 s; the large deadtime is due to the travel speed of the scan devices.

With the help of combined thermal and strain-induced crystallization, it was shown in the literature, that under biaxial strain the crystallites orient with their b-axis normal to the sample plane, whereas the a- and c-axis are oriented randomly. [3, 4, 84] This implies that of course the fiber symmetry no longer holds and that without further assumptions, the crystallinity can only be quantified after rotation around more than one axis, requiring a sophisticated experimental design. At presence, the general relation between crystallinity and non-uniaxial strain fields is unknown.

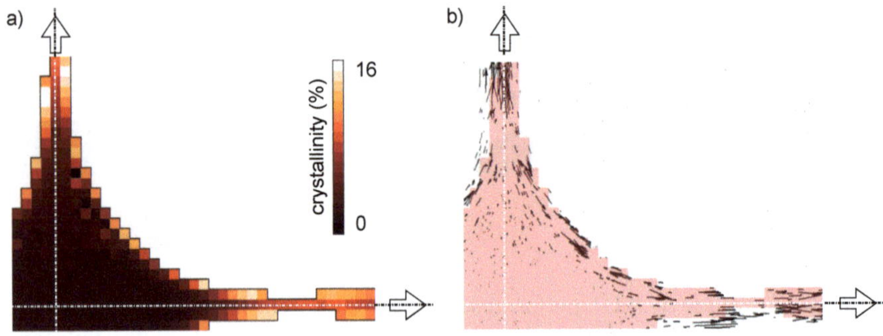

Fig. 15 Crystallinity in an equibiaxially stretched unfilled NR sample. One quadrant of the cruciform sample is shown. Horizontal and vertical displacements are 35 mm each. a) degree of crystallinity, b) orientation of the crystallites (the length of the arrows is proportional to the crystallinity).

4 Kinetics of Strain-Induced Crystallization

Considering that most rubber parts, e. g. tires, are predominantly subjected to dynamic cyclic loads, not only the equilibrium structure, but also the transient structures and the rate of structure formation are important. The mechanical characterization of rubber materials under dynamic conditions is done routinely; however the structural characterization presents a challenge, since it requires experimental techniques with very high time resolution. Owing to the increase in brilliance of third generation synchrotrons by roughly three orders of magnitude over second generation sources, exposure times could be reduced from the second range to the millisecond range over the last 15 years. Along with the sources, the capabilities of X-ray detectors also made significant progress over the last decade. Since 2005, pixel detectors like the Pilatus series developed by the Paul Scherrer Institute (Switzerland) are commercially available and have become more affordable over the years. The Pilatus 300K detector features a readout time of 3 ms, as opposed to several seconds in classic CCD detectors. The readout process does not add any noise to the signal, such that good signal-to-noise ratios can be obtained at very short exposure times, and multiple frames can be combined without accumulating noise. The MiNaXS beamline at Petra III (Hasylab, Hamburg) provides the powerful computer infrastructure needed to process data rates as high as 140 MB s^{-1} over more than a minute. Taking advantage of all these capabilities, WAXD experiments with a pattern acquisition rate of more than 140 fps were performed, corresponding to exposure times of 4 ms. [14]

To tackle the SIC kinetics, two approaches were followed: a) In analogy to pressure jump or temperature jump techniques applied in the study of chemical reaction kinetics, the governing variable of SIC, the strain, was changed at a rate faster than the crystallization rate, and then the system was observed to return to equilibrium over time at constant strain. b) In order to mimic real-life loading conditions more closely, the sample was subjected to a dynamic cyclic strain at a frequency of 1 Hz and the crystallization process was followed with a time resolution capable of resolving each cycle individually as well as following the trend over several cycles. The experiments were made on unfilled rubber (NR), filled rubber (20 phr N234 carbon black, NR-F) and synthetic isoprene rubber (IR).

4.1 Tensile Impact Experiments

Tensile impact experiments were carried out on a specifically designed tensile machine, stretching the sample (fig. 3 a), tensile geometry) to a prescribed strain within less than 10 ms. The synchronicity between the tensile machine and the WAXD patterns was ensured by a shadow on the detector, which would move as the tensile machine moved.

Provided the strain step was sufficiently large, crystalline diffraction peaks were usually observable on the very first pattern after the strain step, i. e. less than 10 ms after completion of the strain step (fig. 16 a). The required strain step heights were 400 % for unfilled NR, 300 % for filled NR and 570 % for IR. Only in a limited region of strains slightly below these values, was crystallization observed to set in after a longer time (fig. 16 b). The crystallinity directly after the strain step is already considerable, around half of the final crystallinity Φ_f after 60 s. Long-time experiments show that beyond 60 s, the crystallinity can be considered constant (fig. 17). After the strain step, crystallization proceeds in an exponential manner and gradually approaches the final level Φ_f, as shown in fig. 18.

The most popular model for the description of crystallization kinetics is the Avrami model:

$$\Phi_{\text{fit}}(t) = \Phi_f \left(1 - \exp\left(-(kt)^{m+\delta}\right)\right). \tag{2}$$

Fitting the experimental data by the Avrami equation yields exponents $m+\delta$ below 0.5, which is incompatible with the physics behind the model. The Avrami model postulates a two step process, consisting of nucleation and growth. It predicts an exponent $m+\delta$ for a linear growth of the crystallites in

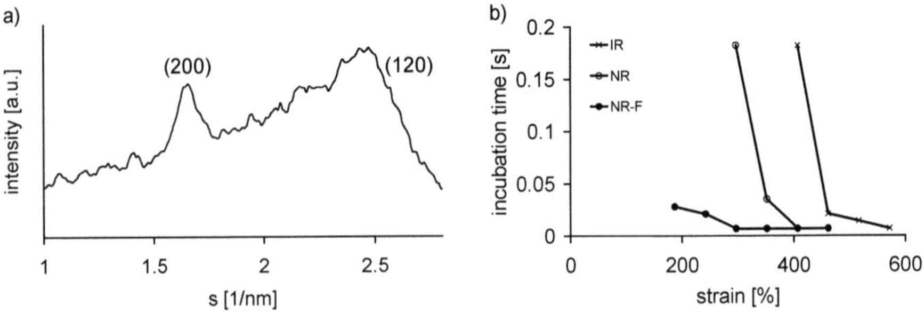

Fig. 16 a) Equatorial diffraction curve of NR, obtained from the first pattern after completion of the strain step. The first signs of (200) and (120) crystalline peaks are clearly visible. b) Incubation times after strain steps in different materials. The incubation time is defined as the time period from completion of the strain step to the detection of the first crystalline peaks.

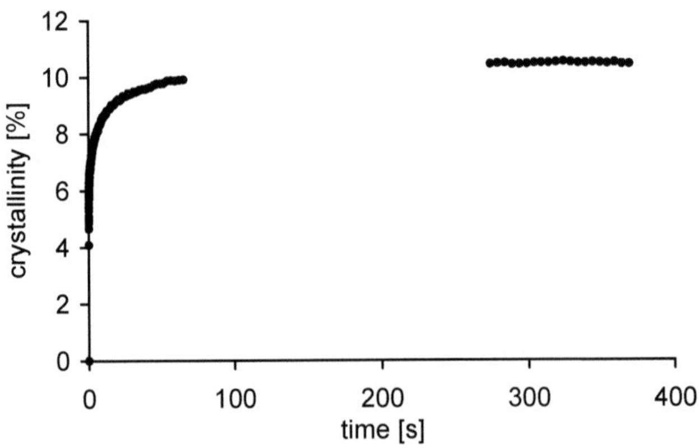

Fig. 17 Crystallinity Φ vs. time in NR for a strain step of 410 %. After 60 s, the crystallinity can be considered stable.

m dimensions, with $\delta = 0$ if the nucleation rate is 0 and $\delta = 1$ if the nucleation rate is constant (and greater than 0). [66] The mismatch between the Avrami model and the experimental data of the SIC kinetics does not come as a surprise, since the Avrami model was intended for thermal crystallization. In fact, as Mandelkern pointed out in 2004, the current understanding of the SIC mechanism is still in its infancy and a lot of work, both theoretical and experimental, remains to be done. [68] Other points disfavoring the use of the Avrami equation are different morphologies (spherulitic vs. fibrillar), different size scales (crystallite sizes are only about 10 nm in SIC), anisotropy due to the strain (reduced diffusion transversal to the strain) and the presence

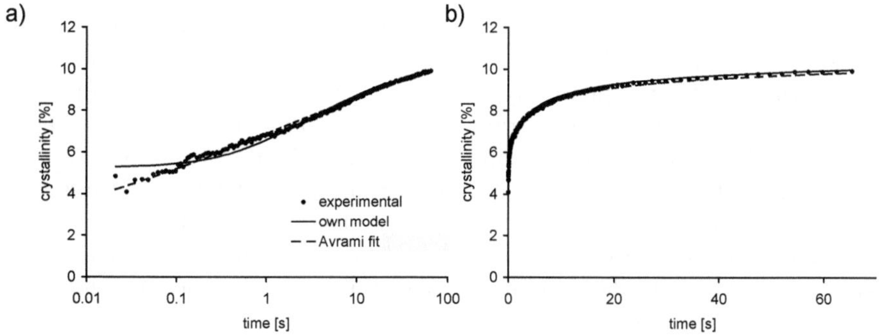

Fig. 18 Crystallinity Φ vs. time in NR for a strain step of 410 % with fits acc. to eq. 4 (solid line) and Avrami fit (dashed line, $m + \delta = 0.18$): a) logarithmic time scale, b) linear time scale

of crosslinks (rendering large scale diffusion impossible). Other mechanisms suggested in the literature include an instantaneous coil-to-stretch transition, originally put forward by De Gennes and later modified for the application to networks by Hsiao. [21, 104] He proposed that in networks the transition takes place in chain segments between crosslinks. In fact, the crystallites observed in SIC are smaller than the distance between crosslinks. In line with this model, we postulate that the crystallization mechanism consists of two steps: First an instantaneous nucleation and second a growth phase. The growth rate is proportional to the number of crystallizable chain segments to the power of n, where n describes the propensity of amorphous chains towards crystallization:

$$\frac{d\Phi}{dt} = \left(k\left(\Phi_{\mathrm{f}} - \Phi\right)\right)^n \tag{3}$$

Integration gives

$$\Phi(t) = \Phi_{\mathrm{f}} - \left((n-1)\,k^n t + \left(\Phi_{\mathrm{f}} - \Phi_0\right)^{1-n}\right)^{\frac{1}{1-n}} . \tag{4}$$

Φ_0 is the crystallinity after the nucleation step, t is the time and k is a proportionality constant. The quality of the fit is demonstrated in fig. 18 a). The crystallinity after nucleation, Φ_0, and the final crystallinity, Φ_{f}, are plotted over strain in fig. 19. The exponent n was found to be close to 4 in all cases. Despite being so far off equilibrium after the strain step, the material approaches almost the same final state as in quasistatic experiments, i. e. the crystallinity measured one minute after the strain step, Φ_{f}, is close to, but slightly lower than, the crystallinity values obtained from quasistatic tensile tests, Φ_{qs} (fig. 19 b).

Application of the Scherrer equation to the diffraction data yields an increase in crystallite size over time in the a and b crystallographic directions, whereas no growth takes place in the c direction, which corresponds to the tensile direction (fig. 20 a)). This can be explained by the arrangement of the crystallites. Similar to shish structures formed in flow-induced crystallization, the fibrils in SIC are apparently made up of small individual crystallites of roughly 10 nm in size, which arrange in rows to form the fibrils. This means that growth can only take place in the radial a and b directions. SEM observations (fig. 20 b)) suggest that the fibrils are around 20 to 50 nm in diameter, which is significantly larger than the sizes derived by the Scherrer equation. The reason for the discrepancy could be that either fibrils are made up of several crystallites in their cross section, or the crystalline core is surrounded by oriented amorphous material.

The melting kinetics are much faster than the crystallization kinetics. After step-like unloading (within less than 10 ms) of any of the materials under investigation, the crystalline peaks disappeared instantaneously on the experimental accessible time scale, i. e. within approximately 10 ms. This is exemplified for IR in fig. 21.

4.2 Dynamic Cyclic Experiments

Dynamic cyclic experiments were carried out on a selfmade electrodynamic tensile machine at a frequency of roughly 1 Hz. Fig. 22 a) shows crystallinity over time for a cyclic experiment performed on unfilled NR. First, for $t < 0$ s, the sample was held at a constant strain of 365 % and a crystallinity of almost 6 % was observed. After starting the cyclic experiment (cycling between 0 % strain and 365 % strain) at $t = 0$ s, the crystallinity did not reach the same level as during the initial quasistatic part. Instead, the maximum crystallinity remained below 2 %. This can be ascribed to the finite crystallization kinetics described in the previous section. Because in each cycle the sample resides above the SIC onset strain for only a fraction of a second, time is too short to

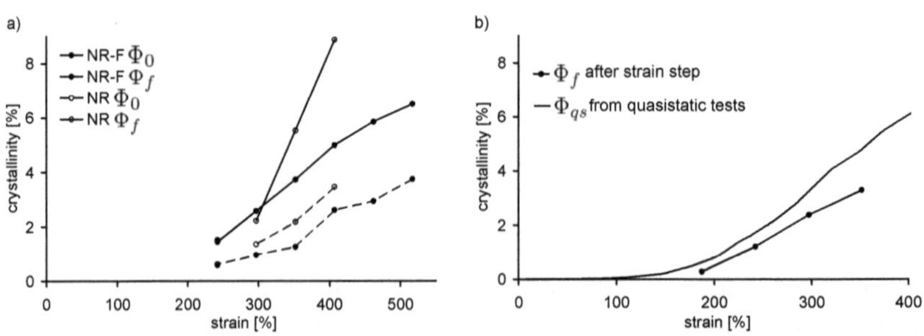

Fig. 19 a) Initial crystallinity Φ_0 (dashed lines) and final crystallinity Φ_f (solid lines) vs. strain step height for NR (open symbols) and NR-F (filled symbols). b) Final crystallinity after step strain Φ_f in comparison with crystallinity Φ_{qs} obtained from quasistatic experiments in NR-F.

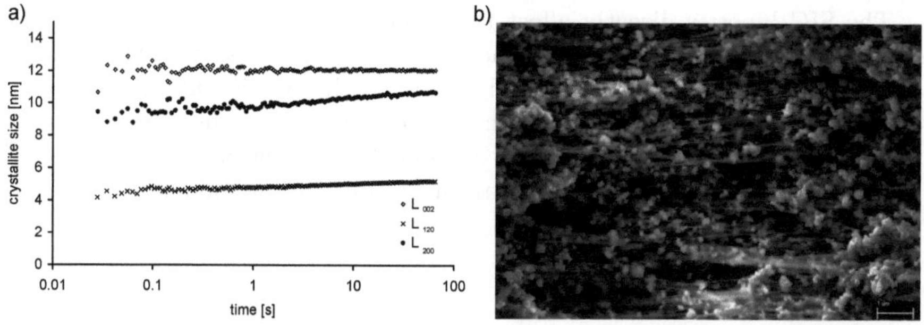

Fig. 20 a) Crystallite size L_{hkl} vs. time after a strain step of 410 % in NR. b) SEM micrograph of highly stretched NR filled with 150 phr N772 carbon black.

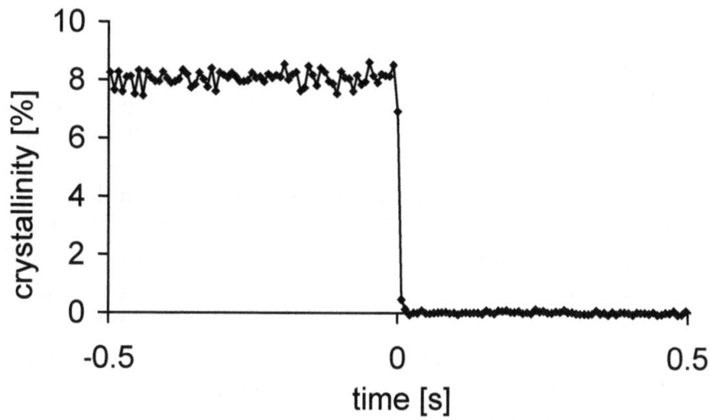

Fig. 21 Crystallinity vs. time in IR after steplike unloading from 500 % to 0 % at $t = 0$ s. The crystallites disappear instantaneously. The time between two consecutive points is 7 ms.

let the crystallinity reach the quasistatic level. Upon unloading, the sample becomes amorphous, as expected. No accumulation of crystallinity occured over time when the sample was extensively cycled.

Not only the maximum strain but also the minimum strain determine crystallinity. This is shown in fig. 22 b), where the sample was cycled between 290 % and 365 % strain - the same maximum strain as in the previous experiment. The crystallinity ranges around 4 %, which is still less than the quasistatic level of 6 %, but above the previous cyclic level of 2 %, when the sample was unloaded completely. Again, no accumulation of crystallinity over the number of cycles was observed. Hence these results disprove the hypothetical accumulation of crystallinity around a crack tip under non-relaxing loading conditions.[93]

The SIC kinetics directly affect the mechanical behavior of NR. For instance, the disparity between the enormous resistance to static crack growth and the reduced resistance under dynamic conditions, is one of the consequences of the crystallization kinetics. In more general terms, the SIC kinetics add an unusual time-dependent component to the mechanical behavior of NR, which should also find its expression in constitutive models.

5 Filler Orientation

In order to facilitate the understanding of the behavior of carbon black in a deformed matrix, it is advisable to start with model fillers to circumvent the problems associated with the evaluation of anisotropic carbon black scattering patterns. The advantage of scattering experiments on geometrically simple scattering entities is that their scattering patterns can be calculated analytically. Further criteria for the selection of model fillers are their size, which should be similar to the size of carbon black aggregates, such that the structural changes can be observed at similar scattering angles and that the deformation of the matrix occurs on a comparable length scale. Moreover, a good scattering contrast between filler and matrix is desirable.

Given these requirements, nanosized aluminum oxide Al_2O_3 (Nanophase Nanodur) was chosen to represent spherical particles and aragonite (Schaefer Kalk), a form of precipitated calcium carbonate, was chosen to represent high aspect ratio cylindrical particles. [13] The beauty of these fillers is that they are monocrystalline substances, meaning their orientation can independently be followed by WAXD. Scanning electron micrographs of the raw fillers and after incorporation into the natural rubber matrix are shown in fig. 23. The filler particles are sufficiently well dispersed and only few agglomerates are visible. The samples were crosslinked with dicumyl peroxide to avoid the

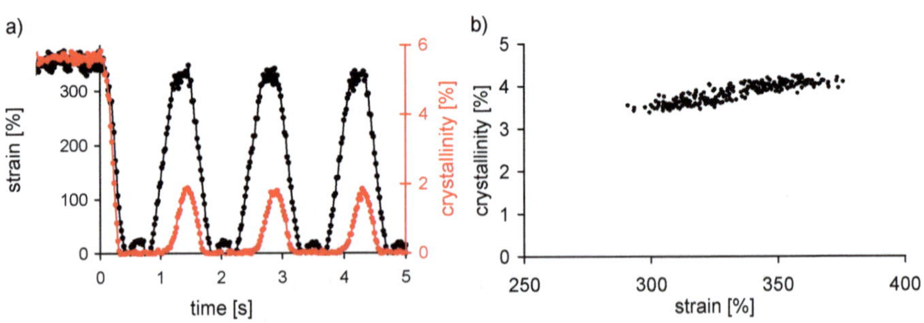

Fig. 22 a) Crystallinity (red) and strain (black) over time, NR cycled between 0 % and 365 % strain. b) Crystallinity vs. strain, NR cycled between 290 % and 365 % strain.

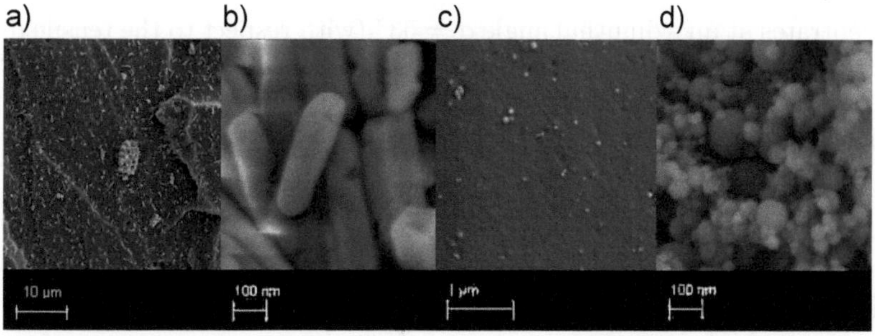

Fig. 23 Scanning electron micrographs of a) NR with 27 phr aragonite (scale bar 10 μm), b) aragonite powder (scale bar 100 nm), c) NR with 10 phr Al_2O_3 (scale bar 1 μm), d) Al_2O_3 powder (scale bar 100 nm) [13]

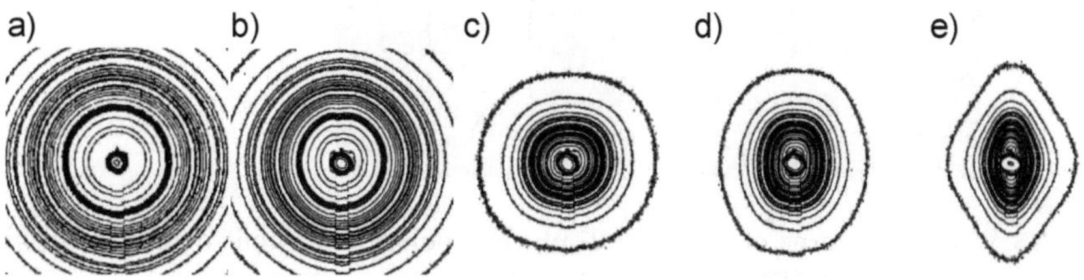

Fig. 24 USAXS patterns of NR with 10 phr Al_2O_3: a) undeformed, b) 500 % strain; NR with 27 phr aragonite: c) undeformed, d) 150 % strain, e) 500 % strain. The maximum scattering vector q is 0.12 nm^{-1}. Stretching direction is horizontal [13].

scattering of zinc oxide, which contributes to the SAXS and USAXS signals since its size is in the same range as that of the fillers. [79]

The USAXS patterns of NR with 10 phr Al_2O_3 are shown in fig. 24 a) and b). As expected, the pattern remains isotropic under strain due to the spherical geometry of the filler. The same conclusion can be drawn from the WAXD patterns, where the Al_2O_3 rings have a constant intensity over the azimuthal angle. In contrast to this, the high aspect ratio aragonite filler orientates under strain. This is reflected in the anisotropy of the USAXS patterns (fig. 24 c-e)). The iso-intensity lines turn from isotropic circular to ellipsoidal and finally, at large strains, assume a lozenge shape. The orientation is fully reversible when the load is removed.

In order to interpret these patterns, the information obtained from WAXD is particularly instructive. Under deformation, the aragonite diffraction rings turn into arcs, and increasingly narrow down at certain azimuthal angles, designating an orientation of the particles along the tensile direction (fig. 25). The orientation of the (111) plane is analyzed in fig. 26. Due to the position of the (111) plane in the crystallograpic lattice, the diffraction intensity concentrates at an azimuthal angle $\alpha = 53°$ (with respect to the tensile direction) when the long axis of the cylinders is aligned with the tensile direction. The orientational distribution of the diffraction intensity can be well fit by a Maier-Saupe function:

$$f_{\mathrm{MS}}(\alpha) = \frac{1}{z} \exp\left(m \cos^2(\alpha)\right), \qquad (5)$$

where $\frac{1}{z}$ is a normalization factor. The orientation parameter m increases almost linearly with strain, as shown in fig. 26 b).

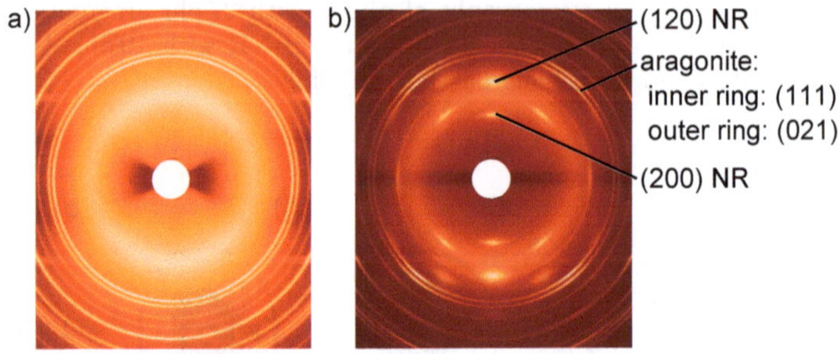

Fig. 25 WAXD patterns of NR with 27 phr aragonite. a) without deformation, b) at 550 % strain. Stretching direction is horizontal.

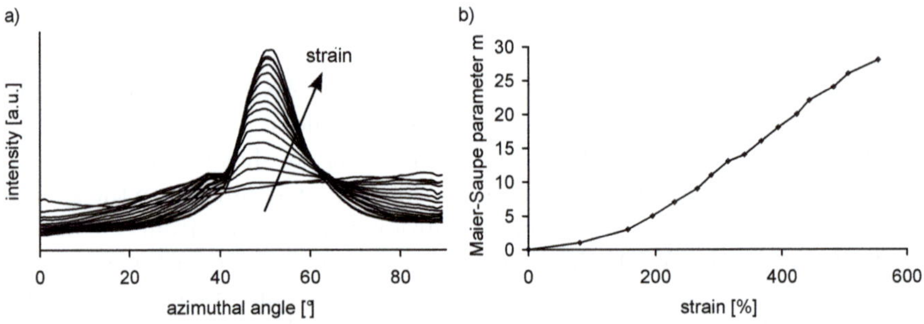

Fig. 26 a) Intensity of the (111) peak of aragonite over the azimuthal angle α in NR filled with 27 phr aragonite. The strain increases from 0 % to 550 %. The kink at 40° is an artifact due to a blind area at the boundary between two panels of the detector. b) Maier-Saupe orientational distribution parameter derived from a) with eq. 5 [13].

With this information one can compute the USAXS patterns if the geometry of the filler particles is known. Assuming a Gaussian size distribution of the cylinder length l, one obtains

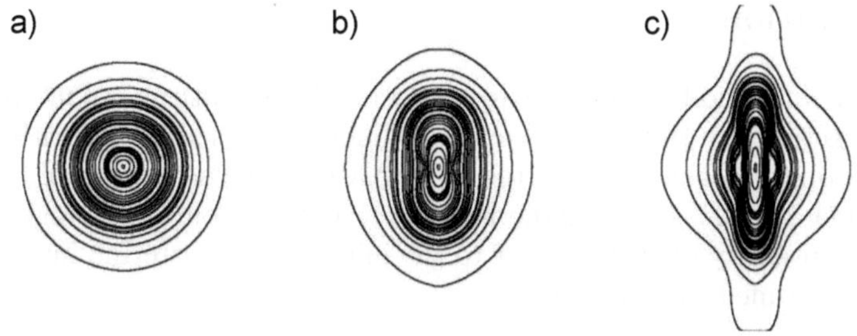

Fig. 27 Simulated scattering patterns acc. to eq. 8 for rubber filled with cylindrical particles: a) no deformation, b) 150 % strain, c) 500 % strain. The maximum scattering vector q is $0.12\,\mathrm{nm}^{-1}$. Stretching direction is horizontal [13].

$$D(l) = \frac{1}{\sigma_l\sqrt{2\pi}} \exp\left(\frac{-\left(l-\bar{l}\right)^2}{2{\sigma_l}^2}\right) \tag{6}$$

with the standard deviation σ_l and an average length $\bar{l}$. From SEM images, the length was taken as $\bar{l} = 600\,\mathrm{nm}$ with a standard deviation $\sigma_l = 100\,\mathrm{nm}$. The aspect ratio a was kept fixed at 6. The size distribution was cut off at $\bar{l} \pm 3\sigma_l$.

The form factor of a cylinder, f_{cyl}, as a function of azimuthal angle α and scattering vector $q = 4\pi\frac{\sin\Theta}{\lambda}$ is given by

$$f_{\mathrm{cyl}} = k\frac{a\sin\left(q\frac{l}{2}\cos(\alpha)\right) J_1\left(q\frac{l}{2a}\sin(\alpha)\right)}{q^2l^2\cos(\alpha)\sin(\alpha)}, \tag{7}$$

where J_1 is the first order Bessel function of the first kind and k is a constant. [39, 63] Considering the moderate volume fraction of the aragonite filler (27 phr correspond to 7.5 vol %), it is well justified to approximate the structure factor as unity and take the scattering function of the polydisperse oriented ensemble simply as the sum of all inidividual contributions:

$$I_{\mathrm{cyl}} = \int_{\alpha=0}^{\pi} \left((1-\phi) f_{\mathrm{MS}}(\alpha) + \phi\right) f_{\mathrm{cyl}}(l,\alpha)^2 d\alpha, \tag{8}$$

where ϕ is the isotropic, unoriented portion of the particles, e.g. due to agglomeration. ϕ was taken as a fitting parameter and kept fixed at 0.2. The scattering patterns obtained from eq. 8 are depicted in fig. 27. The qualitative agreement with the experimental patterns (fig. 24) is obvious.

6 Cavitation

Cavitation and void formation[4] can be studied by a multitude of methods, as listed in section 1.5. But only few of these offer a bulk sensitivity with a good spatial resolution. Thus we resorted to USAXS, taking advantage of the work by Zhang et al., [117] deriving the void fraction ϕ_{void} from the change in total scattering intensity Q with respect to the initial total scattering intensity Q_0 of the undeformed sample: [44]

$$\frac{Q}{Q_0} = 1 + \left(\frac{\phi_{\mathrm{r}}{\rho_{\mathrm{r}}}^2 + \phi_{\mathrm{f}}{\rho_{\mathrm{f}}}^2}{\phi_{\mathrm{r}}\phi_{\mathrm{f}}\left(\rho_{\mathrm{r}} - \rho_{\mathrm{f}}\right)^2} - 1 \right) \phi_{\mathrm{void}} \tag{9}$$

Here, ρ_{r} is the scattering length density of the rubber phase ($8.756 \times 10^{10}\,\mathrm{cm}^{-2}$), ρ_{f} is the scattering length density of the filler phase ($15.26 \times 10^{10}\,\mathrm{cm}^{-2}$ for carbon black) and ϕ_{r} and ϕ_{f} are the volume fractions of the rubber and filler phase, respectively. The total scattering intensity Q is taken with respect to a unit volume and is normalized with the intensity of the incoming beam. Fiber symmetry is assumed in the integration over the reciprocal space.[5] Integration is carried out only over the experimentally accessible range of scattering vectors, i. e. no extrapolation to lower scattering angles was done.

We confirmed that in unfilled natural rubber, no increase in total scattering intensity could be detected in the USAXS regime[6] and that this scattering regime is insensitive to crystallinity. The absence of cavities is in line with the literature, reporting constant volume for unfilled rubbers under deformation. Likewise, in moderately carbon black-filled rubbers (up to 30 phr) and in rubbers filled with non-reinforcing fillers (calcium carbonate, aluminum oxide), no cavitation was observed. Only in SBR and NR filled with at least 40 phr carbon black, an increase in total scattering intensity was found (fig. 28 e)). The increase occurs over a broad range of scattering vectors, hinting at a broad size distribution of cavities and rendering the evaluation of void fraction relatively insensitive to the exact choice of the region of scattering angles over which the integration is performed to obtain the total scattering intensity. Investigating the local structure around a crack tip by scanning USAXS[7] with a spatial resolution of 100 µm showed that cavitation occurs in a small zone directly in front of the crack tip (fig. 28 a-d)). The maximum

---

[4] The terms *cavitation* and *void formation* are often used interchangeably, however some authors prefer to use *voids* to describe stable hollow regions in thermoplastics and other glassy materials, whereas *cavities* is used in the context of elastomers for hollow regions growing in an unstable manner. [29]

[5] Taking the tensile axis as symmetry axis s_3, fiber symmetry is expressed as $s_1 = s_2 = s_{12}$.

[6] experiments at the BW4 beamline at DESY, $0.003\,\mathrm{nm}^{-1} < s < 0.04\,\mathrm{nm}^{-1}$

[7] experiments at the MiNaXS beamline at DESY, beam size 25 µm, $0.005\,\mathrm{nm}^{-1} < s < 0.16\,\mathrm{nm}^{-1}$

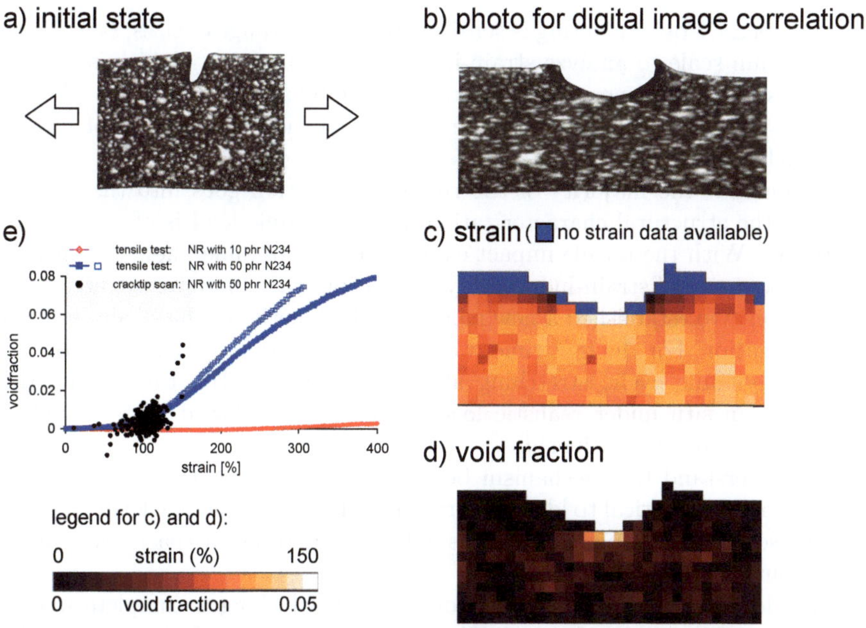

Fig. 28 Scanning USAXS performed on a strained tensile sample with a crack. The sample consists of carbon black reinforced (50 phr N234), peroxide vulcanized (1 phr) natural rubber. a) Photograph of the sample under very moderate strain. The speckle pattern serves for the strain field evaluation by digital image correlation. Arrows indicate the stretching direction. b) Photograph of sample under high strain, the crack has propagated. c) Strain field analysis (strain along the tensile direction). d) Void fraction obtained by scanning USAXS. Each pixel represents 100 µm × 100 µm. e) Plot of void fraction over strain obtained from simple tensile experiments and from the scan.

detected void fraction was around 0.05, but this value may vary with the step size of the scan. The magnitude of the void fraction is comparable to the values observed in a bulk tensile sample under uniaxial strain (fig. 28 e)). Similar values were also obtained for the same material under equibiaxial load.

7 Conclusion

This chapter conclusively illustrated that in order to gain a fundamental understanding of the mechanical behavior of reinforced rubbers, it is essential to characterize the structure in situ under mechanical load. X-ray scattering is one of the few methods that can afford this; it further boasts representative quantitative results and bulk sensitivity. Thanks to the brilliance of the latest synchroton sources, also fast phenomena can be studied as well as local phenomena. X-ray scattering is sensitive to a broad range of sizes, starting at the sub-nm scale to analyze strain-induced crystallization - which is critical for the self-reinforcement of natural rubber, going all the way to the USAXS regime, giving new insights into the filler orientation on the scale of several hundred nanometers.

Given that the majority of the rubber production goes into the tire industry, the structural characterization under dynamic load is of utmost importance. With the tensile impact experiments we obtained new insight into the early stages of strain-induced crystallization with an unprecedented time resolution. We established a new model for the kinetics of the strain-induced crystallization. For the first time, we could follow the crystallization directly during dynamic cyclic stretching at a frequency of 1 Hz. Elucidating the structure in situ under realistic loading conditions allowed us to establish structure-property relationships.

To understand the mechanism behind tear fatigue in strain-crystallizing elastomers, it is critical to identify structural changes at the crack tip on small length scales. WAXD cracktip scans with an unparalleled spatial resolution contributed to this understanding.

A field with a longstanding history, but still lacking a complete understanding, is the study of the reinforcement mechanisms of fillers, especially highly structured fillers like carbon black. Using model fillers, we established a model that explains the interplay between external deformation and the filler orientation on a local scale. This remains to be carried over to industrially relevant fillers.

The current opinion in the literature on the role and magnitude of cavitational processes is very diverse. We confirmed the existence of cavities in certain filled rubbers and for the first time quantified the cavitation in the vicinity of a crack tip.

Despite all the progress made over the last years, a complete quantitative understanding, allowing the development of novel physically motivated constitutive models to predict the mechanical, fatigue and fracture behavior of filled rubbers, is still in the future.

Acknowledgement. Portions of this research were carried out at the light sources DORIS and PETRA III at DESY, a member of the Helmholtz Association (HGF). We would like to thank J. Perlich, S.V. Roth, A. Buffet, A. Rothkirch and G. Santoro for assistance in using the BW4 and MiNaXS beamlines.

References

1. Acken, M.F., Singer, W.E., Davey, W.: Industrial & Engineering Chemistry 24(1), 54–57 (1932)
2. Adriaensens, P., Pollaris, A., Vanderzande, D., Gelan, J., White, J.L., Kelchtermans, M.: Macromolecules 33(19), 7116–7121 (2000)
3. Alam, M.M., Asano, T.: Journal of Macromolecular Science, Part B 47(4), 754–764 (2008)

4. Alam, M.M., Osaka, H., Asano, T.: Journal of Macromolecular Science, Part B: Physics 47(2), 317–337 (2008)
5. Andrews, E.H.: Proceedings of the Royal Society of London. Series A, Mathematical and Physical Sciences 270(1341), 232–241 (1962)
6. Andrews, E.H.: Journal of the Mechanics and Physics of Solids 11(4), 231–242 (1963)
7. Andrews, E.H.: Proceedings of the Royal Society of London. Series A, Mathematical and Physical Sciences 277(1371), 562–570 (1964)
8. Avnir, D., Biham, O., Lidar, D., Malcai, O.: Science 279(5347), 39 (1998)
9. Bale, H.D., Schmidt, P.W.: Physical Review Letters 53(6), 596–599 (1984)
10. Bayraktar, E., Isac, N., Bessri, K., Bathias, C.: Fatigue & Fracture of Engineering Materials & Structures 31(2), 184–196 (2008)
11. Beurrot, S., Huneau, B., Verron, E.: Proceedings of the VII. European Conference on Constitutive Models for Rubber, pp. 23–28 (2011)
12. Botet, R., Jullien, R.: Phase Transitions: A Multinational Journal 24(2), 691–736 (1990)
13. Brüning, K., Schneider, K., Heinrich, G.: Journal of Polymer Science Part B: Polymer Physics 50(24), 1728–1732 (2012)
14. Brüning, K., Schneider, K., Roth, S.V., Heinrich, G.: Macromolecules 45(19), 7914–7919 (2012)
15. Bunn, C.W.: Proceedings of the Royal Society of London. Series A. Mathematical and Physical Sciences 180(980), 40 (1942)
16. Candau, N., Chazeau, L., Chenal, J.M., Gauthier, C., Munch, E., Rochas, C.: Polymer 53, 2540–2543 (2012)
17. Chenal, J.M., Gauthier, C., Chazeau, L., Guy, L., Bomal, Y.: Polymer 48(23), 6893–6901 (2007)
18. Cho, K., Gent, A.N.: Journal of Materials Science 23(1), 141–144 (1988)
19. Christensen, R., Hoeve, C.: Rubber Chemistry and Technology 43, 1473–1481 (1970)
20. Dalal, E.N., Taylor, K.D., Phillips, P.J.: Polymer 24(12), 1623–1630 (1983)
21. De Gennes, P.G.: Journal of Chemical Physics 60(12), 5030–5042 (1974)
22. Domurath, J., Saphiannikova, M., Ausias, G., Heinrich, G.: Journal of Non-Newtonian Fluid Mechanics 171, 8–16 (2012)
23. Dorfmann, A., Fuller, K.N.G., Ogden, R.W.: International Journal of Solids and Structures 39(7), 1845–1861 (2002)
24. Dunning, D.J., Pennells, P.J.: Rubber Chemistry and Technology 40, 1381 (1967)
25. Dupres, S., Long, D.R., Albouy, P.A., Sotta, P.: Macromolecules 42(7), 2634–2644 (2009)
26. Ehrburger-Dolle, F., Hindermann-Bischoff, M., Livet, F., Bley, F., Rochas, C., Geissler, E.: Langmuir 17(2), 329–334 (2001)
27. Falconer, K.J.: Fractal geometry. Wiley & Sons (1990)
28. Feder, J.: Fractals. Plenum Press (1988)
29. Fond, C.: Journal of Polymer Science Part B: Polymer Physics 39(17) (2001)
30. Gent, A.N.: Rubber Chemistry and Technology 63(3), 49–53 (1990)
31. Gent, A.N., Lindley, P.B.: Rubber Chemistry and Technology 31(2), 393–394 (1958)
32. Gent, A.N., Lindley, P.B.: Proceedings of the Royal Society of London. Series A. Mathematical and Physical Sciences 249(1257), 195–205 (1959)

33. Gent, A.N., Park, B.: Journal of Materials Science 19(6), 1947–1956 (1984)
34. Gent, A.N., Tompkins, D.A.: Journal of Applied Physics 40, 2520 (1969)
35. Goebel, J.C., Tobolsky, A.V.: Macromolecules 4(2), 208–209 (1971)
36. Hainsworth, S.V.: Polymer Testing 26(1), 60–70 (2007)
37. Harbour, R.J., Fatemi, A., Mars, W.V.: Fatigue & Fracture of Engineering Materials & Structures 30(7), 640–652 (2007)
38. Harwood, J.A.C., Mullins, L., Payne, A.R.: Journal of Applied Polymer Science 9(9), 3011–3021 (1965)
39. Hayter, J.B., Penfold, J.: The Journal of Physical Chemistry 88(20), 4589–4593 (1984)
40. Herd, C.R., McDonald, G.C., Hess, W.M.: Rubber Chemistry and Technology 65, 107 (1992)
41. Hiratsuka, H., Hashiyama, M., Tomita, S., Kawai, H.: Journal of Macromolecular Science, Part B 8(1-2), 101–126 (1973)
42. Huneau, B.: Rubber Chemistry and Technology 84(3), 425–452 (2011)
43. Hyeon-Lee, J., Beaucage, G., Pratsinis, S.E., Vemurys, S.: Langmuir 14(20), 5751–5756 (1998)
44. Jánosi, A.: Monatshefte für Chemie/Chemical Monthly 114(4), 383–387 (1983)
45. Kakavas, P.A., Chang, W.V.: Journal of Applied Polymer Science 45(5), 865–869 (1992)
46. Kaneko, S., Frederick, J.E., McIntyre, D.: Journal of Applied Polymer Science 26(12) (1981)
47. Katz, J.R.: Naturwissenschaften 13(19), 410–416 (1925)
48. Kawahara, S., Takano, K., Yunyongwattanakorn, J., Isono, Y., Hikosaka, M., Sakdapipanich, J.T., Tanaka, Y.: Polymer Journal 36(5), 361–367 (2004)
49. Kawai, H.: Rheologica Acta 14, 27–47 (1975)
50. Klüppel, M.: Advances in Polymer Science 164, 1–86 (2003)
51. Koga, T., Hashimoto, T., Takenaka, M., Aizawa, K., Naoya Amino, O., Nakamura, M., Yamaguchi, D., Koizumi, S.: Macromolecules 41(2), 453–464 (2008)
52. Krigbaum, W.R., Dawkins, J.V., Via, G.H., Balta, Y.I.: Journal of Polymer Science Part A-2 Polymer Physics 4(3), 475–489 (1966)
53. Kroon, M.: Mechanics of Materials 42(9), 873–885 (2010)
54. Kumar, P., Fukahori, Y., Thomas, A.G., Busfield, J.J.C.: Journal of Polymer Science Part B: Polymer Physics 45(23), 3169–3180 (2007)
55. Layouni, K., Laiarinandrasana, L., Piques, R.: Compressibility induced by damage in carbon black reinforced natural rubber. In: Proceedings of the Third European Conference on Constitutive Models for Rubber, September 15-17, p. 273. Taylor & Francis, London (2003)
56. Le Cam, J.B.: Rubber Chemistry and Technology 83, 247 (2010)
57. Le Cam, J.B., Huneau, B., Verron, E., Gornet, L.: Macromolecules 37(13), 5011–5017 (2004)
58. Le Cam, J.B., Toussaint, E.: Macromolecules 41(20), 7579–7583 (2008)
59. Le Cam, J.B., Toussaint, E.: Mechanics of Materials 41(7), 898–901 (2009)
60. Le Cam, J.B., Toussaint, E.: Macromolecules 43(10), 4708–4714 (2010)
61. Lee, D.J., Donovan, J.A.: Rubber Chemistry and Technology 60(5), 910–923 (1987)
62. Lindley, P.B.: International Journal of Fracture 9(4), 449–462 (1973)
63. Lindner, P., Zemb, T.: Neutrons, X-rays, and Light: Scattering Methods Applied to Soft Condensed Matter. Elsevier North-Holland (2002)

64. Luch, D., Yeh, G.S.Y.: Journal of Macromolecular Science, Part B 7(1), 121–155 (1973)
65. Magill, J.H.: Rubber Chemistry and Technology 68, 507 (1995)
66. Mandelkern, L.: Biophysical chemistry 112(2-3), 109–116 (2004)
67. Mandelkern, L.: Crystallization of Polymers Volume 1: Equilibrium concepts, vol. 1. Cambridge University Press (2004)
68. Mandelkern, L.: Crystallization of Polymers Volume 2: Kinetics and Mechanisms, vol. 2. Cambridge University Press (2004)
69. Mars, W.V.: Rubber Chemistry and Technology 82, 51 (2009)
70. Mars, W.V., Fatemi, A.: International Journal of Fatigue 24(9), 949–961 (2002)
71. Mars, W.V., Fatemi, A.: Rubber Chemistry and Technology 76(5), 1241–1258 (2003)
72. Mars, W.V., Fatemi, A.: Fatigue & Fracture of Engineering Materials & Structures 26(9), 779–789 (2003)
73. Mars, W.V., Fatemi, A.: Rubber Chemistry and Technology 77, 391–412 (2004)
74. Martin, J.E.: Journal of Applied Crystallography 19(1), 25–27 (1986)
75. Medalia, A.I., Heckman, F.A.: Carbon 7(5), 567–568 (1969)
76. Mitchell, G.R.: Polymer 25(11), 1562–1572 (1984)
77. Mitchell, J.C., Meier, D.J.: Journal of Polymer Science Part A-2: Polymer Physics 6(10), 1689–1703 (1968)
78. Monaco, A.I.C.T.G., Tonelli, A.E.: Macromolecules 38(4), 1223–1232 (2005)
79. Morfin, I., Ehrburger-Dolle, F., Grillo, I., Livet, F., Bley, F.: Journal of Synchrotron Radiation 13(6), 445–452 (2006)
80. Natta, G., Corradini, P.: Angewandte Chemie 68(19), 615–616 (1956)
81. Nyburg, S.C.: Acta Crystallographica 7(5), 385–392 (1954)
82. Oberdisse, J., Hine, P., Pyckhout-Hintzen, W.: Soft Matter 3(4), 476–485 (2007)
83. Oberdisse, J., Rharbi, Y., Boué, F.: Computational and Theoretical Polymer Science 10(1-2), 207–217 (2000)
84. Oono, R., Miyasaka, K., Ishikawa, K.: Journal of Polymer Science: Polymer Physics Edition 11(8), 1477–1488 (1973)
85. Penn, R.W.: Journal of Rheology 14, 509 (1970)
86. Persson, B.N.J., Albohr, O., Heinrich, G., Ueba, H.: Journal of Physics of Condensed Matter 17(44), R1071–R1142 (2005)
87. Poompradub, S., Tosaka, M., Kohjiya, S., Ikeda, Y., Toki, S., Sics, I., Hsiao, B.S.: Journal of Applied Physics 97, 103529 (2005)
88. Ramier, J., Chazeau, L., Gauthier, C., Stelandre, L., Guy, L., Peuvrel-Disdier, E.: Journal of Materials Science 42(19), 8130–8138 (2007)
89. Rault, J., Marchal, J., Judeinstein, P., Albouy, P.A.: Macromolecules 39(24), 8356–8368 (2006)
90. Reichert, W.F., Hopfenmüller, M.K., Göritz, D.: Journal of Materials Science 22(10), 3470–3476 (1987)
91. Ruland, W.: Carbon 39(2), 323–324 (2001)
92. Ryu, S.R., Sung, J.W., Lee, D.J.: Rubber Chemistry and Technology 85(2), 207–218 (2012)
93. Saintier, N., Cailletaud, G., Piques, R.: Materials Science and Engineering: A 528(3), 1078–1086 (2011)

94. Schmidt, P.W.: Journal of Applied Crystallography 24(5), 414–435 (1991)
95. Schneider, G.J.: Analyse der Struktur von aktiven Füllstoffen mittels Streumethoden. PhD thesis, Universität Regensburg (2006)
96. Schneider, G.J.: The Journal of Chemical Physics 130, 234912 (2009)
97. Schneider, G.J., Göritz, D.: Journal of Applied Crystallography 43(1), 12–16 (2009)
98. Schneider, G.J., Göritz, D.: The Journal of Chemical Physics 133, 024903 (2010)
99. Schneider, G.J., Vollnhals, V., Brandt, K., Roth, S.V., Göritz, D.: The Journal of Chemical Physics 133, 094902 (2010)
100. Shinohara, Y., Kishimoto, H., Inoue, K., Suzuki, Y., Takeuchi, A., Uesugi, K., Yagi, N., Muraoka, K., Mizoguchi, T., Amemiya, Y.: Journal of Applied Crystallography 40, s397–s401 (2007)
101. Shinomura, T., Takahashi, M.: Rubber Chemistry and Technology 43, 1015–1024 (1970)
102. Shinomura, T., Takahashi, M.: Rubber Chemistry and Technology 43, 1025–1035 (1970)
103. Shuttleworth, R.: European Polymer Journal 4(1), 31–38 (1968)
104. Somani, R.H., Yang, L., Zhu, L., Hsiao, B.S.: Polymer 46(20), 8587–8623 (2005)
105. Stribeck, N.: Fibre Diffraction Review 6, 20–24 (1997)
106. Stribeck, N.: X-ray Scattering of Soft Matter. Springer (2007)
107. Takahashi, Y., Kumano, T.: Macromolecules 37(13), 4860–4864 (2004)
108. Toki, S., Sics, I., Ran, S., Liu, L., Hsiao, B.S., Murakami, S., Tosaka, M., Kohjiya, S., Poompradub, S., Ikeda, Y., et al.: Rubber Chemistry and Technology 77, 317 (2004)
109. Tosaka, M., Kawakami, D., Senoo, K., Kohjiya, S., Ikeda, Y., Toki, S., Hsiao, B.S.: Macromolecules 39(15), 5100–5105 (2006)
110. Tosaka, M., Senoo, K., Kohjiya, S., Ikeda, Y.: Journal of Applied Physics 101, 084909 (2007)
111. Tosaka, M., Senoo, K., Sato, K., Noda, M., Ohta, N.: Polymer 53(3), 864–872 (2012)
112. Trabelsi, S., Albouy, P.A., Rault, J.: Macromolecules 35(27), 10054–10061 (2002)
113. Trabelsi, S., Albouy, P.A., Rault, J.: Macromolecules 36(20), 7624–7639 (2003)
114. Trabelsi, S., Albouy, P.A., Rault, J.: Macromolecules 36(24), 9093–9099 (2003)
115. Treloar, L.R.G.: The physics of rubber elasticity, 3rd edn. Oxford Classic Series (1975)
116. Wood, L.A., Bekkedahl, N.: Journal of Applied Physics 17, 362 (1946)
117. Zhang, H., Scholz, A.K., de Crevoisier, J., Vion-Loisel, F., Besnard, G., Hexemer, A., Brown, H.R., Kramer, E.J., Creton, C.: Macromolecules 45(3), 1529–1543 (2012)
118. Zhang, H.P., Niemczura, J., Dennis, G., Ravi-Chandar, K., Marder, M.: Physical Review Letters 102(24), 245503 (2009)

Morphology and Micro-mechanics of Filled Elastomer Blends: Impact on Dynamic Crack Propagation

Hagen Lorenz, Dagmar Steinhauser, and Manfred Klüppel*

Deutsches Institut für Kautschuktechnologie e. V., Eupener Straße 33
D-30519 Hannover, Germany
manfred.klueppel@dikautschuk.de

Abstract. Reinforcement of elastomers by nanoscopic fillers like carbon black (CB) is essentially determined by the structure formed by the filler particles and their interaction with the polymer. Accordingly, morphological investigations and micromechanical models are shown to be key factors for understanding the dynamic crack growth of filler-reinforced elastomers and high performance rubber blends. The non-linear dynamic behaviour is intensively studied and described by the Dynamic Flocculation Model (DFM) which is based on the breakage and re-aggregation of filler clusters. Physically meaningful material parameters characterizing the microstructure of filled elastomers are determined for different polymer and filler types at various filler concentrations. The model is shown to be useful for the interpretation of crack propagation measurements with the Tear Fatigue Analyser. In particular, the strength of filler-filler bonds and the residual set behaviour as well as its relaxation is shown to have a significant impact on fracture mechanical properties. Applications of the model with the Finite Element Method (FEM) focus on the numerical calculation of stress softening in a notched specimen close to the crack tip. Systematic decrease of the Paris-Erdogan coefficient *A* and exponent *B* with increasing content of CB are detected for different rubbers: styrene butadiene (SBR), ethylene-propylene-dine (EPDM) and natural rubber (NR). Thereby, an anomaly seems to take place in the vicinity of the gel point of the filler network where *B* depends on the chosen tearing energy range. The morphology of fracture surfaces is correlated with crack growth rates, giving important hints about the nature of crack growth in dependence of crack velocity and filler content. Beside the different homo-polymers, with only one rubber component, also blends are examined. SBR/EPDM blends show large phase domains which are visible even in the light microscope and a low amount of interphase. At low tearing energy, cracks in the blend propagate faster than in the single rubbers and also crack resistence is not improved by CB. Contrary, SBR/NR blends show higher interphase volume fractions and the resistence

* Corresponding author.

W. Grellmann et al. (Eds.): *Fracture Mechanics & Statistical Mech.*, LNACM 70, pp. 81–128.
DOI: 10.1007/978-3-642-37910-9_3

against dynamic crack propagation is located between those of the single rubbers (unfilled as well as filled). The analysis of CB distribution in the phases and interphases exhibit characteristic differences depending on the compatibility of the respective rubbers which also impacts the fatigue crack propagation behaviour of rubber blends.

Keywords: Elastomers, Fracture Mechanics, Reinforcement, Micro-Mechanical Modelling, Finite Element Method.

1 Introduction: Fracture Mechanics of Reinforced Elastomers

1.1 Reinforcement of Rubber by Active Fillers

Cross-linked rubber may be the main but is in no way the only essential part of elastomer components. On the one hand side, macroscopic structures like steel cord or fabric serve for the mechanical reinforcement. But, only by structures like active fillers that occupy the whole polymer matrix on a nanoscopic level, a fatigue and wear resistant material can be created [1]-[3]. Filler particles in the nanometre scale can have various effects in elastomers: for instance, they increase hardness and serve as phase intermediate agents between non-miscible polymers [4], [2]. It was around the year 1900 that S. C. Mote of India Rubber Gutta Percha & Telegraph Works Co. (England) realized that carbon black (CB) could reduce in particular wear of rubber products significantly. CB consists of aggregates which are made up statistically of a number of "sintered" primary carbon particles. In order to rationalize the mechanism of reinforcement, there were efforts to measure mechanical properties of fillers and to correlate with properties of filled vulcanizates, since the 1940s.

The so-called “mechanical reinforcement” is achieved by fillers, most notably when the particle diameters are below a typical value of about 100 nm [5]. Another important factor is the so-called activity of the filler surface which means essentially the (physical) binding energy with organic molecules. In general, activity increases with decreasing particle size. However, even if the term "reinforcement" is widely used, it is not uniquely defined. Kraus [6] speaks of reinforcement, if viscoelastic and fracture mechanical properties like modulus and strength are influenced positively without impairing the reversible extensibility. Either if dealing with wear or rupture of a component - cracks always play a pivotal role. Active fillers do not only hamper crack propagation but also crack initiation. An experiment giving insight into these processes was carried out by Hess et al. who analyzed microtome cuts and copies of stained SBR vulcanizates with 50 phr CB under an electron microscope. The finer the CB, the higher stresses the specimen survived without aggregates separating from the matrix. With HAF CB (where the primary particle size is about 30 nm) this were about 10 MPa; at 23 MPa, 20% of the aggregates were separated [7]. Measurements of vacuole forming showed similar results: while in a strained vulcanizate with coarse CB (N 880) the volume grew by several percent, HAF showed hardly any volume increase [5].

The reinforcement by active fillers is always connected with an increase of stress relaxation and mechanical losses. Also, it seems that the relaxation is not only a by-product of the large polymer-filler interface (about 90 m^2/g for HAF CB) but also the cause of the reinforcement of fracture mechanical properties. Since many years a correlation is known between energy dissipation and the critical fracture energy [8] as well as abrasion resistance [9]. On the one hand side, the necessary energy for crack propagation is mainly spent for viscoelastic deformation of the material in the vicinity around the crack tip. On the other hand side, the ability of the particles to deviate the crack has to be recognized [9], [10], [3]. A third factor of importance is the residual deformation at the tip of the crack which results in a time-dependent change of the crack or notch radius. How the notch radius controls crack propagation will be discussed below, starting from linear-elastic fracture mechanics. The reinforcement of elastomers by nanoscopic fillers is essentially determined by the structure formed by the filler particles and their interaction with the polymer. Accordingly, beside crack growth and morphological investigations, micromechanical models are key factors for understanding the dynamic crack growth of filler-reinforced elastomers as well as high performance rubber blends. Though, an extensive literature exists today, there is only a partial consensus in a sense that the fundamental mechanism is connected with bonds by which filler particles are connected.

The Dynamic Flocculation Model (DFM) was developed with the aim of describing the reinforcment mechanisms on a microscopic level leading to the strongly non-linear mechanical behaviour. The bonds between aggregated filler particles (clusters) consist of nanoscopic polymer layers, and their behaviour in respondance to time-dependent mechanical loading and temperature affects the overall behaviour of the filled elastomer. On the next higher structural level, this leads to the breakage and re-aggregation of filler clusters. A closer explanation will be given in the 3rd section, followed by an evaluation of micro-mechanical material parameters for various elastomer materials by fitting multi-hysteresis stress-strain cycles to the DFM [11]. We will use these parameters for a discussion of the fatigue crack propagation behaviour obtained for pulsed loading under high severity conditions. Thereby, we will demonstrate that the characteristics of the filler network are closely related to the fracture mechanical properties of filled elastomers.

Since many years there have been efforts of a quantitative characterization and physical understanding of crack propagation and tear resistance in hyper-visco-elastic solids, most of them are based on the fundamental work of Rivlin & Thomas [12], and some of the most recent being [10], [13]. However, there is still a high scientific and technological potential for the development of fracture mechanical models of filler-reinforced elastomers which take into full account the strongly non-linear deformation behaviour [14]-[17]. The micro-mechanical mechanisms of crack initiation and propagation in elastomer materials are still not exactly known, in particular, how these processes start and how they proceed under dynamic loading conditions [17]. An extended knowledge shall lead to a better understanding of fatigue crack propagation under high severity conditions and will allow for a more fundamental, physically motivated prediction of service life properties of high performance elastomer products like tires, conveyor belts, seals, air springs and dampers.

1.2 Linear-Elastic Fracture Mechanics

Here we will focus first on linear-elastic fracture mechanics. Following the concept of Griffith [18], the energy needed to form the crack surfaces has to be released in the form of potential energy from the loaded body when the crack propagates through the body. Here we assume that a small inner crack of length $2c$ is present in a loaded sample as shown in Fig. 1. For crack growth by $\mathrm{d}c$ the relesed elastic energy - $\mathrm{d}W$ must be larger than the surface energy $\mathrm{d}S$ (Griffith Criteria): $\frac{\mathrm{d}(W+S)}{\mathrm{d}c} \leq 0$, provided no external work is done, i.e. the length of the sample is constant. By taking $W = b_0 \frac{\pi . \sigma^2 c^2}{E}$ and $S = 4b_0 c \gamma_{\mathrm{eff}}$ one obtains the instability criteria: $\frac{2\pi}{E}\sigma^2 c \geq 4\gamma_{\mathrm{eff}}$, from which the external critical stress results as $\sigma_{\mathrm{c}} = \sqrt{2E\gamma_{\mathrm{eff}} / \pi c}$. Here γ_{eff} is the effective surface energy which may be much larger than the surface energy obtained from contact angle measurements. This observation can be related to two effects: First, the energy to crack polymer chains is larger than the value obtained from contact angle measurements. Second, a large amount of energy dissipation occurs around the crack tip in viscoelastic materials [19]. The crack propagation criteria, derived from energetic balance, considers the body as a whole and is a necessary condition for unstable crack propagation. However, the energy density at the crack tip must be high enough that chemical bonds can be broken irreversibly and viscoelastic losses can be compensated.

Cracks are initiated, in general, at defects like voids and microcracks which form at non-dispersed agglomerates. These defects act as stress concentrators. Under small deformation, cracks can be described by linear-elastic fracture mechanics - under finite deformations of an elastomer, this can only be regarded as an approximation, because elastic as well as geometric non-linearities appear. For a crack tip of infinitesimal curvature radius, a singularity of stresses results [20]. The stress tensor components depend on the distance r from the tip like: $\sigma_i = \frac{K_I}{\sqrt{2\pi\, r}} f_i(\theta)$, where $f_i(\theta)$ are angular functions and K_I is the stress intensity factor for uniaxial loading (mode I) [21].

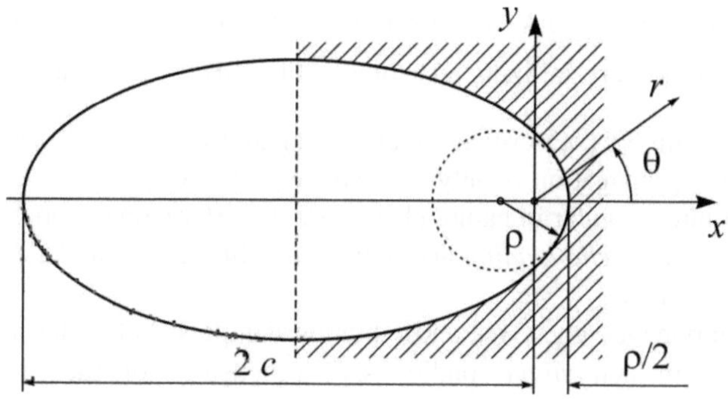

Fig. 1 Elliptical inner crack (notch) with radius of curvature ρ

For a small inner crack of lenght $2c$ under uniaxial loading stress σ, an approximate solution was found by Irwin: $K_I = \sigma\sqrt{\pi\, c}$, which also holds for a crack with finite radius (notch) [22], [23]. Along the symmetry axis (θ = 0) of a notch with curvature radius ρ, the axial stress in a distance r (measured from the centre between curvature centre and apex) is: $\sigma_y = \frac{K_I}{\sqrt{2\pi\, r}}\left[1+\frac{\rho}{2r}\right] = \sigma\sqrt{\frac{c}{2r}}\left[1+\frac{\rho}{2r}\right]$, therefore, the maximum stress at the apex (r = $\rho/2$) is:

$$\sigma_{\max} = 2\sigma\sqrt{\frac{c}{\rho}}\,. \qquad (1)$$

One realizes that failure does not only depend on the intrinsic strength of the material but, most of all, on the geometry of existing cracks: long and sharp cracks are the most critical concerning the attainable strength. As we will see later, the failure criterion at the crack tip as for the specimen as a whole is controlled by mechanical energy. When stress is expressed by the stored energy density in linear-elastic approximation: $\sigma = \sqrt{2Ew}$, according to eq. (1) we see that the relation of maximum to global energy $w_{\max}/w$ density is proportional to c/ρ - similar to a lever rule.

The rate of crack propagation $\frac{\mathrm{d}c}{\mathrm{d}t}$ in a specific elastomer is controlled (beside temperature etc.) by the (elastic) energy release rate $T_{\mathrm{el}} = \left.\frac{\mathrm{d}W}{\mathrm{d}S}\right|_l$ at constant length l of the specimen, also called *tearing energy*. As Rivlin & Thomas showed, crack propagation does not depend on the global specimen geometry explicitly but only on T_{el} and on the geometry of the crack tip [12]. T_{el}, in turn, depends on the global geometry and loading conditions, but is independent of the material.

On the one hand, crack propagation can be analyzed in statically loaded specimens. In non-crystallizing elastomers like SBR, $\frac{\mathrm{d}c}{\mathrm{d}t}$ generally increases continuously with T_{el} (sub-critical crack propagation). In crystallizing elastomers like NR, crack propagation is only seen above a relatively high load, the process itself being quite irregular in nature [17], [19], [10]. Here, on the other hand, crack propagation under *cyclic loading* and defined amplitude of ε, and T_{el} respectively, shall be investigated.

Many components, e.g. tyres and vibration dampers, fail under a many times repeating load. In such *fatigue* conditions, the propagation of cracks is generally of crucial importance. Compared to crack propagation expected at static loading, an additional progress can result during the increasing flank of the load cycle [19], [24]. An increment dc of the crack length can amount to some nano- or micrometers and only result in visible crack progress after a number n of many thousands of cycles. Crack propagation depends on many experimental factors, as static preload or idle time after a load pulse, which are hardly examined systematically until today. As a rule, a dependency of crack growth dc during a load cycle on the maximum of T_{el} is found, which is characterized by a power law according to Paris & Erdogan [25]:

$$\frac{\mathrm{d}c}{\mathrm{d}n} = A T_{\mathrm{el}}^{B} . \tag{2}$$

The cycle duration plays a much smaller role on the crack increment. Because the relation is independent of specimen geometry, eq. (2) is a constitutive law where the parameters A and B vary in dependence of the investigated material.

In the following we want to examine structure-property-relationships responsible for crack propagation in *unfilled* elastomers. The dependence on filler content will later be analyzed systematically. In particular, the connection between Paris-Erdogan parameters and parameters of the microstructure-based model for the filled elastomers shall be regarded. In qualification, it shall be stated that eq. (2) is valid only over a growth rate range of a few decades. Below a critical value of $T_{\mathrm{el}} = G_0$ (in absence of ozone which attacks the main chains chemically) there is no detectable crack growth. For various elastomers a critical value of $G_0 \approx 50\ \mathrm{J/m^2}$ is found which can be tracked down to the chemical binding strength of the chains and the molecular structure of the vulcanizate [19].

The crack resistance is also dependent on the cross-linking system. Short and stable C-C bonds rather raise the sensitivity to cracking while, on the other side of the spectrum, polysulfidic S_8-bonds have a tendency to prevent cracking. The reason seems to be related to the lower strength of the latter: when a bond breaks, macromolecules are left undamaged, but fewer restraining conditions exist for them implying higher extensibility [24]. The effective cross-link density decreases, sliding movements between molecules and relaxation intensify. The crack front is not one-dimensional, there is rather a process zone which stretches into the volume around the tip.

The relaxation of stress can be described by a scaling function of time. However, because cracking is rather controlled by energy than stress and because of the complex stress state at the crack tip, not the same scaling exponents are found in the rate dependence of the tearing energy. Some approaches exist which allow for a correlation of relaxation time spectra with the rate dependence of tearing energy: [10], [13].

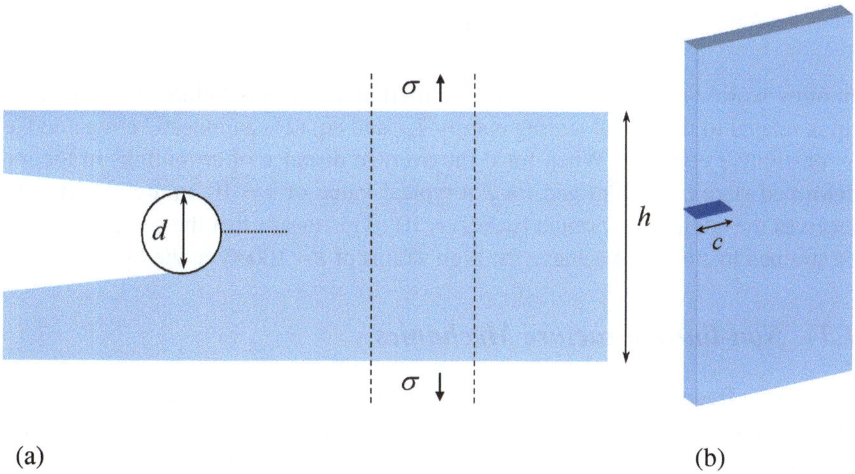

Fig. 2 (a) Scheme of crack tip in a pure shear sample; (b) SEN specimen with implanted crack (undeformed)

For a model notch with a half-circle tip of diameter d, the energy release rate (tearing energy) is [26]:

$$T_{el} = w_t \, d. \tag{3}$$

Therein, w_t denotes an averaged energy density over the edge of the notch ground. In this area of the rubber component the largest deformation occurs of $\varepsilon \approx 10$ where polymer chains are ultimately extended. When the notch propagates to the right in Fig. 2 (a), the stored energy is not completely dissipated but serves to deform the new notch ground. Normally, there will be hysteretic losses, when the time of deformation is smaller than the largest relaxation times. The minimum energy release rate necessary for low-velocity cracks results from the dissociation energy E_{diss} of the weakest C-C single bonds and the number of bonds per volume ν_{bond}. Inserting into eq. (3) gives: $G_0 \approx \nu_{bond} E_{diss} d$. Lake & Thomas [19] assumed the values: $\nu_{bond} = 2.4 \cdot 10^{28} \text{m}^{-3}$, $E_{diss} = 3.3 \cdot 10^{-19}$J and $d = 5 \cdot 10^{-9}$m, from which is follows that $G_0 \approx 40$ J/m^2.

The tearing energy of a *single edge notched* (SEN) strip specimen shown in Fig. 2 (b) can be calculated from the elastic energy density w_{el} in the homogeneous deformation region ahead of the tip:

$$T_{el} = 2k\, w_{el}\, c, \tag{3a}$$

where $k = \frac{\pi}{\sqrt{\lambda}}$ is a coefficient of order one[1]. From eqs. (3) and (3a) a relation between energy density in the notch ground and in the homogeneous deformation region derived:

$$w_t = 2kw_{el}\frac{c}{d}. \tag{4}$$

In other words: for a given value of the tip diameter d, the failure criterion at the tip is related to the global tearing energy T_{el}, and eq. (2) can indeed be regarded as a constitutive equation. When for d the average distance of cross-links in the undeformed state $d \approx 10^{-8}$m and for c a typical value of $c \approx 10^{-3}$m are chosen, one receives the additional formula $(w_t/w_{el}) \approx 10^5$. This means that the specimen has to be strained to $\varepsilon \approx 3\%$ to achieve the high strains of $\varepsilon \approx 1000\%$ at the tip.

1.3 Non-linear Fracture Mechanics

The plastic fracture theory mostly comes back to the non-linear Ramberg-Osgood material law between strain and stress: $\varepsilon = \sigma + \alpha\sigma^n$; therein, ε and σ are the values of strain and stress related to the respective yield values, n is the "hardening exponent" (which actually expresses the softening). With this assumption, stress and strain field of a crack can be calculated in the case of *loading*, also called Hutchinson-, Rice- & Rosengren- (HRR-)field [27]. The dependence of stress on distance r from the tip is: $\sigma \sim r^{-1/(n+1)}$ and of strain: $\varepsilon \sim r^{-n/(n+1)}$. In the case of rubber, the exponent n depends on whether low ($n \approx 2$) or high ($n < 1$) deformations shall be described. Similar approaches [28] exist also for large elastic deformation, also regarding incompressibility and altered geometry. However, these loose their applicability when polymer chains approach their finite extensibility near the tip. Also, stress softening has not been considered yet. In all cases, at least the energy density should show a singularity of the form $w \sim r^{-1}$, the distance r given in the reference configuration attached to the material [28]. This corresponds to eq. (4): the maximum energy density at the notch ground is inversely proportional to the notch radius.

For plastic materials, the Ramberg-Osgood approach is not valid in the *unloading* case, because then stress follows an elastic straight line, and the energy density has to be calculated from stress: $w = \frac{1}{2E}\sigma^2$. However, plastic flow leads to

[1] This relation is valid only if c is small relative to width and length of the specimen.

a *residual deformation* which causes a *rounded-off crack tip* after unloading. Finite element analysis with elasto-plastic material behaviour accordingly show a non-singular stress field [29].

In particular in filled elastomers, a similar effect in connection with *hysteresis* leads to a residual (set) strain which is smaller than that occuring in thermoplastics. Additionally, the residual strain is time-dependent, which has an effect on dynamic crack propagation. As a result of the round-off, the tearing energy necessary to propagate the crack is increased, according to eq. (3) and assumed that w_t has to reach a critical value. The crack may slow down or even come to rest. When the crack is unloaded, a sharp tip forms again, and crack propagation can take place in the next loading cycle. This mechanism gives an explanation for the increased crack growth under cyclic as opposed to static loading.

Finally, the effect of stress softening on crack propagation shall be discussed, whereby, only elastic properties but no inelastic phenomena (hysteresis) shall be considered. When a cracked specimen is loaded cyclically, stress softening increases towards approaching the singularity, and the strain amplification factor decreases towards unity when $r \rightarrow 0$. The maximum stress in the cycle should depend on the maximum strain, as attained at a location with radius r, in a manner similar to an ordinary tension experiment. As known from literature, at high strains the loading cycle of elastomers filled with active CB shows a more or less linear stress-strain curve [6], [30]. This stress-strain curve is followed only once in the first loading cycle, in particular, when the crack propagates. During unloading and further loading cycles without crack progress, every local material area follows a different curve, where the slope decreases as the tip is approached. Due to stress softening, the energy density w, as determined from the area under the unloading curve, accordingly increases less than quadratic with maximum strain. At a singularity of the form $w \sim r^{-1}$, this means that macroscopic strain rises over-proportionally with stress when $r \rightarrow 0$. But, because of decreasing amplification factor, this does not hold for the microscopic matrix strain. Consequently, no statement concerning the effect of stress softening on crack propagation can be derived. Also, stress softening is only caused by filler-reinforcement and is always coincided with hysteresis and its effect cannot be isolated in practice.

2 Experimental

For experimental investigations a sample pool has been prepared based on two types of polymers with variable type and amount of fillers. The composites have been mixed in an industrial type, intermeshing mixer (Werner & Pfleiderer GK 5 E) at 50 rpm. Vulcanization of the samples was performed semi-efficiently with sulfur and accelerator (CBS and DPG) at 160°C in a heat press up to 90% of the vulcameter torque maximum (t_{90} time). As reinforcing fillers, variable amounts of carbon blacks (N234, N550), differing in activity have been used. The basic polymers were a solution-styrene-butadien rubber with high vinyl content (S-SBR; VSL 5025-0), an amorphous ethylen-propylen-dien rubber (EPDM) and natural rubber (NR). All samples were compounded with the processing additives stearic acid and ZnO and protected against aging by IPPD. The ingredients are summarised in Tab. 1.

Table 1 List of ingredients (in phr*) of the rubber samples used

sample name	S-SBR 5025-0/4041	NR CV 50	EPDM Keltan 512	carbon black				x-link / accelerator			anti-aging
				N234	N550	ZnO	stear. acid	S	CBS	DPG	IPPD
S-SBR	100					3	1	1.7	2.5		1.5
NR		100				3	1	1.7	2.5		1.5
EPDM			100			3	1	1.7	2.5	1.5	1.5
S20N2	100			20		3	1	1.7	2.5		1.5
S40N2	100			40		3	1	1.7	2.5		1.5
S60N2	100			60		3	1	1.7	2.5		1.5
N20N2		100		20		3	1	1.7	2.5		1.5
N40N2		100		40		3	1	1.7	2.5		1.5
N60N2		100		60		3	1	1.7	2.5		1.5
E20N2			100	20		3	1	1.7	2.5	1.5	1.5
E40N2			100	40		3	1	1.7	2.5	1.5	1.5
E60N2			100	60		3	1	1.7	2.5	1.5	1.5
S60N5	100				60	3	1	1.7	2.5		1.5
N60N5		100			60	3	1	1.7	2.5		1.5
E60N5			100		60	3	1	1.7	2.5	1.5	1.5
Blends:											
SBR-EPDM0	50		50			3	1	1.7	2.5	1.5	1.5
SBR-NR0	50	50				3	1	1.7	2.5		1.5
S-E60N2N	50		50	60		3	1	1.7	2.5	1.5	1.5
S-E60N2M	50		50	60		3	1	1.7	2.5	1.5	1.5
S-NR60N2	50	50		60		3	1	1.7	2.5		1.5

* per hundred rubber.

The quasi-static mechanical testing of the vulcanized samples was performed with a tensile tester Zwick 1445. The uniaxial hysteresis cycles in the small and medium strain regime were obtained with dumbbells by applying five times repeated cycles with successively increasing load (discontinuous damage mode). In order to avoid dynamic contributions, the stretching velocity was chosen to be small (20 mm/min), corresponding to a strain rate of $\partial\varepsilon/\partial t \approx 10^{-2}\ s^{-1}$.

Dynamic fracture mechanical investigations were performed at notched strip samples of size 100 x 15 mm with a Tear Fatigue Analyzer (TFA) manufactured by Coesfeld, allowing for a fully automatic detection of crack length [31]. Ten specimens can be analyzed simultaneously, the duration between two measurements of the same specimen being about 2 min. Measurements were performed in tension mode with a small off-set stress of 0.1 MPa under pulsed excitation (1 Hz, sinusoidal pulse) at room temperature. The strain amplitude was varied between 10% and 55%. A schematic representation of stress response measured by the TFA is depicted in Fig. 3.

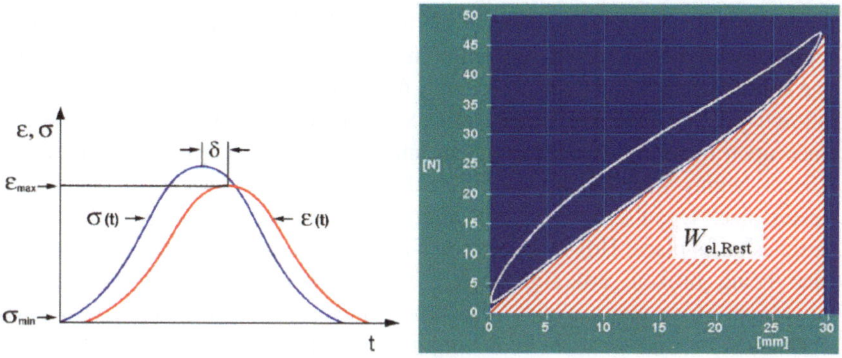

(a) (b)

Fig. 3 Schematic representation of stress response measured by the TFA: (a) harmonic loading pulse, (b) determination of elastic energy from force-displacement curve

Light microscopic oberservation were carried out with a Keyence digital microskope VHX-600 and VH-Z100 UR universal zoom lense at 500x magnification. For electron microscopic examinations the respective surfaces were sputtered with gold. A scanning electron microscope Zeiss EVO MA10 was deployed.

For the validation of FEM simulations, 3D photogrammetric measurements were carried out which yield an area distribution of the strains on the specimen surface. An ARAMIS system from GOM with two CCD cameras (f = 50 mm) was used for this purpose. A single edge notched (SEN) specimen was investigated which had a central hole, 2 mm in diameter, to prevent premature failure under tension. A black/white pattern was applied to the surface using a spraying technique. The tracking of the pattern by image correlation permits the computation of local deformations. The cross-head velocity was the same as in the multihysteresis tests, v = 20 mm/min. The slow movement guaranteed that during the exposure time an object only moved 1/2 pixel or less in the digital image. In a first cycle the strain was raised to 40%, then again to 0% and finally till tensile failure occurred.

3 Dynamic Flocculation Model

3.1 Basic Assumptions

The Dynamic Flocculation Model (DFM) has been introduced by Klüppel [11] in view of explaining the strongly non-linear stress-strain behaviour of filler-reinforced rubbers. Its relevance for describing filler induced stress-softening and hysteresis has been demonstrated in several papers [32-35,41]. This model combines well-established concepts of rubber elasticity with a micro-mechanical approach of dynamic filler flocculation in strained rubbers at different elongations. The stress response of filled rubbers as a function of strain can be derived from the following mesoscopic phenomena. A successive breakdown of filler clusters takes place upon increasing the strain of an unconditioned, virgin rubber sample. This process begins with the largest filler clusters and continues up to a minimum cluster size. Upon decreasing the strain, complete re-aggregation of the filler particles takes place. However, the filler-filler bonds, which are formed again after once being broken, are significantly weaker than in the virgin sample. At subsequent stress-strain cycles of a pre-strained, reinforced sample, two micro-mechanical mechanisms can be distinguished. (i) Hydrodynamic reinforcement of the rubber matrix by a fraction of rigid filler aggregates with strong virgin filler-filler bonds, which have not been broken during previous deformations. (ii) Cyclic breakdown and re-aggregation of the remaining fraction of more soft filler clusters with damaged and hence weaker filler-filler bonds. The fraction of rigid (unbroken) filler clusters decreases with increasing strain, while the fraction of soft filler-clusters increases. The mechanical action of the soft filler clusters refers primarily to a viscoelastic effect, since any type of cluster that is stretched in the stress field of the rubber stores energy that is dissipated when the cluster breaks. This mechanism leads to a filler-induced viscoelastic hysteresis contribution to the total stress, which significantly affects the internal friction of the filled rubber samples. Note that this kind of hysteresis response is present also in the limit of quasi-static deformations, where no explicit time dependency of the stress-strain cycles is taken into account.

Accordingly, the apparent stress resulting in the DFM consists of two contributions: (i) the stress of the rubber matrix including hydrodynamic reinforcement, and (ii) the stress of the strained and broken filler clusters. The free energy density of filler reinforced rubber is given by:

$$W(\varepsilon_\mu,\dot{\varepsilon}_\mu,\varepsilon_{\text{min/max}}) = (1-\Phi_{\text{eff}})\,W_{\text{R}}(\varepsilon_\mu,\varepsilon_{\text{min/max}}) + \Phi_{\text{eff}}\,W_{\text{A}}(\varepsilon_\mu,\dot{\varepsilon}_\mu,\varepsilon_{\text{min/max}}), \quad (5)$$

with Φ_{eff} being the effective filler volume fraction of the structured filler particles, e. g. primary carbon black aggregates. The first addend considers the equilibrium energy density stored in the strained rubber matrix, which includes hydrodynamic strain amplification effects varying with the fraction of relatively stiff filler clusters with strong virgin filler-filler bonds. The second addend considers the energy stored in the residual fraction of more soft filler clusters with damaged bonds that are deformed in the stress field of the rubber matrix.

The free energy density of the strained rubber matrix is described by a non-affine tube model of rubber elasticity:

$$W_{\text{R}} = \frac{G_{\text{C}}}{2}\left\{ \frac{\left(\sum_{\mu=1}^{3}\lambda_\mu^2 - 3\right)\left(1-\frac{T_{\text{e}}}{n_{\text{e}}}\right)}{1-\frac{T_{\text{e}}}{n_{\text{e}}}\left(\sum_{\mu=1}^{3}\lambda_\mu^2 - 3\right)} + \ln\left[1-\frac{T_{\text{e}}}{n_{\text{e}}}\left(\sum_{\mu=1}^{3}\lambda_\mu^2 - 3\right)\right]\right\} + 2G_{\text{e}}\left(\sum_{\mu=1}^{3}\lambda_\mu^{-1} - 3\right). \quad (6)$$

The first bracket term of eq. (6) considers the inter-chain junctions, with an elastic modulus G_{c} proportional to the density of network junctions. The second addend is the result of tube constraints, whereby the tube constraint modulus G_{e} is proportional to the entanglement density of the rubber. The parenthetical expression in the first addend corresponds to a non-Gaussian extension of the tube model taking into account the finite chain extensibility of the polymer network. The finite extensibility parameter is chosen as the ratio $n_{\text{e}}/T_{\text{e}}$ with n_{e} being the number of statistical chain segments between two successive entanglements and T_{e} is the trapping factor characterizing the portion of elastically active entanglements. Since T_{e} increases with the number of cross-links in the system ($0 < T_{\text{e}} < 1$), the finite extensibility parameter decreases with increasing cross-linking density.

The presence of rigid filler clusters, with bonds in the virgin, unbroken state of the sample, give rise to hydrodynamic reinforcement of the rubber matrix. This is specified by the strain amplification factor X, which relates the external strain ε_μ of the sample to the internal strain ratio λ_μ of the rubber matrix:

$$\lambda_\mu = 1 + X\,\varepsilon_\mu \quad (7)$$

In the DFM, the strain amplification factor depends on the pre-conditioning of the sample and gives rise to the well known stress softening effect of filler reinforced rubbers. In the case of pre-conditioned samples and for strains smaller than the previous straining ($\varepsilon_\mu < \varepsilon_{\mu,\max}$), the strain amplification factor X is independent of strain and determined by $\varepsilon_{\mu,\max}$: $X = X(\varepsilon_{\mu,\max})$. For the first deformation of virgin samples, it depends on the external strain $X = X(\varepsilon_\mu)$. For fractal clusters, $X(\varepsilon_{\mu,\max})$ or $X(\varepsilon_\mu)$ can be evaluated by averaging over the size distribution of rigid clusters in all space directions. In the case of pre-conditioned samples, this yields:

$$X_{\max} = X(\varepsilon_{\min}, \varepsilon_{\max}) = 1 + c\Phi_{\text{eff}}^{\frac{2}{3-d_f}} \left[\int_0^{x_{1,\min}} x^{d_w - d_f} \phi_1(x)\, \mathrm{d}x + \int_{x_{1,\min}}^{\infty} \phi_1(x)\, \mathrm{d}x \right] \tag{8}$$

Here, c is a constant of order one, Φ_{eff} is the effective filler volume fraction, ξ_μ is the cluster size, d is the particle size, $d_f \approx 1.8$ is the mass fractal dimension and $d_w \approx 3.1$ is the anomalous diffusion exponent on cluster-cluster aggregation. $\phi(\xi_\mu)$ is the normalized size distribution that can be derived from the Smoluchowski equation of the kinetics of cluster-cluster aggregation of colloids. With the abbreviation $x_\mu \equiv \xi_\mu / d$ and the normalized mean cluster size $x_0 \equiv \xi_0/d$ it reads:

$$\phi(x_\mu) = \frac{4x_\mu}{x_0} \exp\left(-\frac{2x_\mu}{x_0} \right) \qquad \mu = 1,2,3 \tag{9}$$

We point out that with this distribution function the integrals in eqs. (8) and (10) can be solved analytically.

The second addend in eq. (5) considers the energy stored in the substantially strained filler clusters. It delivers a filler induced hysteresis due to cyclic stretching, breakdown and re-aggregation which is described by an integral over the soft filler clusters in stretching direction ($\dot{\varepsilon}_\mu > 0$) with strain dependent upper boundary:

$$W_A(\varepsilon, \dot{\varepsilon}_\mu, \varepsilon_{\text{min/max}}) = \sum_{\mu}^{\partial \varepsilon_\mu / \partial t > 0} \frac{1}{2} \int_{x_\mu(\varepsilon_{\mu,\max}; \varepsilon_{\mu,\min})}^{x_\mu(\varepsilon_\mu; \varepsilon_{\mu,\min})} G_A(x)\, \varepsilon_{A,\mu}^2(x, \varepsilon_\mu)\, \phi_\mu(x)\, \mathrm{d}x \tag{10}$$

where the dot denotes time derivative. The sum over stretching directions with ($\dot{\varepsilon}_\mu > 0$) implies that clusters store energy by being stretched and re-aggregate upon contraction. G_A is the elastic modulus and $\varepsilon_{A,\mu}$ is the strain of the soft filler clusters in spatial direction μ. The dependency of these quantities on cluster size ξ and external strain ε_μ can be derived from basic micro-mechanical considerations about elasticity and fracture mechanics of tender filler clusters imbedded into a strained rubber matrix. This also allows for a specification of the strain-dependent integral boundaries $\xi_\mu = \xi_\mu(\varepsilon_\mu)$ and $\xi_{\mu,,\min} = \xi_{\mu,,\min}(\varepsilon_{\mu,\max})$.

For obtaining reasonable fits, a further stress contribution, the set stress σ_{set} is added. In the case of quasistatic, uniaxial deformations this yields for the stress-strain cycles:

$$\sigma_1(\varepsilon, \varepsilon_{\text{min/max}}, T) = (1 - \Phi_{\text{eff}})\, \sigma_{R,1}(\varepsilon, \varepsilon_{\text{min/max}}, T) + \Phi_{\text{eff}}\, \sigma_{A,1}^{up/down}(\varepsilon, \varepsilon_{\text{min/max}}, T) + \sigma_{\text{set}}(\varepsilon_{\text{min/max}}, T) \tag{11}$$

which also depend on temperature T. The set stress σ_{set} takes into account that elastomers show a certain inelastic behaviour also known as *setting*. After the material has been strained to a maximum at $\varepsilon_{\max}$, $\varepsilon_{\min}$, respectively, and then again to $\varepsilon = 0$, a residual stress is measured. This so-called set stress essentially relaxes and vanishes at large times but has to be taken into account in normal deformation experiments - especially with filled rubbers. The residual set stress σ_{set} is the stress required to get the material back into the un-deformed state. In the present time-independent form of the DFM the corresponding stress contribution has to be described by a term separate from elastic potentials. The free fitting parameter, $s_{set,0}$, stands for the set stress at $\varepsilon_{max} = 1$ (while $\varepsilon_{min} = 0$). A more detailed physical description of the DFM and the various experimental tests of the model can be found in references [32-35].

3.2 Parameter Fitting

The dynamic flocculation model offers the unique capability of obtaining micro-structural physical parameters of the polymer- and filler network that are hardly or not at all accessible experimentally. Of course, no exact quantitative values should be expected, because of the necessary simplifications inherent in the model. Here, we want to verify if the model parameters show a comprehensible behaviour under systematic variation of polymer, type and content of filler. Where it is possible, the quantitative values shall be evaluated.

For unfilled rubbers, values for G_c, n and $s_{set,0}$ were obtained from fits to simple stretching data by putting $\Phi_{eff} = 0$. This implies a constant hydrodynamic amplification factor $X_{max} = 1$ and a cluster stress equal to zero. The estimated G_e values were: $G_e = 0.59$ MPa (EPDM), S-SBR: $G_e = 0.32$ MPa (S-SBR), which

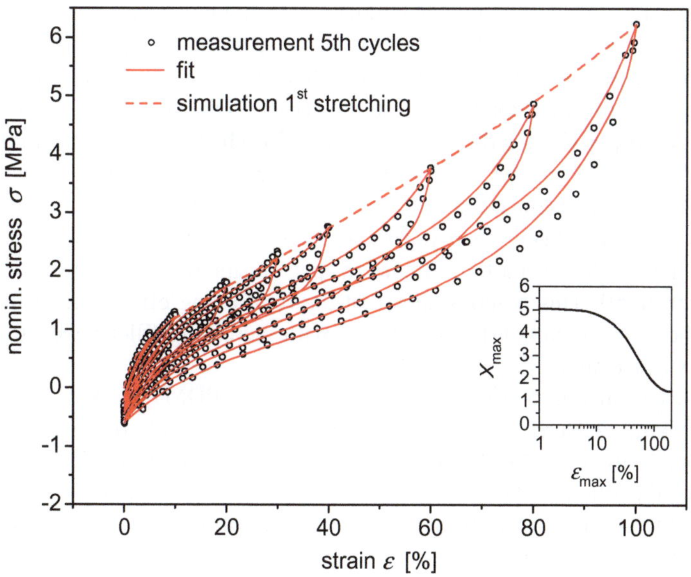

Fig. 4 Uniaxial stress-strain measurements (5th cycles) in tension of EPDM with 60 phr N243 (symbols) and fit with the DFM (lines); The inset shows the strain amplification factor X_{max} vs. maximum strain ε_{max}

agrees fairly well with theoretical values $G_e = 0.3$ MPa for S-SBR and $G_e = 0.6$ MPa for EPDM obtained from dynamic measurements of the plateau modulus.

For the filled rubbers, stress-strain cycles were measured by stretching the rubber sample repeatedly (five times) to a certain maximum strain value which was then increased, successively (5, 10, 20, 30, 40, 60, 80, 100%). Fits with the DFM were obtained with the measured fifth cycles which were considered to be equilibrium cycles. An example of a fit for an EPDM samle with 60 phr N243 is shown in Fig. 4. Obviously, the fit is in fair agreement with the experimental data. The dashed line indicates a simulation of the first stretching of a virgin sample which results from a decrease of the strain amplifcation factor X_{max} (ε) with increasing strain $\varepsilon = \varepsilon_{max}$. The strain amplification factor X_{max} calculated with eq. (8) is shown in the inset of Fig. 4 as a function of maximum strain ε_{max}.

For the fits of the filled samples the modulus of topological constraints G_e was not varied but kept constant as obtained for the unfilled samples: $G_e = 0.3$ MPa for S-SBR and $G_e = 0.6$ MPa for EPDM. The fitting parameters are summerized in Figs. 5 and 6. The amount of N234 is 20, 40 and 60 phr, while for N550 only 60 phr is used to see the effect of CB. From Fig. 5 (a) it strikes that the effective filler ratio Φ_{eff} is more or less independent of the rubber type, but expectedly increases with the structure of the filler particles. Obviously, for a given filler type (N234) there is an almost linear relation between Φ_{eff} and Φ with a constant factor ~1.9 reflecting the structure of the filler particles, i.e. its deviation from spherical shape. A similar value of $\Phi_{eff} \approx 0.31$ for $\Phi = 0.23$ of N550 was found by Morozov et al. [36] They developed a computer simulation technique where three-dimensional filler clusters were constructed following a specially designed algorithm and evaluated the amount of bound and occluded rubber. Also Wang, Wolff & Tan [37] derived a formula relating Φ_{eff} to the crushed DBPA number and filler mass density ρ_f: $\Phi_{eff} = \Phi\,(\rho_f \cdot DBPA(24M4) + 1)/1.59$ which yields comparable values for N550 and N234 to the fitted values in Fig. 5 (a). The average related cluster size x_0, Fig. 5 (b), reflects the state of the filler structure after a possible continuous network has been destroyed by a small deformation. x_0 increases with Φ, but not higher than 9 particle diameters per cluster of N234 and even smaller values for N550. This demonstrates that the reinforcing effects result from quite small filler clusters which become larger with increasing filler concentration and specific surface area.

As seen from Fig. 5 (c), G_c increases linearly (EPDM) or over-proportionally (S-SBR) with Φ. Here, we look at the elastic modulus only of the rubber matrix which needs to be scaled by a factor of $1-\Phi_{eff}$ in order to calculate the elastic energy of the filled rubber, as in eq. (5). If a fraction of active filler particles is added to the polymer, this results in higher values of the cross-link modulus G_c as compared to the cross-link modulus for the unfilled rubber given by $G_c = A_c \nu_c \cdot k_B T$. Here, ν_c denotes the number of mechanically effective network chains, related to chemical cross-links, per unit volume of the rubber matrix, A_c is a microstructure factor that considers the fluctuation of network junctions, T is temperature and k_B is the Boltzmann constant. Accordingly, for filled rubbers G_c is affected by

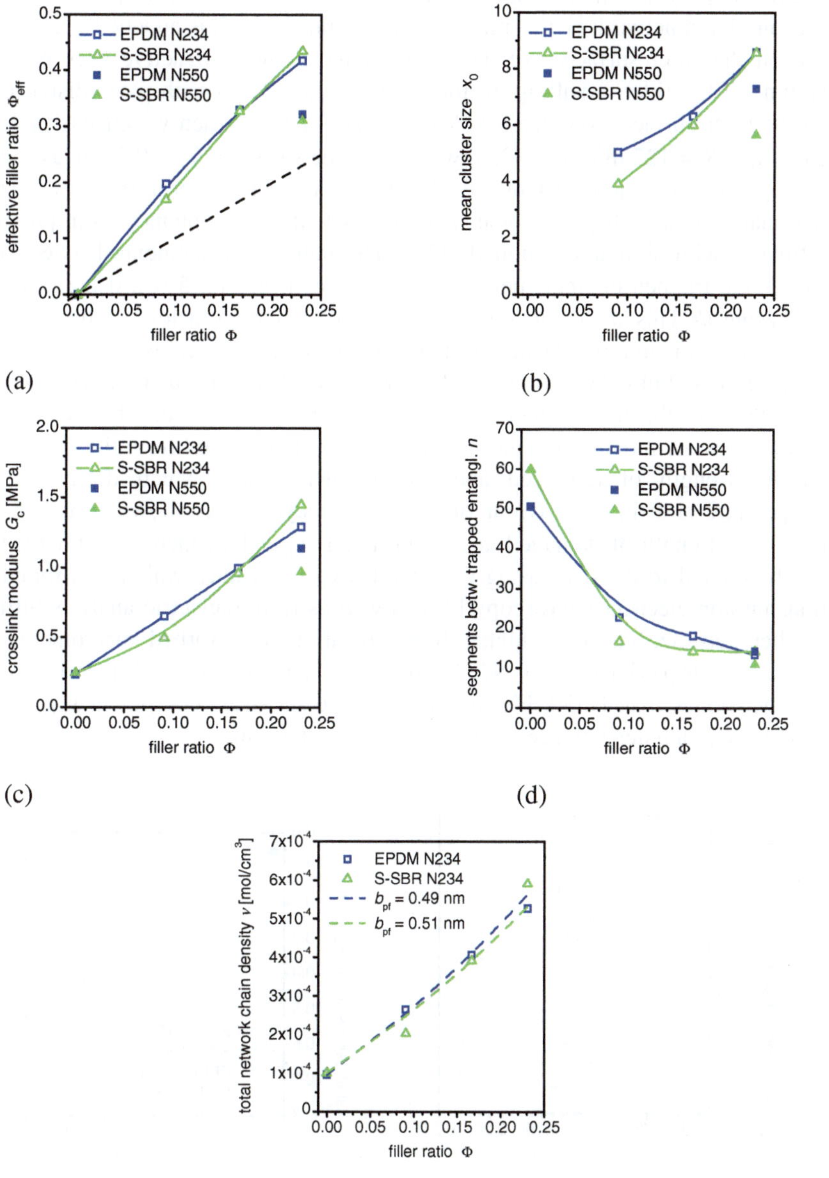

(e)

Fig. 5 Systematic variation of fitting parameters of the DFM related to the filler and polymer structure at varied filler content for EPDM and S-SBR composites: (a) effective filler ratio Φ_{eff}, dashed line: $\Phi_{\mathrm{eff}} = \Phi$; (b) mean cluster size x_0; (c) crosslink modulus G_c; (d) number of segments between two successive trapped entanglements n; (e) total network chain density $\nu = A_c\nu_c + \nu_f\ \Phi/(1\text{-}\Phi)$, as calculated from the fitted G_c values with N234 (symbols); the dashed lines are calculated from $\nu_f = \rho_f\ S_f\ /\ b_{pf}^2$ with the given values b_{pf} of mean contact distance of adsorption sites on the CB surface (see text).

additional polymer-filler couplings, $G_c = (A_c \nu_c + \nu_f\ \Phi/(1-\Phi))\cdot k_B T$, taking into account that a network chain splits into two chains when it is attached to an active site on the filler surface. We denote ν_f as the number of active sites per unit volume of the filler, resulting in filler-induced cross-links with mean distance b_{pf} on the filler surface, whereas it holds: $\nu_f = \rho_f\ S_f\ /\ b_{pf}^2$. When we choose $\rho_f = 1.8$ g/cm^3 and $S_f = 125$ m^2/g for N234, we receive the values of $b_{pf} \approx 0.5$ nm, as shown in Fig. 5 (e). These calculated distances between adsorption sites appear reasonable though they are somewhat smaller than those obtained from previous estimates with similar compounds [38]. The number of segments between two successive trapped entanglements $n_e/T_e \equiv n$, shown in Fig. 5 (d), decreases with increasing Φ. This can be understood because the trapping factor T_e which characterizes the fraction of elastically active entanglements increases due to filler-induced cross-links. We point out that the segment number n_e is independent of the filler since the entanglement density of the rubber is not affected by the filler.

The possibly most important material parameters obtained from the fits are the tensile strength of damaged and virgin filler bonds, which determine the temperature and time behaviour of stress-strain cycles. As shown in Fig. 6 (a), both, the strength of damaged and virgin bonds s_d and s_v, increase with Φ. This can be related to the decreasing gap size between particles which was found by transmission electron microscopy [39] as well as from dielectric analysis [40]. A smaller gap size causes a higher immobilization of adsorbed macromolecules between filler particles. Also, N234 is the more active CB and will therefore yield stronger bonds than N550. Fig. 6 (b) shows the set stress coefficient $s_{set,0}$ which exhibits a strong dependence on the total filler surface area.

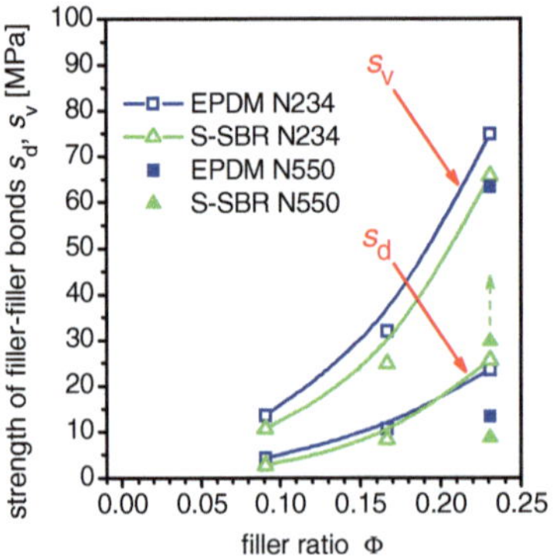

(a)

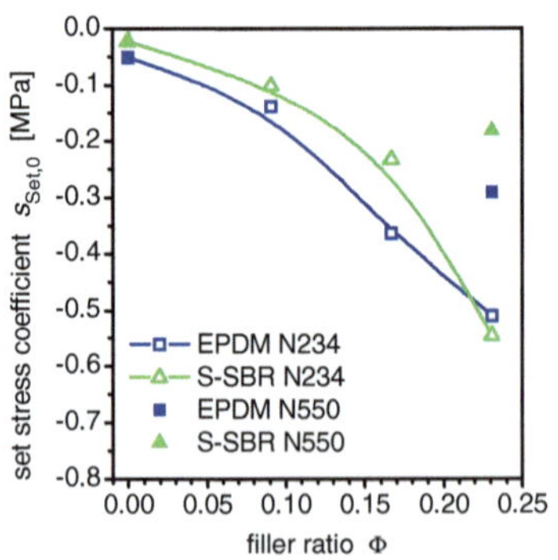

(b)

Fig. 6 Systematic variation of model parameters at varied filler content: (a) strength of virgin and damaged bonds, s_v and s_d, respectively; (b) set stress coefficient $s_{set,0}$.

3.3 FEM Simulation of Stress Softening Close to the Crack Tip

The dynamic flocculation model has been implemented into the Finite-Element-Method (FEM) via the concept of representative directions. In a first step, the model has been implemented into the FE system MSC.MARC by using the subroutine HYPELA2. This requires the second Piola-Kirchhoff stress tensor and a Lagrangean material tangent with respect to the right Cauchy-Green tensor. In contrast to MSC.MARC the FE software ABAQUS uses the subroutine UMAT which requires the Cauchy stress tensor and a different material tangent. The ABAQUS implementation was done by converting the output of the subroutine HYPELA2 into the required formulation. More details are found in Ref. [41]. The following FEM simulation was done with ABAQUS using hybrid linear continuum elements. We demonstrate the evolution of the amplification factor and the spatial distribution of strains in comparison to the experiment.

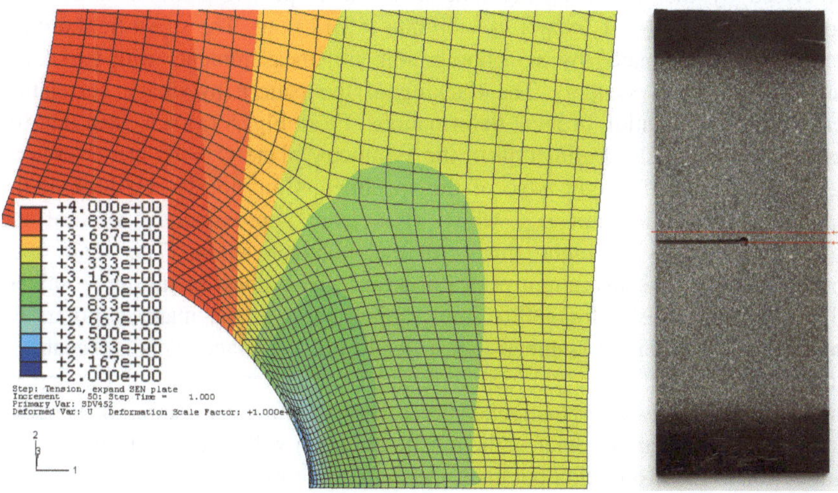

(a) (b)

Fig. 7 SEN tensile specimen of EPDM filled with carbon black, E60N5: (a) FEM simulation of stress softening expressed by the averaged amplification factor $<X_{max}>$ for a nominal tensile stain of $\varepsilon_y = 0.5$; (b) photograph of prepared specimen for experimental evaluation of the strain field by photogrammetry, stains are evaluated along the indicated paths Y=0 and Y=2 (comp. Fig. 8).

Fig. 7 shows the calculated stress softening of a SEN tensile specimen of carbon black filled EPDM as expressed by the averaged amplification factor $<X_{max}>$ for a nominal tensile stain of $\varepsilon_y = 0.5$. The FEM model comprised the upper half of the specimen, symmetry conditions were also given in the flat plane of the specimen forcing a symmetric deformation in the normal z-direction. The used material parameters were obtained solely from multihysteresis uniaxial tensile tests with E60N5. Fig. 7 (a) shows the strain amplification factor X_{max} averaged over the 50 representative directions, at a nominal tensile stain $\varepsilon_y = 0.5$. Prior to loading $<X_{max}>$ is ubiquitously equal to 4.0. It decreases during the test, especially in the vicinity of the notch ground. This happens due to increasing stress which leads to breakdown of rigid filler clusters. Because the decrease of $<X_{max}>$ corresponds to permanent stress softening, it can be regarded as a kind of damage variable. The smallest values of $<X_{max}> \approx 2.0$ are found at the notch ground near the symmetry plane normal to the Y-direction where, simultaneously, the highest stresses are found.

For experimental validation of the calculated strain fields, a SEN specimen was prepared for photogrammetric investigations. A photograph of the unstretched sample is shown in Fig. 7 (b). For stretching the sample, it was fixed at its ends between screwed clamps and a nominal axial strain $\varepsilon_y = \Delta l/l_0$ (l_0 = 80 mm) was prescribed by the tensile testing machine. The respective nominal stress was measured: $\sigma_y = F/A_0$ ($A_0 = b_0 \times h_0$ = 40 mm × 2.14 mm). The strain field around the notch was estimated by 3D photogrammetric measurements with two CCD cameras (ARAMIS system). The tracking of the pattern by image correlation permits the computation of local deformations. From Fig. 8 (a) it is seen that the deformation along the X-direction in the symmetry plane ($Y = 0$) increases with the distance from the right edge of the specimen. Consciously, we look at the maximum principal strain ε_{max} which rises from approximately 80% of the nominal strain up to eight times of the nominal strain. The main axis corresponding to the maximum principal strain (the highest of the three values) will follow the tensile force flux line running through the material point. By approximation, it can be assumed that it lies in the X-Y-plane. This is, because a tensile load is applied in y-direction (space-fixed Euler system of the testing machine), and the two perpendicular directions will show contraction or negative strain. When the material rotates, the tensile direction will gradually turn from the Y-direction to the X-direction but hardly to the Z-direction.

Generally, fair agreement of the simulation with measured values of ε_{max} were found. In the simulation, higher strains were reached, but this is because of the much finer discretization in the FEM model compared to the photogrammetric measurement. The agreement is also seen for a path along Y = 2 mm where ε_{max} runs through a maximum near the notch ground. It is remarkable that complex stress-strain states induced by the notch are reproduced using a parameter set which was obtained from fitting of uniaxial tension tests only. Finally, the simulated nominal stress is very close to the experiment which shows stress softening, hysteresis and set behaviour, Fig. 8 (b) (the experiment stops at $\varepsilon_y = 0.6$ because of tensile failure).

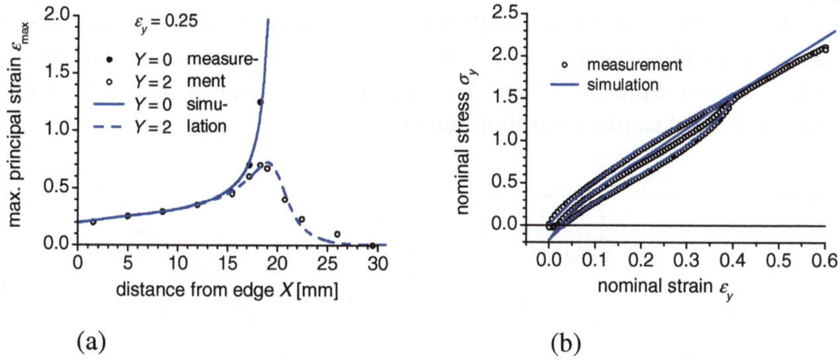

Fig. 8 Comparison of measurements (symbols) and FEM simulations (lines) of a SEN tensile specimen, CB-filled EPDM, E60N5; (a) maximum principal strains ε_{max} along two horizontal paths Y=0 and Y=2 (comp. Fig. 7 (b)); (b) nominal stress as a function of the strain history

4 Stress Relaxation of Unfilled and Filled Elastomers

Here, the relaxation of stress at constant stain shall be regarded. The measured stress-time curves were adapted with power laws of the form:

$$\sigma(t) = \sigma_\infty + bt^{-\beta}. \tag{12}$$

The equilibrium stress σ_∞, as a result of the entropy-elastic behaviour of the rubber matrix, depends on ε and T. Characteristic exponents β were determined for the up cycle, where the stress relaxes downward, (S-SBR: $\beta \approx 0.16$) and for the down cycle, where the stress relaxes upwards, (S-SBR: $\beta \approx 0.28$). Filled specimens were pre-conditioned with ε_{max} = 100%, far above the experimental deformation range of $\varepsilon = 0 \ldots 40\%$, in order to guarantee that the inner structure of the material is not influenced by the loading during the experiment.

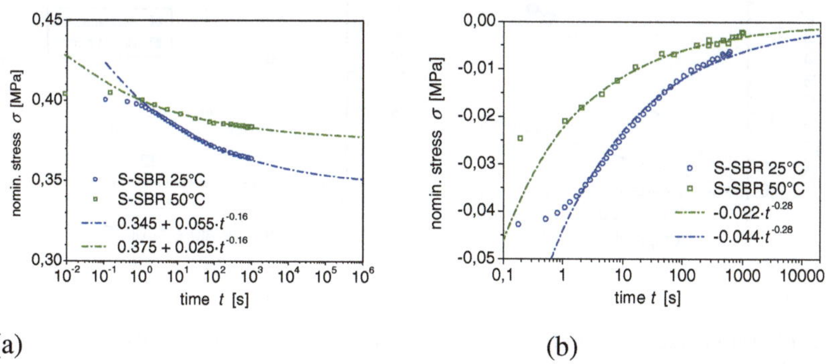

Fig. 9 Stress relaxation for unfilled S-SBR at 25°C and 50°C (symbols) and fits with eq. (12) (lines); (a) up cycle, $\varepsilon = 40\%$, $\beta = 0.16$; (b) down cycle, $\varepsilon = 0\%$, $\beta = 0.28$

Fig. 9 shows measurements for unfilled S-SBR at 25°C and 50°C. Fig. 10 shows the relaxation for filled S-SBR for the same temperatures and also at 80°C, because the reinforcement admitted the specimens to survive the long loading period of 1000 s without rupture. Measurements, not shown here, with EPDM specimens showed similar relaxation functions.

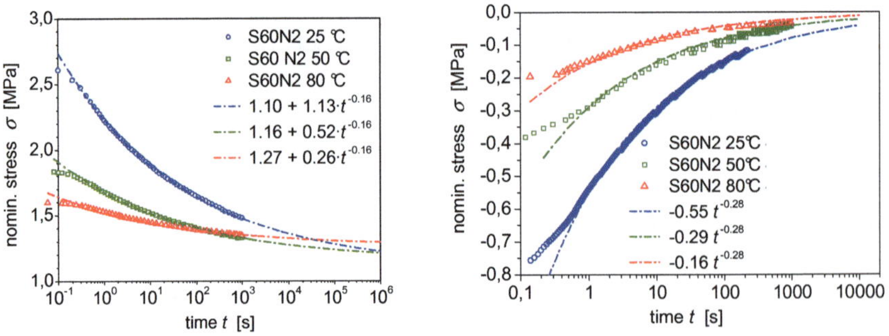

(a) (b)

Fig. 10 Stress relaxation for filled S-SBR, 60 phr N 234, at 25°C, 50°C and 80°C (symbols) and fits with eq. (12) (lines); (a) up cycle, $\varepsilon = 40\%$, $\beta = 0.16$; (b) down cycle, $\varepsilon = 0\%$, $\beta = 0.28$

Fig. 11 (a) shows the expected linear dependence of the equilibrium stress σ_∞ reflecting the entropy elastic behaviour of the rubber matrix, which often cannot be measured because the dependence is overshadowed by relaxation effects showing an opposite temperature dependence. It can only be observed with a correct extrapolation of stress for infinite times t_∞, which means that eq. (12) is the

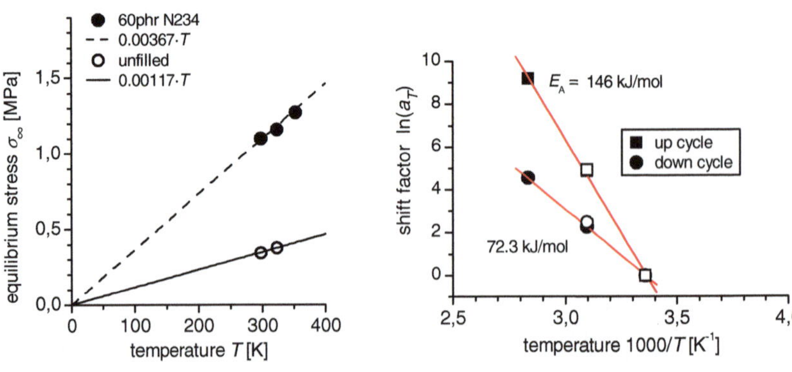

(a) (b)

Fig. 11 Temperature dependence of relaxation parameters for S-SBR, unfilled (open symbols) and filled with 60 phr N 234 (closed symbols): (a) equilibrium stress σ_∞ during up cycle; (b) shift factor a_T in an Arrhenius-plot.

correct relaxation function. Fig. 11 (b) demonstrates that the relaxation measurements exhibit a time-temperature superposition according to an Arrhenius behaviour of the shift factor a_T. The shift factor was calculated from the pre-factor b of the power law eq. (12), because the simple relation holds: $\sigma(a_T t) - \sigma_\infty = a_T^{-\beta}(\sigma(t) - \sigma_\infty)$. However, different apparent activation energy values result for the up cycle (E_A = 146 kJ/mol) und down cycle (E_A = 72 kJ/mol).

Further, the effect of cross-linking on relaxation, in particular of set stress at ε = 0%, was examined. From Fig. 12 it is seen that the relaxation strength is influenced by the filler as well as by the degree of cross-linking. However, the form of the relaxation (power law exponent b) is almost unchanged. Only during the conditioning process (indicated by "kondit" in Fig. 12) an additional part of relaxation appears. It is remarkable that also at half of the normal amount of cross-linking agents no plastic deformation is observed - that means: no (quasi-static) flow but only an increase of the dynamic hysteresis takes place.

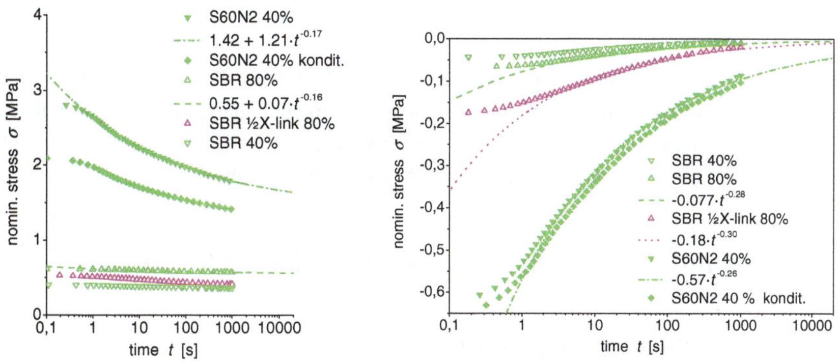

(a) (b)

Fig. 12 Stress relaxation for unfilled S-SBR (open symbols) at different cross-linker concentration (green: normal amount; margenta: halve amount) and filled S-SBR with 60 phr N 234 (closed symbols); (a) ε = 40%, 80% as indicated (up cycle); (b) ε = 0% (dopwn cycle); lines are fits with eq. (12)

The relaxation of set stress at ε = 0% follows a simple time dependence, because it shows no residual stress for infinitely long time, σ_∞ (ε = 0) = 0. Fig. 13 (a) shows the time evolution in dependence on maximum stress ε_{max}, holt for t = 1000 s at the end of the up cycle. With increasing maximum stress a systematic increase of the set stress is observed. Fig. 13 (b) demonstrates that for unfilled and filled elastomers as well as at different cross-linking degrees a linear dependence of set stress (here taken after 1 s at ε = 0) on maximum stress ε_{max} at the end of the up cycle is found. Compared with unfilled S-SBR, an increase of filler as well as a reduction of cross-linking leads to an increase of the slope, however, a combination of both yields no further increase. At larger negative values of set stress a saturation is observed, no values below $\sigma_{set} \approx$ -0.7 MPa are measured (for ε_{max} = 40%). However, it has to be taken into account that the time between $\varepsilon = \varepsilon_{max}$ and $\varepsilon_{max} = 0$ can already be much larger than 1 s.

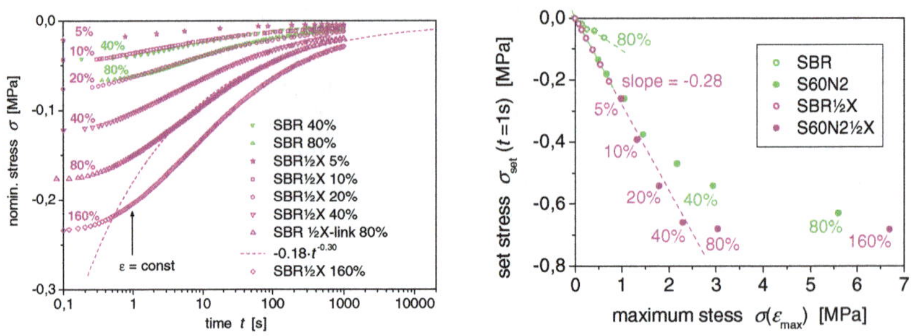

(a) (b)

Fig. 13 (a) Stress relaxation of the set stress at $\varepsilon = 0\%$ for unfilled S-SBR with different cross-linker concentration (green: normal amount; margenta: halve amount) for different pre-stretchingvalues ε_{max} as indicated; (b) set stress at $\varepsilon = 0\%$ after 1 s for unfilled S-SBR (open symbols) and filled S-SBR with 60 phr N 234 (closed symbols)

5 Dynamic-Mechanical Analysis of Blend Morphology

5.1 Theoretical Background

In this section we will present a dynamic mechanical analysis technique concerning the morphology of filled rubber blends giving information about the amount of interphase and the distribution of filler between the phases of the blends listed in Tab. 1. Following the general principle of additivity of energy, the total stored or dissipated energy of a dynamically excited rubber blend system is the sum of all energy contributions of the different phases. The stored and dissipated energy per unit volume can be expressed by the storage and loss modulus, respectively, multiplied by the square of the mean strain. Therefore, the complex modulus of the blend is given by

$$G^*_{blend} = \sum_i G^*_{\mathrm{P},i} \phi_i \left(\frac{\gamma_{0,i}}{\gamma_0} \right)^2 , \tag{13}$$

where $G^*_{\mathrm{P},i}$, ϕ_i, $\gamma_{0,i}$ are the modulus, the volume fraction and the averaged strain, respectively, of the i-th blend phase and γ_0 is the external strain of the blend. The modulus $G^*_{\mathrm{P},i}$ of the single phases can be measured by dynamic-mechanical analysis of the single, unblended rubbers. Unfortunately, the averaged strain of the i-th phase is generally difficult to obtain, since it depends on the relative stiffness of the components and the morphological structure of the interpenetrating phase network. Instead, it has been shown that a second order approximation can be used to estimate the loss modulus of a blend [42] [43]:

$$G''_{blend} = \sum_i G''_{\mathrm{P},i}\left(a_i + b_i\phi_i + c_i\phi_i^2\right). \tag{14}$$

The coefficients a_i, b_i, c_i can be obtained by a fitting procedure to the measured data of blends with different blend ratios.

Besides the two phases consisting of polymer A and polymer B (for a blend system with two polymers), a third phase, namely the interphase, can contribute to the loss modulus. The polymers are separated in domains at which border a thin layer of a mixture of both components exists. The interphase depends strongly on the polymer interactions. In general, the amount of interphase increases with increasing compatibility of the polymers [43]. In order to take the interphase into account, it has been assumed that the contribution of the interphase to the modulus of the blend system changes linearly with the volume fraction of the interphase. The amount of interphase depends on the contact probability of the two phases, which can be estimated by the product of the volume fraction of the both polymer components ϕ_A and ϕ_B. Thus, the interphase volume fraction is given by $\phi_{IPH} \sim \phi_A\phi_B$. Furthermore, it has been assumed that the interphase consists to equal parts of both polymers and that the modulus of the pure interphase would be given by $G''_{IPH} = \left(G''_A + G''_B\right)/2$.

So far, unfilled rubber blends have been considered. Adding fillers such as carbon black, the rubber matrix is reinforced by the filler particles. This leads to an increase in the loss modulus. The maximum of the loss modulus for a filler-reinforced single rubber can be empirically described by a linear expression [42] [43]:

$$G''_{\max} = G_0 + \alpha\phi_{\mathrm{F}}, \tag{15}$$

where G_0 is the maximum value of the loss modulus of the unfilled rubber and ϕ_{F} is the filler volume fraction. Broadening of the single rubber glass transition $G''(T)$ due the filler is considered by a spline fitting procedure, as described below.

Combining eq. (14) for the loss modulus of the unfilled rubber blend with eq. (15) for the filler-reinforced modulus of a single rubber, the temperature-dependent loss modulus of filled rubber blend is given by

$$G''_{blend}(\phi_i, \phi_{F,i}, T) = \sum_i A_i(\phi_i, \phi_{F,i}) G''_{P,i}(T), \tag{16}$$

where

$$A_i(\phi_i, \phi_{F,i}) = \left(1 + \alpha_i \frac{\phi_{F,i}}{\phi_{F,i} + \phi_i}\right)\left(a_i + b_i(\phi_{F,i} + \phi_i) + c_i(\phi_{F,i} + \phi_i)^2\right). \tag{17}$$

$\phi_{F,i}$ is the filler volume fraction in the *i*-th phase relative to the total composite. Note that the filler is not equally distributed over all phases. The phases are loaded with different amount of filler, depending on the affinity of carbon black to the polymer in each phase and also on the mixing procedure [44] [43]. In eq. (17) it has been taken into account that the total amount of phase i is increased by adding the filler, leading to the phase volume fraction that is the sum $\phi_{F,i} + \phi_i$ of the volume fraction of the polymer component and the volume fraction of the filler inside this phase. Furthermore, the filler volume fraction to describe the filler reinforcement effect of the modulus for each phase (first bracket) has been normalized to the corresponding phase. If all coefficients a_i, b_i, c_i and α_i are known and the modulus of unblended single rubber $G''_{P,i}$ and blended rubber G''_{blend} are measured, equation (4) can be used to determine the filler amount in the different phases.

5.2 *Carbon Black Distribution between the Blend Phases*

The temperature-dependent loss modulus during the glass transition of filled rubber is shown exemplarily for S-SBR in Fig. 14 (a). As for all other investigated polymers, the glass transition temperature, indicated by the temperature at the maximum of G", is independent of carbon black amount and is for all filled samples the same as for the unfilled rubber. Thus, there is no influence of the filler on the glass transition as reported elsewhere [45]. The maximum of the loss modulus increases with larger filler content. Furthermore, the glass transition gets broader, indicating a broader distribution of relaxation times. This has been explained by the temperature-dependent localization of polymer chain modes close to the strongly attractive filler surface [40] [46] [47].

The increase in the maximum is plotted versus the carbon black volume fraction for all polymers in Fig. 14 (b). As can be seen, a linear expression can describe the dependence of the loss modulus on filler volume fraction. This justifies the usage of eq. (15) in order to describe the effect of filler on the loss modulus. The maximum of the loss modulus G_0 obtained for $\phi_F = 0$ and the slope α are summarized in Tab. 2. NR has the largest modulus, followed by S-SBR, and EPDM has the smallest. Furthermore, it has been found that the larger the modulus of the pure rubber is, the larger is also the slope.

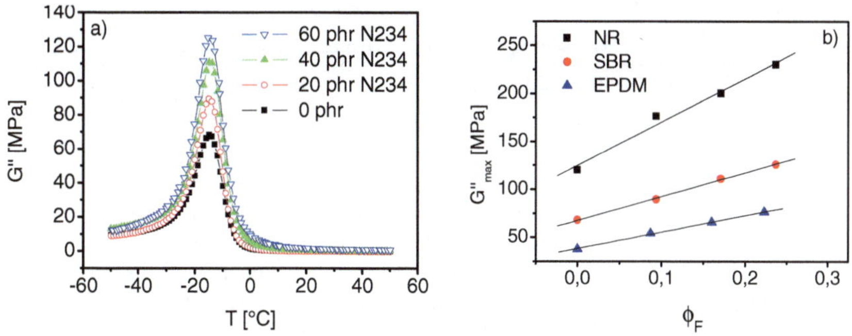

Fig. 14 (a) Loss modulus during the glass transition of S-SBR filled with different amounts of carbon black N234; (b) maximum value of the loss modulus versus filler volume fraction for different rubber systems filled with N234

Table 2 Maximum value of the loss modulus G_0 for unfilled rubber and slope α of eq. (15)

	G_0 [MPa]	α [MPa]
NR	125	453
SBR	67	248
EPDM	38	171

The loss modulus of the unfilled rubber blend NR/SBR is shown in Fig. 15 (a) in comparison to the unblended homopolymers NR and SBR. The blend has two separated glass transitions, indicating the immiscibility of the both polymers. Thus, the both polymer components are separated in phase domains. The maxima of the loss modulus are slightly shifted apart from the ones of the unblended rubbers towards each other. This can be explained by a partial solubility of a small amount of short chains in the other rubber component. Both maxima of the blend with blend ratio 50:50 are not equal to the half of the maxima of SBR and NR, respectively. The maximum at small temperatures is clearly above half of the NR maximum, and the other maximum is clearly below half of the SBR maximum. Thus, the modulus of the blend depends non-linear on blend ratio. The strong non-linearity has been investigated and discussed with varying blend ratio and different blended polymers in [42] [43]. It has been shown that the dependence on the blend ratio can be described by the empirical eq. (14). This is demonstrated in Fig. 15 (b) for the NR/SBR blends which is adapted from [43]. The coefficients a_i, b_i, c_i are obtained from the fitted lines to the data points.

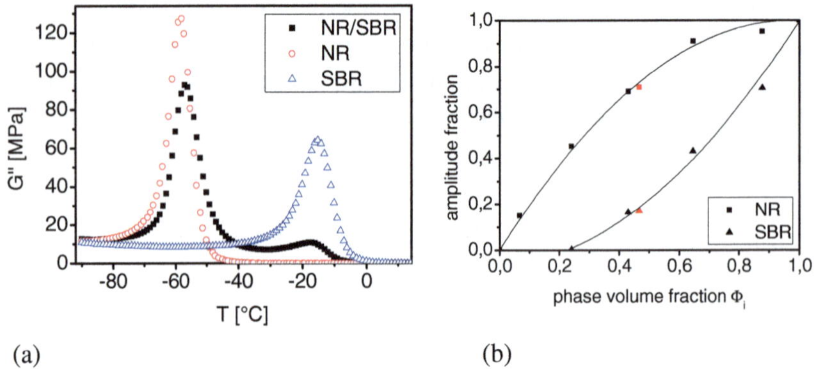

Fig. 15 (a) Loss modulus of the unfilled rubber blend NR/SBR with blend ratio 50:50 compared to the unblended rubbers. (b) Amplitude contributions from the NR- and SBR-phases vs. phase volume fraction for different blend ratios (symbols) and fits with eq. (14) (lines) as adapted from [43]. The two symbols depicted in red are new data points corresponding to the amplitudes shown in (a).

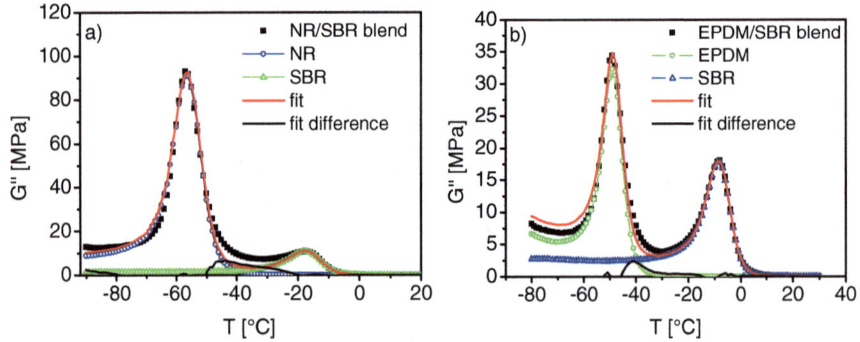

Fig. 16 Two-phase fit of unfilled blends for (a) NR/SBR and (b) EPDM/SBR in order to separate the interphase

In order to separate the contribution from the different polymer phases, the rubber blend has been fitted by the sum of two fit functions for both polymers. The spline parameters used here are the amplitude, the broadness and the temperature shift of the loss maximum. The fits are seen in Fig. 16 (a) for NR/SBR and Fig. 16 (b) for EPDM/SBR. The fits are reasonable in the regions of both glass transitions, but larger deviations, in particular for NR/SBR, are found in the region between the maxima. This can be explained by an additional phase, namely the interphase, which reaches its maximum in this region. The difference between the two-phase fit and the measured data has been taken for the construction of the interphase. For later usage, the signal has been slightly smoothed by fitting it with an empirical function. The results are summarized in

Tab. 3. The interphase is relatively large for NR/SBR and delivers a significant contribution to the overall loss modulus of the blend. For EPDM/SBR, the contribution of the interphase is not that much pronounced but is still not negligible, in particular in the further analysis of filled rubber blends. This can be related to the difference of solubility parameter $\Delta\delta = 0.77\ (J/cm^3)^{1/2}$ for NR/SBR and $\Delta\delta = 1.55\ (J/cm^3)^{1/2}$ for EPDM/SBR [43].

Table 3 Volume fraction of different components in the unfilled rubber blends

	ϕ_{NR}	ϕ_{SBR}	ϕ_{EPDM}	$\phi_{\text{interphase}}$
SBR/NR	0.465	0.465	-	0.07
EPDM/SBR	-	0.46	0.50	0.04

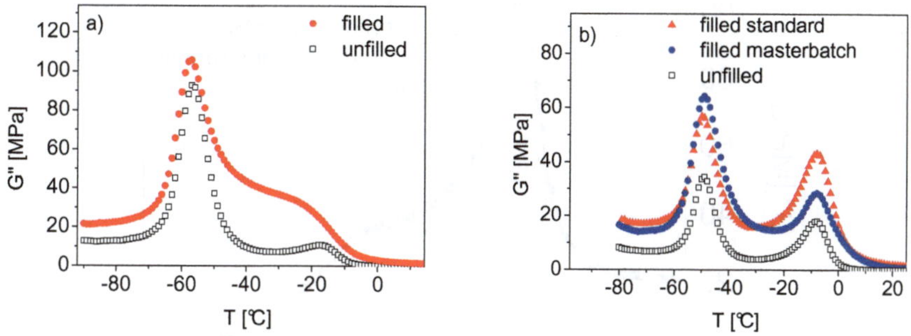

Fig. 17 Loss modulus during the glass transition of unfilled and filled rubber blends with 60 phr N234; (a) NR/SBR; (b) EPDM/SBR for different blending procedures, as indicated (see text)

The loss modulus of rubber blends filled with carbon black is shown in Fig. 17 (a) for NR/SBR and Fig. 17 (b) for EPDM/SBR. Just as for the unblended systems, the filler causes an increase of the maxima of loss modulus. For NR/SBR, it is striking that the rise in the modulus is extremely large in the region between the two maxima in which the interphase is located. This indicates high filler loading of the interphase. In contrary, both maxima increase strongly for EPDM/SBR, showing that the filler here is mainly located in the domains of the pure polymers. For this blend system, two mixing procedures have been used: first blending the two polymers and then adding carbon black (here named as standard procedure) and adding carbon black to EPDM and thereafter blending with S-SBR (master batch procedure). The two mixing procedures lead to different loss moduli as seen in Fig. 17 (b). The maximum at small temperatures, the contribution from the EPDM phase, is larger for the master batch procedure than for the standard procedure, whereas the maximum at large temperatures, contribution from the SBR phase, is smaller. This indicates that the filler content is larger in EPDM and smaller in SBR for the masterbatch than for the standard procedure.

In order to quantify the filler distribution in the blends, the measured curve has been separated into the contributions of the single phases. For this, the filled rubber blend has been fitted by the sum of three functions for the two polymers and the interphase, which have been determined in the fitting procedure of the unfilled rubbers (see previous section) and are assumed not to be altered by the filler. The spline parameters used here are the amplitude and the broadness of the loss maxima. The glass temperatures have been kept the same as for the phase contributions of the unfilled rubber blend. The 3-phase fit is shown exemplarily for NR/SBR in Fig. 18. The filler volume fractions in the different phases have been evaluated from the separation procedure by using eqs. (16) and (17). The results are summarized in Tab. 4.

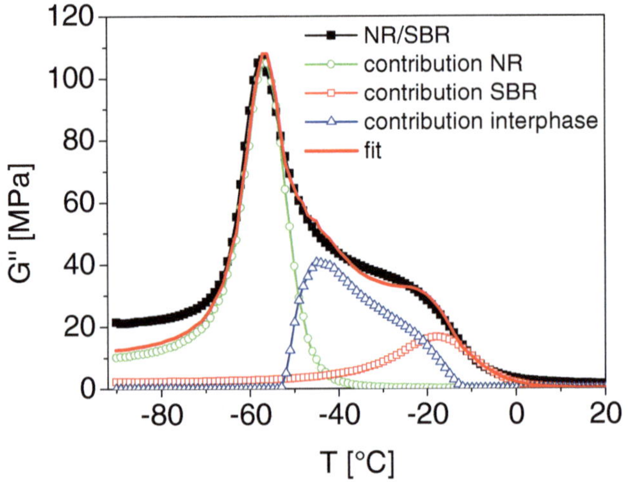

Fig. 18 Three-phase fit for filled NR/SBR in order to separate the contribution of the different phases to the loss modulus of the blend

For the NR/SBR blend, most of the carbon black is found in the SBR phase, while much less is found in the NR phase. This can be attributed to the high affinity between the carbon black surface and styrene units of the SBR which both contain conjugated π-electron rings. But, on the other hand, the NR has a high viscosity because of high molecular weight, and this will reduce the diffusion of carbon black aggregates significantly. Furthermore, more carbon black is inside the interphase than in the NR phase. Since the amount of interphase is much less than of the single polymer phases (see Tab. 3), the filler volume fractions relative to the corresponding phase have been calculated. Here, it can be seen that the interphase is highly loaded. Thus, the carbon black in NR/SBR is preferably located in the interphase.

Table 4 Filler distribution in the rubber blends: the filler volume fraction denoted by $\phi_{F,i}$ is relative to the total sample and the one denoted by $\tilde{\phi}_{F,i}$ is relative to the corresponding phase *i*.

	$\phi_{F,NR}$	$\phi_{F,SBR}$	$\phi_{F,EPDM}$	$\phi_{F,iph}$	$\tilde{\phi}_{F,NR}$	$\tilde{\phi}_{F,SBR}$	$\tilde{\phi}_{F,EPDM}$	$\tilde{\phi}_{F,iph}$
NR/SBR	0.03	0.11	-	0.08	0.07	0.23	-	0.61
EPDM/ SBR standard	-	0.15	0.06	0.02	-	0.29	0.14	0.38
EPDM/ SBR masterbatch	-	0.08	0.08	0.05	-	0.17	0.18	0.62

For the EPDM/SBR blend fabricated by the standard procedure, the differences in the filler volume fractions related to the phases are much less than in the NR/SBR blend. Here, the interphase is loaded slightly higher than the SBR phase. Less carbon black is located in the EPDM phase than in SBR, showing the higher affinity of the carbon black to SBR. Using the masterbatch procedure at which the carbon black was firstly batched with EDPM, the carbon black is almost equally distributed in both polymer phases. Interestingly, the interphase is again highly loaded, comparable to the NR/SBR blend. Obviously, a carbon black transfer has taken place from the EPDM to the more preferred SBR phase and a large amount of carbon black got stuck at the border of the phase domains.

6 Crack Propagation under Fatigue Conditions

In order to examine crack propagation under fatigue conditions, measurements in the Tear Fatigue Analyzer (TFA) were carried out. Before going into the crack propagation measurements themselves, it is appropriate to display the nature of crack propagation at selected examples.

6.1 Crack Growth Rates and Morphology

First, a cut of about 1 mm in length was made with a sharp razor blade. During the first few hundred cycles, mostly a relatively fast crack propagation is observed. This is connected with the very smooth starting crack front. This front is then

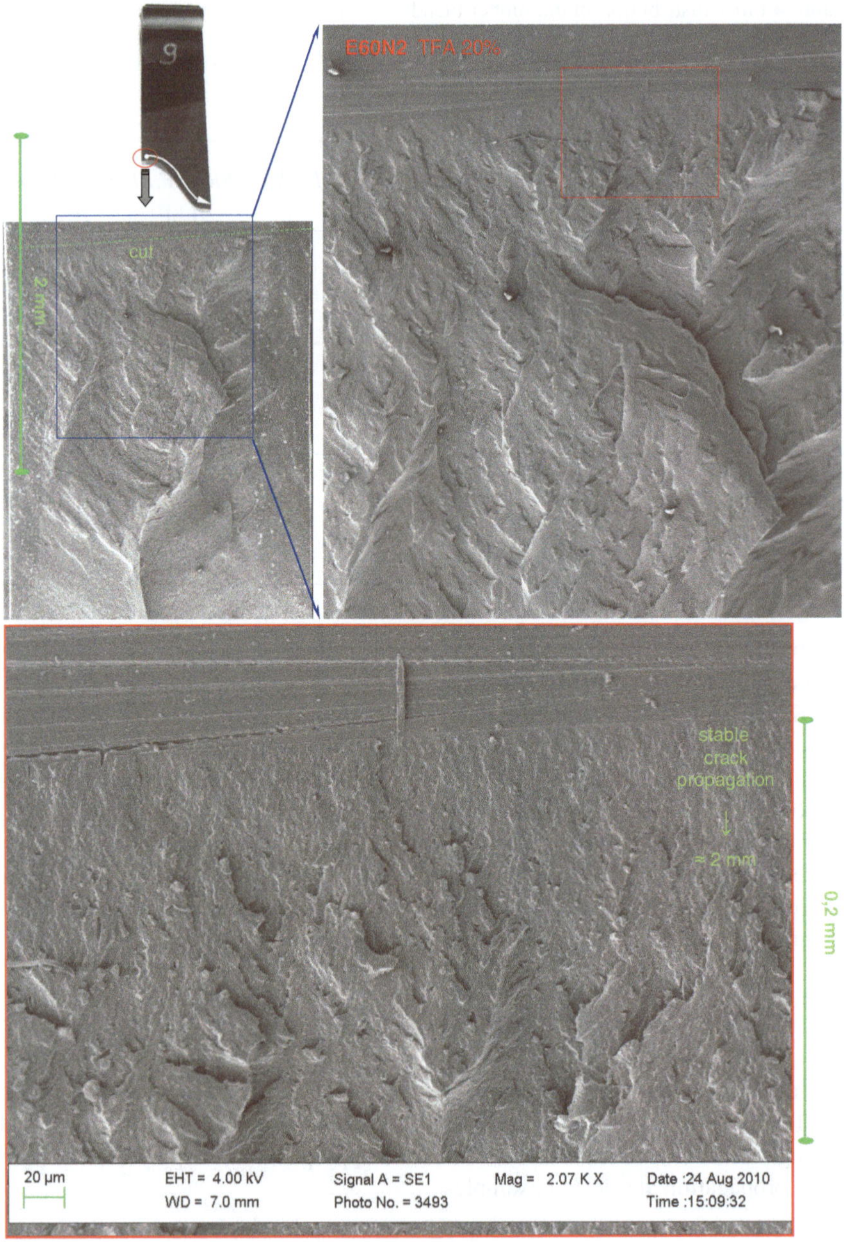

Fig. 19 REM micrograph of a crack surface from a TFA measurement at a SEN specimen (top left) of CB filled EPDM at different magnifications. The transition from smooth to rough with increasing crack length becomes apparent.

fragmented over a path of a few hundred μm, which is recognized by a very rough crack surface (Fig. 19, see also [24]). On the other hand, the "effective notch radius" will change with beginning crack propagation. Therefore, the crack propagation necessarily has to by analyzed only after a corresponding "running distance". Fig. 19 shows that the crack front does not run in a straight direction, as assumed in theory. Rather, at the μm-scale neighbouring crack facets seem to run on different levels and, after a certain running distance, to melt forming a common crack front. From single overlapping material regions which can also be seen, crack splitting could be expected. Obviously, the real crack surface is larger than that calculated using Euclidean geometry, $S = 2b_0c$; actually the calculation had to be carried out via fractal exponents of the surface, as determined in recent works [48].

From the TFA values of crack length, a number of regions or regimes of the crack propagation can be distinguished. After the initial phase, described above, the (half) contour length c (of the unloaded specimen) grows more or less linearly with the number of cycles n, and the velocity shows a slight increase according to the increase of c and T_{el}, respectively. The final phase is characterized by a kind of singularity. This can be described by eq. (2), if for the tearing energy eq. (3a) is substituted, resulting in the following differential equation:

$$\frac{dc}{dn} = A\left(\frac{2\pi}{\sqrt{\lambda}} w_{el}\right)^B c^B. \tag{18}$$

This makes clear that crack propagation in a SEN specimen results from a self-amplifying mechanism, because the velocity of change of crack length is over-proportional to the latter, and experimental values of the exponent B lie in the range between 1 und 5. Starting from a cut length c_0 and a constant energy density w_{el}, the differential equation can be solved for c, explicitly, and integrated:

$$c(n) = \left[c_0^{-(B-1)} - (B-1)\,A\left(\frac{2\pi}{\sqrt{\lambda}} w_{el}\right)^B n\right]^{-\frac{1}{B-1}}. \tag{19}$$

The validity of this equation is demonstrated for the example of unfilled S-SBR at ε_{max} = 25% and harmonic pulse frequency of f = 1 Hz in Fig. 20. The chosen parameters c_0 = 1.4 mm and w_{el} = 0.066 N/mm^2 are taken directly from the TFA results, while A = 0.00122 mm/cycle and B = 3.508 result from the plotting of $\log(\frac{dc}{dn})$ over log T_{el} (Paris plot, see bolow). A second example of a CB-filled EPDM, E60N2 at $\varepsilon_{max} = 20\%$ is depicted in Fig. 21 (a).

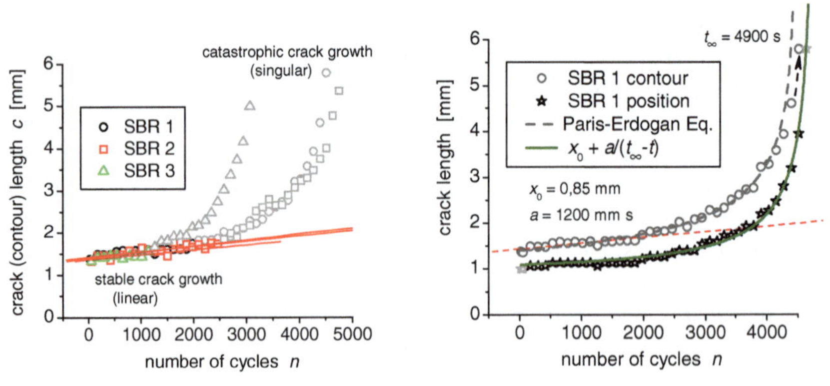

(a) (b)

Fig. 20 Kinematics of crack propagation of three similary loaded samples of unfilled S-SBR, $\varepsilon_{\max} = 25\%$, $f = 1$ Hz; (a) half counter length c (symbols), linear regressions (lines); (b) sample 1: half counter length c (circles) and simulation with eq. (19) (dashed line); position of the crack tip x (stars) and simulation with eq. (20) (solid line)

The kinematics of the crack can also be described by a phenomenological relation between n or time t, respectively, and the crack position x - but, only when the crack propagates perpendicularly to the loading direction and does not deviate from this plane. This is the case for unfilled S-SBR, see Fig. 20. For EPDM filled with N 234, in contrast, the cracks in all specimens turn after a certain crack progress, see Fig. 21 (a). For the crack position as a function of time we choose:

$$x(t) = x_0 + \frac{a}{t_\infty - t}, \tag{20}$$

i.e. a hyperbola diverging at $t = t_\infty$. From this, the velocity v along the crack path as a function of x can be derived:

$$v = \frac{\mathrm{d}x}{\mathrm{d}t} = \frac{1}{a}(x - x_0)^2, \tag{21}$$

This means that v should approximately follow a parabola along the crack path whose apex is situated at $x = x_0$; a is a free fitting parameter. For S-SBR and EPDM, unfilled as well as filled, a value of $x_0 = 0.85$ mm is found; the parameters a and t_∞ depend on the material and amplitude of loading, these are given in Figs. 20 and 21.

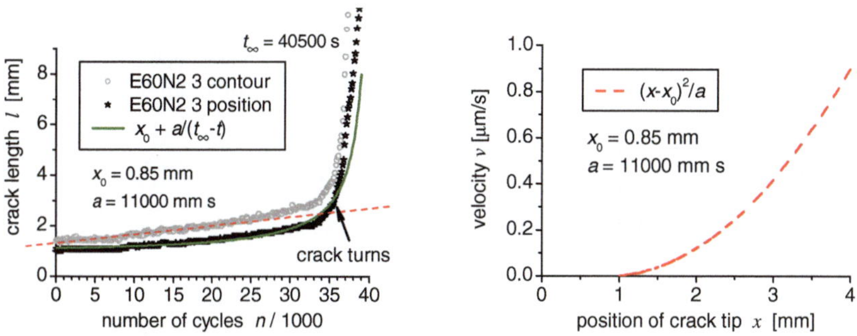

(a) (b)

Fig. 21 Kinematics of crack propagation of CB filled EPDM (E60N2), ε_{max} = 20%, f = 1 Hz, (a) half counter length c (circles) and position of the crack tip x (stars), Simulation with equ. (20) (solid line); (b) crack velocity vs. position

In this way, the crack surface structure, as observed under the microscope, can be connected with a crack velocity $v(x)$ at the respective position. For E60N2 (ε_{max} = 20%, Fig. 19), the crack surface becomes more and more rough till $x \approx 1.5$ mm. This takes place at very low (average) velocities of $v < 50$ nm/s, which still corresponds to several network chain lengths per cycle (these, however, are only averaged values - it can not be seen from the measurements if sections of the crack front are arrested for a longer period and move rather in jumps). When the crack length approaches $x \approx 2$ mm, a transition to a smoother surface is observed. At $x \approx$ 3 mm and velocities of $v > 400$ nm/s, the surface becomes almost flat, while now a beginning deviation of the propagation direction is recognized. The following catastrophic crack propagation takes places at velocities of several µm/s and leads to a surface that is smooth down to the lower µm range. Only occasional "defects" of approx. 10 µm size are to be seen which probably originate from ZnO particles or CB agglomerates. Also to mention: if crack velocity, instead according to position x, is determined according to the half contour length c (see: Figs. 20 and 21), somewhat higher velocity values result, but this has no qualitative effect.

6.2 Analysis of TFA Measurements According to Paris and Erdogan

In order to determine the material-specific parameters of the Paris-Erdogan equation (2), an elaborate statistics for a number of different amplitudes with 3 (S-SBR) or 4 (EPDM) equally loaded specimens each is necessary. Two examples are depicted in Fig. 22. Despite of statistical variation, it shows a faster crack propagation with increasing cut length c_0, as expected from theory. The crack growth rate was analyzed at selected sections of the curves for which also the averaged crack length and energy density were determined. The single results were assembled for growth-rate tearing-energy curves (Paris plots).

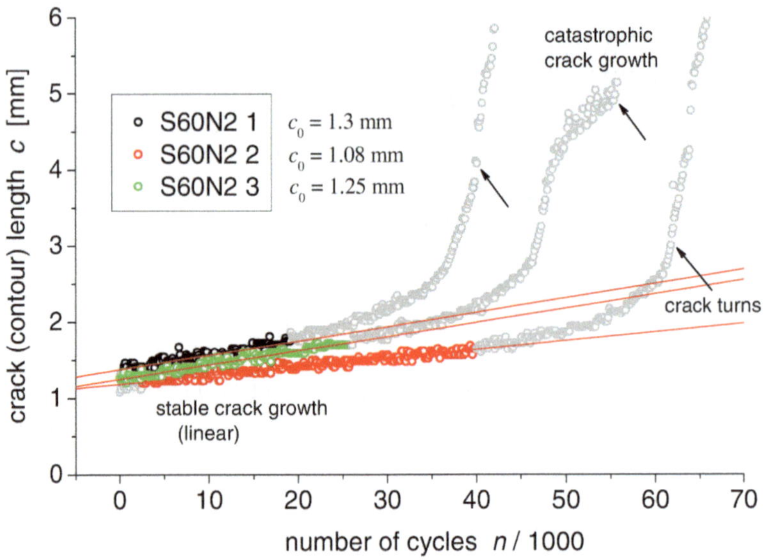

(a)

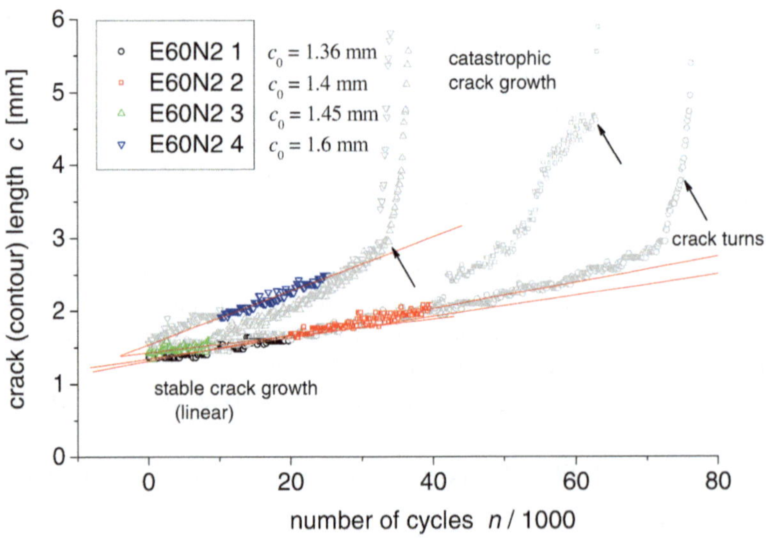

(b)

Fig. 22 TFA-results of crack propagation for rubber samples filled with 60 phr N234, (a) S-SBR (S60N2) and (b) EPDM (E60N2) with different initial crack length at ε_{max} = 20% and f = 1 Hz (symbols); The initial crack propagation is described by linear regression (lines)

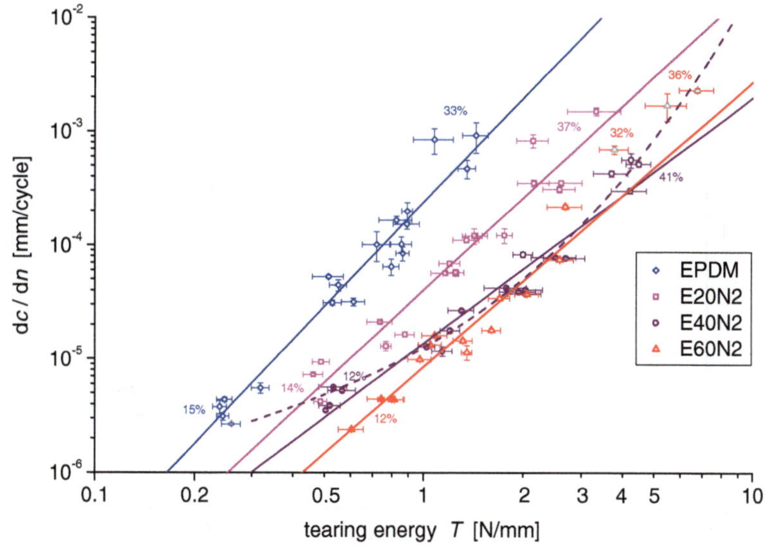

(a)

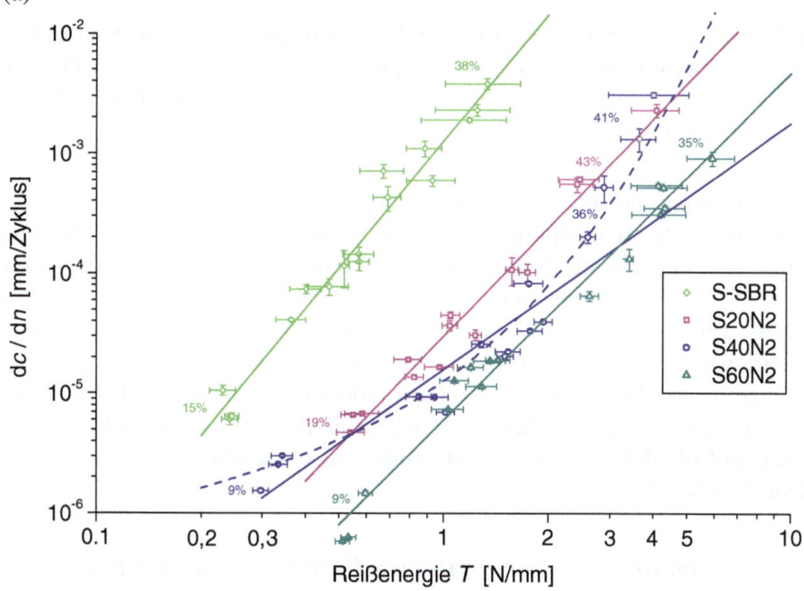

(b)

Fig. 23 Paris plot for unfilled and filled (N234) rubber samples from TFA measurement (symbols), (a) EPDM and (b) S-SBR; the strain amplitude varies between $\varepsilon_{\max} = 9\%$ ... 43%, linear regression (solid lines)

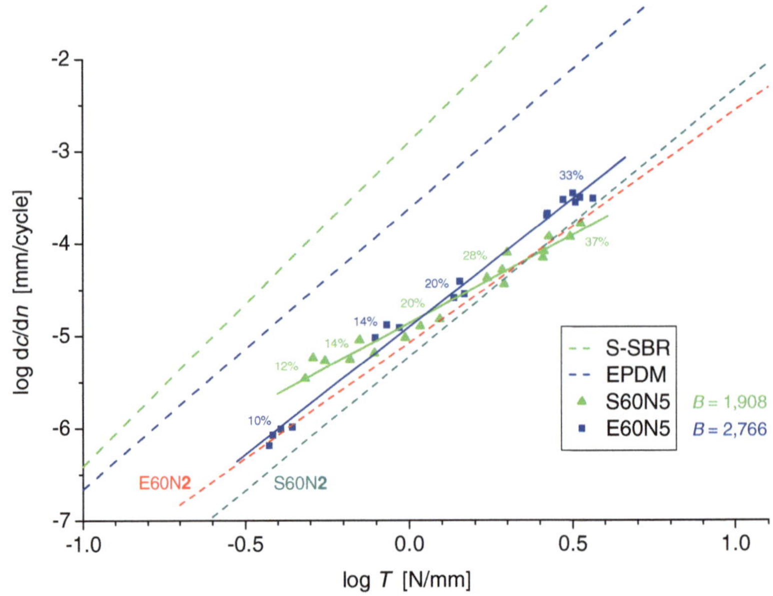

Fig. 24 Paris plot for S-SBR and EPDM filled with 60 phr N550 (S60N5, E60N5) from TFA measurement (symbols); the strain amplitude varies between ε_{max} = 10% ... 37%, linear regression (solid lines); for comparison, plots for unfilled S-SBR and EPDM and filled with N234 (S60N2, E60N2) are given (dashed lines)

Linear regression of the logarithmic plots are depicted in Fig. 23 for varied N234 content. At high amplitudes, often the growth rate could only be determined from a few data points, because many specimens showed crack deviation or failure after some hundreds of cycles. The remaining measurements have been analyzed (grey symbols) but were not included in the regression, because the statistics would then only consider specimens with low growth rate and therefore would be biased. Fig. 24 shows the Paris plots for vulcanizates filled with N550. As with N234, a higher potential of CB reinforcement is seen with S-SBR in relation to EPDM. In average, N234 counteracts crack propagation in some more extend than N 550.

6.3 *Correlation of Microstructure Parameters with Fracture Mechanical Properties*

TFA measurements with CB and silica-filled S-SBR and EPDM have already been carried out in [49]. It could be demonstrated that the structure of the filler network is connected closely with the fracture-mechanical characteristics. This shall be regarded in the following for the vulcanizates examined here, whereby it is still to be dealt with the change of the Paris plots as a function of polymer and filler as well as the underlying mechanisms in more detail.

Both for the SBR and the EPDM compounds (Fig. 25) the same tendency results with rising content of N 234: The factor *A* decreases, which goes back to the reinforcement effect. The exponent *B* decreases likewise with rising filler content, which indicates an increase of the tearing resistance at high tearing energy. Since an ideal-elastic material would exhibit an infinitely high slope the log-log plot, the drop of *B* can be explained first of all by the increasing visco-elasticity. However, a comparison between CB and silica-filled S-SBR [49] shows a contrary trend: the same quantity of slica (chemically bound to the rubber) causes a clearly smaller hysteresis than N 339. The tearing resistance of the silica vulcanizate is comparably high, *B* is however significantly smaller. Thus, the exponent can not be attributed completely to visco-elasticity, therefore in the following the influence of the filler network on the tearing resistance shall be examined more closely.

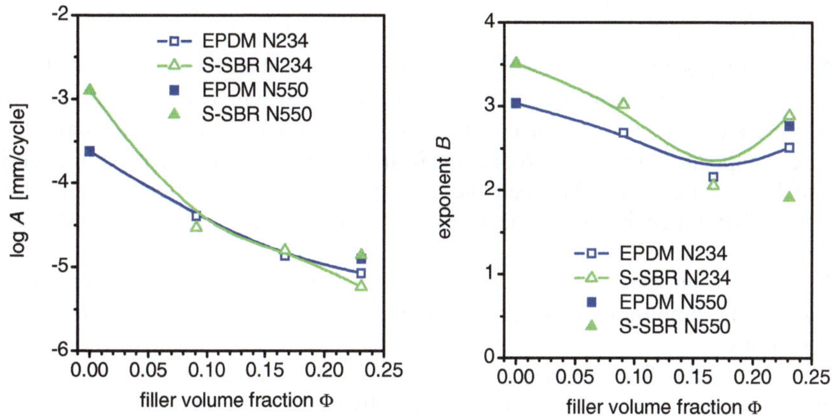

(a) (b)

Fig. 25 Paris-Erdogan parameters for S-SBR und EPDM as a function of filler type and filler content (N 234 und N 550): (a) prefactor *A*; (b) exponent *B*

At about 40 phr of N 234 a maximum of *B* occurs, this is apparently connected with the gel point $\Phi^* \approx 0.16$ of the CB network which is reached at this filler content [50]. At a closer view, however, at smaller amplitude a higher growth rate is measured, than it would correspond to the Paris Erdogan relationship, so that the exponent *B* increases with increasing tearing energy. This is seen for S-SBR as well as EPDM with each 40 phr N 234 (dashed lines in Fig. 23). In former investigations [49] a variation of N 339 content in S-SBR was carried out, where a minimum of *B* was also seen at 40 phr. But, an increase of *B* with rising T_{el} was not observed, because only measurements with tearing energies up to $T_{el} \approx 1$ N/mm could be realized (due to the high cycle frequency of 10 Hz), but the increase only appears above $T_{el} \approx 2$ N/mm. The increase of *B* from 40 to 60 phr CB was explained by the temperature increase near the crack tip (flash temperature) due to the increased energy dissipation at rising filler content (larger loss angle) as well as rising tearing energy. This appears reasonable because the heating of the specimen by cyclic deformation can not be neglected at the chosen pulse frequency of 10 Hz. However, in our case of 1 Hz the thermal heating is much smaller. Also, the determined value of B is found to be influenced by the range of T_{el} in which the slope is analyzed in the Paris plot. A possible reason for the deviation from a linear dependence in the log-log plot could result from the stress softening (Payne effect) which increases significantly above the gel point.

On reaching the gelling point (percolation threshold), a pervasive filler network forms. This manifests itself primarily in a significantly higher storage modulus G'. The Payne effect — specifically the drop in G' brought about by the breakup of the network as the deformation amplitude increases — occurs above all in the case of small elongations of up to $\varepsilon_{max} \approx 10\%$. If the load on a specimen in the area of homogeneous strain ahead of the crack tip has a smaller amplitude, this means that G', the value crucial for the measured energy density w_{el}, has not dropped as much as directly at the tip of the crack, where much higher elongations prevail. The effect of this is that the crack process zone is more markedly deformed than it would have been without stress softening. This implies a higher energy density for the same stress level. Of course, this applies only for the mean value across the entire composite. Due to the reduction in the reinforcement factor, elongation is more likely to decrease between the clusters. It is solely within the broken-up clusters that much higher elongations are attained. This ultimately leads to a higher crack rate at small amplitudes, as the TFA measurement on the elastomers filled with 40 phr carbon black shows. The effect should also occur at higher filler loadings, the amplitude at which the breakup occurs shifts, however, to smaller elongations.

As far as the filler influence is concerned, the relevant literature discusses various mechanisms in addition to the generally elevated hysteresis. These include crack tip blunting & branching [19,24]. Crack branching represents an effective reinforcement mechanism in that not only a larger crack surface forms but, above all, the deformations in the process zone are distributed over a larger area. This temporarily impedes crack propagation and ultimately leads to a fissured fracture plane. To be sure, crack branching is seldom in evidence for the SBR and EPDM compounds under study, and then only in the case of 60 phr N234. It plays a greater role in the facture plane morphology of crystallizing elastomers like NR. Applying electron microscope analysis, Hess & Ford observed cracks in vulcanizates filled with carbon black, running directly along the surface of the particles [51]. A characteristic of markedly stretched vulcanizates is the detachment of small polymer areas prior to failure, in particular of particles of a lesser size. There is hardly any observation of carbon black aggregate fracturing. Such a pattern might also manifest itself in the vicinity of the crack tip, although observations of this kind are unknown in the relevant literature. This gives rise to the idea, at least, that cavities (microcracks) form ahead of the tip of the crack, eventually merging to constitute a crack. This might be a mechanism giving an explaination for the complex fracture plane morphology as well as for fissuring of a crack.

To date there has been little discussion of the influence of the filler-filler bond on crack propagation. As the crack can hardly run through the much firmer filler particles, it has to work its way around them. In the case of a sufficiently high filler concentration, its path inevitably crosses the filler-filler bonds. Above the glass temperature the immobilization of the rubber on the filler surface results mainly in an increase in the relaxation or retardation times. This means that for the given load, stretching the chain up to its eventual elongation takes more time than in the case of an unfilled elastomer. It implies that the strength of the bonds becomes correspondingly temperature- and time-dependent. In the preceding section this was determined on the same filled systems at room temperature and constant elongation rate by means of parameter adaptation with the DFM. As filling load increases, s_v and s_d climb more steeply: in the case of an increase of 20 phr N234, the strength values rise by a factor of 2 to 2.5. The mean cluster size x_0 also rises and clusters start to overlap. This means that the crack either has to destroy filler-filler bonds or make a bigger "detour" around the clusters. Parallel to this, there is a shift in the Paris plots in Fig. 23 to lower crack growth rates. If one ignores the curves in the case of 40 phr, whose slopes deviate more markedly, a 2-to-3 factor drop of $\frac{\mathrm{d}c}{\mathrm{d}n}$ (for given crack energy) is evident with a rise of 20 phr in black carbon content. Similarly, there is an increase in T_{el} with the given crack rate. Admittedly, at higher filler loadings, the factor increases for crack energy are smaller than the increases in strength values.

For N550, no concentration series was prepared, so that in Fig. 24 solely the vulcanizates filled with 60 phr are compared with the unfilled vulcanizates. For the filled systems, the factor A is $\approx 10^{-5}$ in each case; the exponent B varies in the range 2-to- 3 range (Fig. 25). In the case of S-SBR with 60 phr N550 (S60N5), the relative small vaue of B (= 1.91) is striking, i.e. relatively pronounced crack growth small crack energy values and relatively weak growth in the case of high energy values. By way of a concrete comparison, for T_{el} = 3 N/mm, this means a reduction in the growth rate by a factor of ≈ 300 in comparison to unfilled S-SBR, which, however, is elongated considerably more due to the smaller modulus. But even with the same elongation of 38%, there is still a reduction by a factor of ≈ 20. Ref. [52] contains a further discussion on the influence of microstructure on dynamic crack growth.

6.4 Dynamic Crack Growth in Rubber Blends

The mechanism of crack growth in rubber blends under dynamic excitation is of high technical interrest because many rubber goods, as e.g. tire treads, consist of two or more blended rubbers delivering improved mechanicval properties due to symbiotic iteractions between the blend components. Paris plots of the unfilled SBR/EPDM and SBR/NR blends in comparison with the homo-polymers are shown in Fig. 26. It becomes obvious that for both systems no significant improvement is observed. For both systems the resistence against dynamic crack propagation is roughly located between those of the single rubbers. At low tearing energy, cracks in the SBR/EPDM blend propagate even faster than in the single rubbers which can be related to the weak interphase. SBR/EPDM blends are quite uncompatible and show large phase domains which are visible even in light microscopic images. The evaluation of blend morphology in Section 5b demonstrates that these systems also exhibit a low amount of interphase. For the unfilled SBR-NR blend the resistance against crack propagation is significantly better compared to the SBR/EPDM blends, but the crack propagation rate is roughly still one order of magnitude larger than for the pure NR.

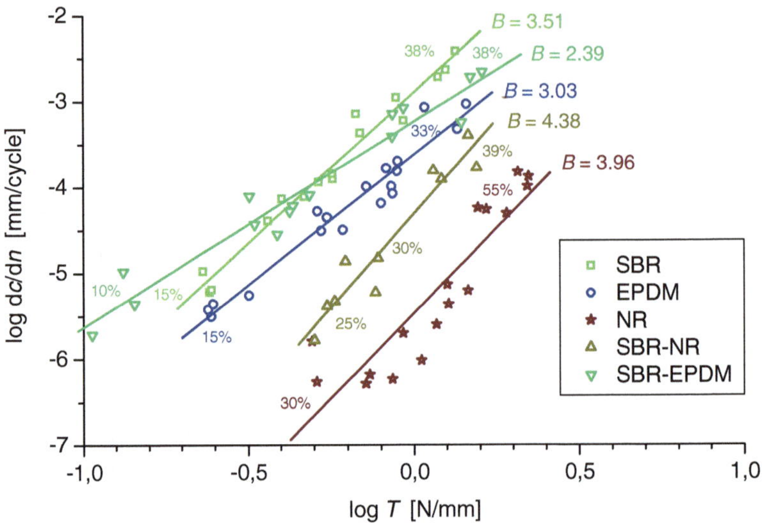

Fig. 26 Dynamic crack propagation behaviour of the unfilled SBR/EPDM and SBR/NR blends in comparison with the homo-polymers

The Paris plot in Fig. 27 shows the dynamic crack propagation behaviour of the SBR/EPDM blends (unfilled and filled with 60 phr N234) in comparison with the unfilled and filled homo-polymers. Surprisingly, it is found that for the SBR/EPDM blend crack resistence is not improved by adding CB. It appears that the weak interphase in these systems can not be reinforced by the addition of filler. For the filled SBR/EPDM blend system, two mixing procedures have been used: first blending the two polymers and then adding carbon black denoted as SE60N2N (standard procedure) and adding carbon black to EPDM and thereafter blending with S-SBR which is denoted SE60N2M (master batch procedure). Fig. 27 demonstrates that an improvement is also not reached by applying the special batching procedure during mixing which increases the amount of CB in the interphase. Obviously, the unsufficient compatibility of the two polymers implies that independent of reinforcing mechanisms the crack propagates quite fast through the weekest structure of this blend system, i.e. the thin interphase with a low level of chain interpenetration.

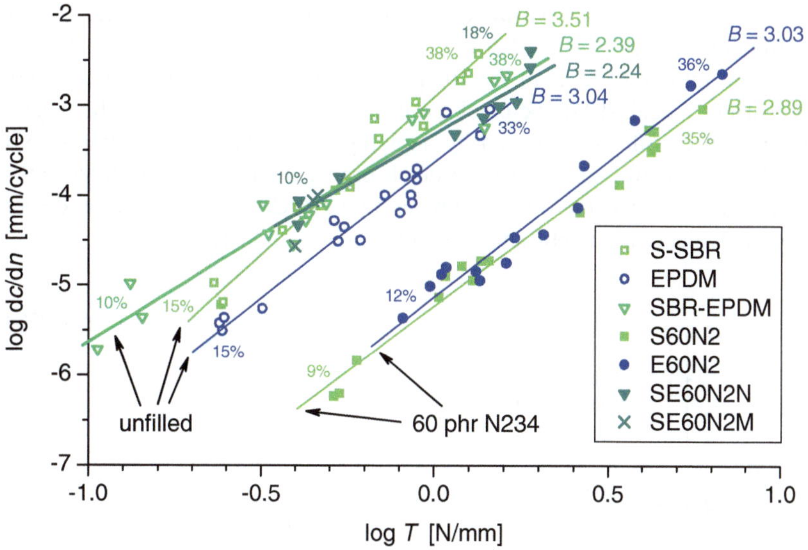

Fig. 27 Dynamic crack propagation behaviour of the SBR/EPDM blends (unfilled and filled with 60 phr N234) in comparison with the unfilled and filled homo-polymers. For the filled blend, samples prepared with two different mixing procedures SE60N2N and SE60N2M are shown.

The Paris plot in Fig. 28 shows the dynamic crack propagation behaviour of the SBR/NR blends (unfilled and filled with 60 phr N234) in comparison with the unfilled and filled homo-polymers. Obviously, the resistence against dynamic crack propagation of the unfilled blend is located right between those of the single rubbers. Roughly speaking, the crack propagation rate of the blend is about one order of magnitude larger than that of pure NR but also one order of magnitude lower than that of the pure SBR. In Section 5b it was demonstrated that the SBR/NR blends show significantly higher interphase volume fractions compared to the EPDM/SBR blends due to the higher compatibility between NR and SBR (Tab. 3). This indicates that the interphase is sufficiently strong to prevent the cracks propagating though the blend boundary between both phases. Instead, the crack seems to propagate through the single phases whereby the NR-phase acts as a crack blocking area slowing down the mean propagation rate. A similar picture is observed for the filled NR/SBR blend. Again, the resistence against dynamic crack propagation of the filled blend is located between those of the single filled rubbers indicating that the crack is not propagating along the interphase but through the rubber phases. But for the filled systems the crack propagation rate lies much closer to the very low value of the NR, specially in the range of low tearing energies. Accordingly, for the filled systems there seems to be an additional synergetic effect of the highly filled interphase (compare Tab. 4) which hinders the propagation of the crack more efficient than the interphase of the unfilled NR/SBR blend. However, we point out that according to Tab. 4 the distribution of carbon black between the phases is rather inhomogeneous with a quite low amount of filler in the NR-phase. In previous studies it was found that the resistance against fatigue crack propagation increases if a more homogeneous distribution of filler is achieved by applying special batching technologies [44]. For that reason we expect that the crack propagation rate of the filled NR/SBR blend can be further improved by preparing a NR/CB masterbatch, i.e. dispersing the carbon black first in NR and then blending it with SBR.

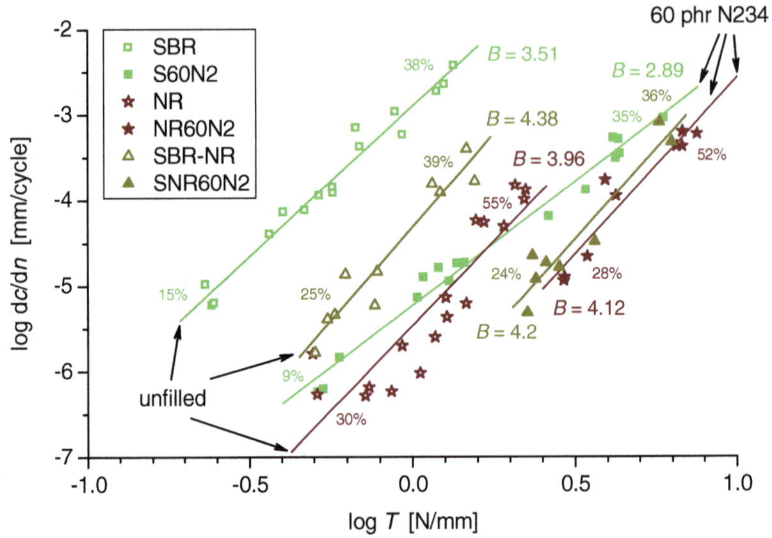

Fig. 28 Dynamic crack propagation behaviour of the SBR/NR blends (unfilled and filled with 60 phr N234) in comparison with the unfilled and filled homo-polymers

7 Conclusions

Reinforcement of elastomers by means of nanoscale fillers owing to polymer-filler interaction as well as their influence on the non-linear dynamic behavior has been discussed. A novel analysis of stress relaxation shows that qualitatively, an S-SBR filled with carbon black exhibits the same relaxation behavior as the unfilled material. It has been shown that even a variation in crosslinking yielded, essentially, nothing more than a modification of the relaxation strength. The stress (minus an entropically elastic final value) decreases with a power function of time. Earlier studies have already found that this is the case also for unfilled elastomers. The findings corroborate that the stress relaxation of filled elastomers proceeds in the same way. This is confirmed by the fact that the final stress arrived at an (indefinitely) long time is proportional to temperature reflecting the entropic elasticity of the matrix. The power function exhibits the property that a scaling of the stress (relaxation intensity) corresponds to a scaling on the time axis. It is possible to interpret this as a characteristic of the filler-filler bonds, which are composed of the matrix material but exhibit retarded dynamics overall. There are, however, still other possible interpretations as well. It must thus be left to future studies to definitively clarify the physical state of the bonds.

The stress relaxation behavior offers further interesting insights regarding the crack propagation characteristics of filled elastomers. The relationship between the elastic energy density present directly at the tip of the crack and causing a breakup of the polymer chain, on the one hand, and the "external" energy density, on the other, is similar to the relationship between the length of the crack and the radius of the tip, resembling a lever principle. In the case of a blunt tip (larger radius), the result is a higher crack energy. There are strong indications that setting effectively triggers a blunting of the crack tip in much the same way as plastic deformation, which vanishes over time upon retraction, however. The setting intensity has been shown to increase with a reduction in crosslinking, although this cannot be equated with a plastic behavior as it disappears over indefinitely long times. The relaxation of the radius of the crack tip also provides an explanation for the known fact that crack propagation for the given crack energy is essentially a function of cycle count and less a function of the load period.

The study presented here expands the "dynamic flocculation model" (DFM) by pointing up practical applications for component calculations. The formulated model describes the rubbery-elastic behavior, the filler-induced hydrodynamic reinforcement, stress softening, hysteresis and setting behavior as well as temperature- and time-dependency. It takes into account physical occurrences at the micromechanical level, especially the fracture and re-aggregation of filler clusters. Determination of the material parameters involved uniaxial multi-hysteresis testing. It was possible to show that the material parameters occurring in the model characterize the microstructure and can be interpreted as physical variables whose numerical values lie within a relatively narrow range. Polymer and filler variation exemplified the systematic behavior of the parameters: the effective filler volume fraction determined exhibits a roughly linear relationship to the filler fraction volume of carbon black N 234, with a constant of proportionality of ≈ 1.9, reflecting the structure of the filler particles i.e. their deviation from the spherical shape. The reinforcement of the rubber matrix was shown further to be effected by relatively small clusters that manifest themselves as correspondingly stiff. Their size increases with increasing filler loading and specific surface area, with the mean cluster size not attaining any more than approximately ten particle diameters, however. The coupling of polymer chains with the active filler surface increases the mechanically effective crosslinking density of the matrix. A mean distance between adsorption sites of ≈ 0.5 nm was determined. The strength of the filler-filler bonds increases with rising filler loading, as regards both virgin and damaged bonds. This can be related to the decreasing gap distance between adjacent filler particles, effecting a stronger immobilization of the adsorbed macromolecules and thus stronger bonds. Relaxation measurements corroborated the hysteric character of setting, with the relaxation strength of the setting stress dependent only on the immediately preceding maximum stress, but not on preconditioning.

Alongside finite element simulation of a tensile specimen notched on one side, it was also possible to successfully achieve validation by means of comparison with corresponding measurements, with the material parameters (in this case, an EPDM with carbon black filler) derived solely from uniaxial tensile testing. Applications of the model were presented, especially in the area of fracture mechanics, both for the calculation of stress softening in a notched specimen and for the interpretation of crack propagation measurements. A tear fatigue analyzer was used for these measurements, which were evaluated with regard to changes in the logarithmic Paris plots, considering crack propagation rate over crack energy. The power law exponent B was found to decrease systematically with rising filler loading, with an aberration occurring in the vicinity of the gelling point where the exponent depends on the crack energy range selected. It declines in the direction of low crack energies. This can be explained by the breakup of the filler network occurring initially at higher filler loadings. There was also found to be a correlation between the filler-filler bond strengths, as determined by model parameter adaptations, and the readings for crack resistance.

Beside the different homo-polymers, with only one elastomer component, also rubber blends have been investigated. A dynamic-mechanical analysis of CB distribution in the blend phases and interphase has been performed. The CB distribution exhibits characteristic differences reflecting the compatibility of the respective rubbers with CB. Nevertheless, the distribution is not determined by equilibrium thermodynamics alone, but it can also be influenced by applying special mixing procedures (master batch mixing). SBR/EPDM blends show large phase domains which are visible even in the light microscope and a low amount of interphase. At low tearing energy, cracks in the blend propagate faster than in the single rubbers which can be related to the weak interphase. For these systems, crack resistence is not improved by adding CB and an improvement is also not reached by applying a special master batch technology. Contrary, SBR/NR blends show higher interphase volume fractions and the resistence against dynamic crack propagation is located between those of the single rubbers. This is the case for unfilled as well as filled systems, indicating that the crack is not propagating along the interphase.

References

[1] Donnet, J.-B., Bansal, R.C., Wang, M.J. (eds.): Carbon Black: Science and Technology. Marcel Decker Inc., New York Hongkong (1993)

[2] Schuster, R.H.: Verstärkung von Elastomeren durch Ruß - Teil 1: Morphologie & Charakterisierung von Rußen; Wechselwirkung zw. Ruß und Kautschuk; Mechanisches Verhalten gefüllter Vulkanisate, W.d.K (ed.), Grünes Buch Nr. 40 (1989)

[3] Schuster, R.H.: Verstärkung von Elastomeren durch Ruß - Teil 2: Dispersion, Hysterese & Reißverhalten; Zugfestigkeit, Bruchenergie & Abrieb, W.d.K (ed.), Grünes Buch Nr. 41 (1990)

[4] Röthemeyer, F., Sommer, F. (eds.): Kautschuktechnologie. Hanser, München Wien (2001)

[5] Edwards, D.C.: Polymer-filler interactions in rubber reinforcement. J. Matls. Sci. 25, 4175–4185 (1990)

[6] Kraus, G.: Reinforcement of elastomers by carbon black. Fortschr. Hochpolym. Forsch (Adv. Polym. Sci.) 8, 155–237 (1971)

[7] Hess, W.M., Lyon, F., Burgess, K.A.: Einfluss der Adhäsion zwischen Ruß und Kautschuk auf die Eigenschaften der Vulkanisate. Kautschuk Gummi Kunstst. 20(3), 135–141 (1967)

[8] Grosch, K.A., Harrwood, J.A.C., Payne, A.R.: Hysteresis in Polymers and its Relation to Strength. Rubber Chem. Technol. 41, 1157–1167 (1968)

[9] Dannenberg, E.M.: The effects of surface chemical interactions on the properties of filler-reinforced rubbers. Rubber Chem. Technol. 48, 410–444 (1975)

[10] Persson, B.N.J., Albohr, O., Heinrich, G., Ueba, H.: Crack propagation in rubber-like materials. J. Phys.: Condens. Matter 17, R1071–R1142 (2005)

[11] Klüppel, M.: The Role of Disorder in Filler Reinforcement of Elastomers on Various Length Scales. Adv. Polym. Sci. 164, 1–86 (2003)

[12] Rivlin, R.S., Thomas, A.G.: Rupture of Rubber. I. Characteristic Energy for Tearing. J. Polym. Sci. 10, 291–318 (1953)

[13] Klüppel, M.: Evaluation of viscoelastic master curves of filled elastomers and applications to fracture mechanics. J. Phys. Condens. Matter 21, 035104 (2009)

[14] Payne, A.R.: Rubber Chem. Technol. 36, 432 (1963)

[15] Kraus, G. (ed.): Reinforcement of elastomers. Wiley Interscience Publ., N.Y. (1965)

[16] Ferry, J.D.: Viscoelastic Properties of Polymers. J. Wiley & Sons, New York (1971)

[17] Gent, A.N. (ed.): Engineering with rubber, 2. Ausgabe, Hanser, München (2001)

[18] Griffith, A.A.: The phenomena of rupture and flow in solids. Philos. Trans. R. Soc. London A 221, 163–198 (1921)

[19] Lake, G.J., Thomas, A.G.: Thomas, Strength. In: [17]

[20] Irwin, G.R.: Analysis of stress and strain near the end of a crack traversing a plate. J. Appl. Mech. 24, 361–364 (1957)

[21] Freund, L.B.: Dynamic Fracture Mechanics. Cambridge University Press, Cambridge (1998)

[22] Rossmanith, H.-P.: Grundlagen der Bruchmechanik. Springer, Wien (1982)

[23] Ward, I.M., Hadley, D.W.: Mechanical Properties of Solid Polymers. Wiley, New York (1993)

[24] Ellul, M.D.: Mechanical Fatigue. In: [17]

[25] Paris, P.C., Gomez, M.P., Anderson, W.E.: A rational analytic theory of fatigue. The Trend in Engineering 13, 9–14 (1961); Paris, P., Erdogan, F.: A critical analysis of crack propagation laws. Journal of Basic Engineering, Transactions of the American Society of Mechanical Engineers, 528–534 (December 1963)

[26] Thomas, A.G.: Rupture of Rubber. II. The Strain Concentration at an Incision. J. Polym. Sci. 18, 177–188 (1955)

[27] Hutchinson, J.W.: Singular behaviour at the end of a tensile crack in a hardening material. J. Mech. Phys. Solids 16, 13–31 (1968)

[28] Gao, Y.-C., Shi, Z.-F.: The plane stress crack-tip field for an incompressible rubber material. Applied Mathematics and Mechanics 15, 499–506 (1994)

[29] McMeeking, R.M., Parks, D.M.: On Criteria for J-Dominance of Crack-Tip Fields In Large-Scale Yielding. ASTM STP 668, 175–194 (1979)
[30] Batzer, H. (ed.): Polymere Werkstoffe. Band III, Thieme, Stuttgart (1984)
[31] Eisele, U., Kelbch, S.A., Engels, H.-W.: The Tear-Analyzer - A New Tool for Quantitative Measurements of the Dynamic Crack Growth of Elastomers. Kautsch. Gummi Kunstst. 45, 1064–1069 (1992)
[32] Lorenz, H., Meier, J., Klüppel, M.: Micromechanics of internal friction of filler reinforced elastomers. In: Besdo, D., Heimann, B., Klüppel, M., Kröger, M., Wriggers, P., Nackenhorst, U. (eds.) Elastomere Friction. LNACM, vol. 51, pp. 27–52. Springer, Heidelberg (2010)
[33] Klüppel, M., Heinrich, G.: Kautsch. Gummi Kunstst. 58, 217 (2005)
[34] Lorenz, H., Freund, M., Juhre, D., Ihlemann, J., Klüppel, M.: Macromol. Theory Simul. 19, 110 (2010)
[35] Vilgis, T.A., Heinrich, G., Klüppel, M.: Reinforcement of Polymer Nano-Composites. Cambridge University Press, Cambridge (2009)
[36] Morozov, I., Lauke, B., Heinrich, G.: A new structural model of carbon black framework in rubbers. Comp. Mat. Sci. 47, 817–825 (2010)
[37] Wang, M.-J., Wolff, S., Tan, E.-H.: Rubber Chem. Technol. 66, 178–195 (1992)
[38] Luo, H., Klüppel, M., Schneider, H.: Study of Filled SBR Elastomers Using NMR and Mechanical Measurements. Macromolecules 37, 8000–8009 (2004)
[39] Kohjiya, S., Katoh, A., Suda, T., Shimanuki, J., Ikeda, Y.: Polymer 47, 3298–3301 (2006)
[40] Fritzsche, J., Klüppel, M.: J. Phys.: Condens. Matter 23, 035104 (2011)
[41] Freund, M., Lorenz, H., Juhre, D., Ihlemann, J., Klüppel, M.: Int. J. Plasticity 27, 902–919 (2011)
[42] Klüppel, M., Schuster, R.H., Schaper, J.: Rubber Chem. Technol. 72, 91–108 (1999)
[43] Schuster, R.H., Meier, J., Klüppel, M.: Kautsch. Gummi Kunstst. 53, 663–674 (2000)
[44] Meier, J.G., Klüppel, M., Geisler, H., Schuster, R.H.: Schuster. Kautsch. Gummi Kunstst. 58, 587–594 (2005)
[45] Li, C., Wu, J., Zhao, J., Zhao, D., Fan, Q.: Eur. Polym. J. 40, 1807–1814 (2004)
[46] Le Gal, A., Yang, X., Klüppel, M.: J. Chem. Phys. 123, 014704 (2005)
[47] Huber, G., Vilgis, T.A.: Eur. Phys. J. B 3, 217 (1998); Ibid. Kautschuk Gummi Kunstst. 52, 102 (1999)
[48] Horst, T., Reincke, K., Illisch, S., Heinrich, G., Grellmann, W.: Fracture Surface Statistics of Filled Elastomers. Phys. Rev. E 80, 046120 (2009)
[49] Klüppel, M.: The Role of Filler Networking in Fatigue Crack Propagation of Elastomers under High-Severity Conditions. Macromol. Mater. Eng. 294, 130–140 (2009)
[50] Gerspacher, M., Nikiel, L., Yang, H.H., O'Farrell, C.P., Schwartz, G.A.: Flocculation in Carbon Black Filled Rubber Compounds. Kautschuk Gummi Kunstst. 55, 596–604 (2002)
[51] Hess, W.M., Ford, F.P.: Microscopy of Pigment-Elastomer Systems. Rubber Chem. Technol. 36, 1175–1229 (1963)
[52] Lorenz, H.: Mikrostruktur-basierte Modellierung des mechanischen Verhaltens gefüllter Elastomere. PhD-Thesis, TU Dresden (2012)

Linking Mesoscopic and Macroscopic Aspects of Crack Propagation in Elastomers

Thomas Horst[1,2], Gert Heinrich[1,2], Martin Schneider[1], Annegret Schulze[1,3], and Mirko Rennert[1,4]

[1] Leibniz-Institut für Polymerforschung Dresden e. V., Hohe Straße 6, 01069 Dresden, Germany
[2] Technische Universität Dresden, Institut für Werkstoffwissenschaft, 01062 Dresden, Germany
[3] Technische Universität Chemnitz, Institut für Fördertechnik und Kunststoffe, 09107 Chemnitz, Germany
[4] Martin-Luther-Universität Halle-Wittenberg, Institut für Polymerwerkstoffe e. V., Geusaer Straße, 06217 Merseburg, Germany

Abstract. The fracture-mechanical characterization of elastomeric materials is based on a global energy balance. Tearing energy was introduced to characterize the energy required for an infinitesimal increase of surface area during crack propagation. The contribution of various energy dissipation mechanisms during such a process of crack propagation is crucial for the understanding and modification of elastomeric materials with respect to an enhanced service life. Energy balance is reviewed from both a theoretical and experimental point of view, leading on the one hand to possibilities and limits of generalized J-integrals for fracture mechanical characterization of elastomeric materials, and on the other hand to alternative procedures of experimental characterization of crack propagation in elastomers. In order to analyze the influence of viscoelastic material behavior on the crack propagation behavior in elastomeric materials, steady state fields were calculated dependent on the crack velocity. The results indicate a change in the size of the fracture process zone where the defect evolution takes place. Such characteristic length scales of the fracture process zone can be estimated from a statistical analysis of fracture surface topography by means of determining characteristic self-affine roughness exponents. A modeling of the material degradation due to the rupture of polymer chain segments within the fracture process zone was proposed, taking into account the overloading of chains both in the fully stretched state and due to fast loading.

W. Grellmann et al. (Eds.): *Fracture Mechanics & Statistical Mech.*, LNACM 70, pp. 129–165.
DOI: 10.1007/978-3-642-37910-9_4

1 Introduction

The fracture of materials is mostly an undesirable process that reduces the service life of structural components dramatically. Defects and flaws exist in every real material and act as initiators of macroscopic cracks. The behavior of an existent crack in a specimen of a specific material is investigated in fracture mechanics. It is an important branch of material science and much effort has been already made to modify materials with the objective of high fracture resistance at specified requirements with respect to other material properties.

Rubber materials are weakly cross-linked polymeric materials. Due to high internal flexibility, rubbers exhibit large deformations even under weak external forces above the glass transition temperature. These materials are usually filled with hard nano-structured particles in order to enhance different mechanical properties such as strength, hardness, stiffness or abrasion resistance. The resulting multiscale structural hierarchies due to the phase morphology, the distribution and dispersion of filler and the multiscale structure of filler itself influences the fracture mechanical behavior of elastomeric material.

The fracture of materials is a complex process in which multiple length scales are involved and the interaction of processes taking place in different length scales has a huge influence on the way the material breaks. Increased fracture toughness can be realized by modifying the structure of the material in such a way that as many non-catastrophic dissipative processes as possible on various length scales in a spacious volume are induced in the specimen. This results in an increase in the material's capability to absorb energy during crack propagation. Generally, these processes can be traced back to specific polymer-polymer, polymer-filler and filler-filler interactions in the material. The understanding of these interactions with respect to their influence on the fracture process is a prerequisite of modeling the processes taking place in the vicinity of the crack tip.

2 Energy Balance and Fracture-Mechanical Characterization of Elastomers

The susceptibility of rubber to large deformations leads to an elastic crack blunting, i.e. a crack opens up if the specimen is loaded (see fig. 1). However, this crack tip blunting is reversible, unlike the plastic crack blunting in metallic materials. As a consequence, the stress and strain fields in the vicinity of the crack front are known beforehand neither quantitatively nor even qualitatively. Thus, characterization of local fields in the vicinity of the crack front, as it can be done in brittle materials for example, is not appropriate. Therefore, the global energy balance in a specimen undergoing crack propagation as a generalization of Griffith's approach [1] seems to be the sole candidate for a characterization of fracture in such materials [2].

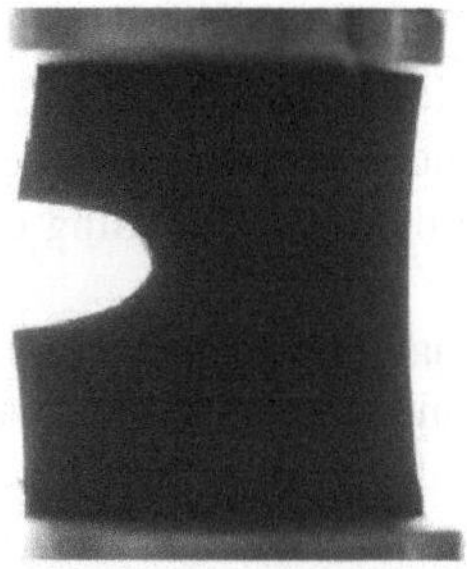

Fig. 1 Elastic crack tip blunting in elastomers

Considering the process of incremental crack propagation in a solid, the global energy balance reads

$$\frac{dW_{ext}}{dA} - \frac{dW}{dA} = \frac{dW_{sep}}{dA} + \frac{dW_{Diss}}{dA} \tag{1}$$

where W_{ext} is the external work and W is the recoverable elastic strain energy, W_{Diss} is the macroscopic remote dissipated energy and W_{sep} is the local work of separation. For the case of plane specimens with a constant thickness b and a straight crack front, the increment of the crack area is $dA = b\,da$ where da denotes the crack length increment. The left-hand side of eq. (1) is called energy release rate [3], the energy released during crack propagation. The right-hand side is called dissipation rate [4-11], also denoted as tearing energy [2] in the case of elastomeric materials, which is understood as the energy needed to advance a fracture plane by one unit area including additional energy losses due to accompanying dissipative processes.

Considering that the external work is zero, i.e. the clamp distance l is constant during the process of crack propagation, the original proposition made by Rivlin and Thomas [2] for the determination of tearing energy

$$T = -\left.\frac{dW}{dA}\right|_{l=const.} = -\left.\frac{dW}{b\,da}\right|_{l=const.} \tag{2}$$

can be derived.

However, the location of the process of energy dissipation in rubber materials is not restricted to the immediate vicinity of the crack front. Energy dissipation mechanisms can be induced in the whole specimen. Moreover, due to the relaxation behavior of the polymer chains [12] and the strain-dependent changes of the filler structure [13], rate-dependent and rate-independent material dissipation processes must be considered. These features complicate the fracture mechanical approach for rubber materials even more.

However, it should be noted that energy dissipation in rubber materials cannot be neglected even in the case of a non-propagating crack. Thus, the dissipation contributions are most crucial in the energy balance, especially with regard to the separation of these contributions, which raises the question of whether the presence of these dissipation mechanisms is only related to the crack propagation or if it is more a consequence of specific loading conditions or if it is influenced by both.

Energy balance can be evaluated from both the experimental and the numerical side. In both cases, the left hand side of eq. (1) will be evaluated. Experimental evaluation is related to the evaluation of energy contributions taken from global loading and unloading curves, whereas numerical evaluation can be either taken from global loading and unloading curves or by determining energy flux integrals, i.e. the J-Integral [14,15] or its generalizations.

2.1 Energy Flux Integrals

Due to the susceptibility of elastomers to large deformations, all considerations should be carried out within the framework of nonlinear continuum mechanics. However, in order to not complicate the issue, the following derivations were done, assuming the small deformation theory. It should be noted that the results derived are transferable to the finite deformation framework.

The fracture of materials can be energetically characterized within the framework of plane continuum mechanics by energy flux integrals [16] such as

$$F(\Gamma) = \int_{\Gamma} \left(w n_1 v_c + \sigma_{ji} n_j \frac{\partial u_i}{\partial t} \right) dL. \tag{3}$$

Eq. (3) can be used to quantify the amount of elastic energy that flows per time unit and thickness unit through a defined contour Γ that moves together with the crack tip without crack deflection, see fig. 2. In eq. (3), w is the elastic energy density, n_j is the outer normal unit vector of Γ, v_c is the crack tip velocity, σ_{ji} is the stress tensor and u_i is the displacement vector of a material point of the

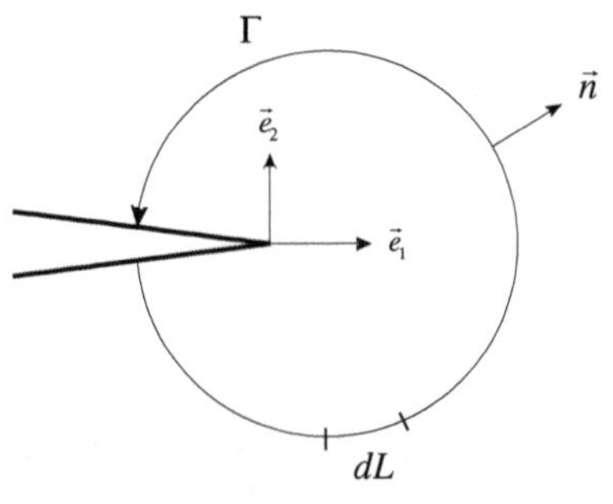

Fig. 2 Definition of a contour Γ encircling the crack tip

continuum. It should be noted that Einstein's summation is used in eq. (3), where repeated indices are implicitly summed over from 1 to 3.

As energy dissipation mechanisms in elastomeric materials are crucial, the material behavior is assumed to be dissipative, i.e. the elastic energy density $w = w(\varepsilon_{ij,}\alpha_i)$ is not only a function of the strain tensor ε_{ij} as an observable variable, but also of a set of internal variables of state α_i [17,18]. These internal variables of state are not controllable and a change in such an internal variable indicates the presence of a dissipative process. If the fields of all state variables within the contour Γ translate invariant with the crack, then the temporal evolution of the observable variable ε_{ij}

$$\frac{\partial \varepsilon_{ij}}{\partial t} = -\frac{\partial \varepsilon_{ij}}{\partial x_1} v_c \tag{4}$$

and the temporal evolution of the internal variables of state α_i

$$\frac{\partial \alpha_i}{\partial t} = -\frac{\partial \alpha_i}{\partial x_1} v_c \tag{5}$$

are fully determined by the corresponding gradient in the direction of crack propagation and by the crack tip velocity. In this case, eq. (3) yields

$$F(\Gamma) = J\, v_c \tag{6}$$

where

$$J = \int_\Gamma \left(w n_1 - \sigma_{ji} n_j \frac{\partial u_i}{\partial x_1} \right) dL \tag{7}$$

is the J-integral [14,15] characterizing the change in elastic energy per unit thickness due to an infinitesimal translation of characteristic fields within the region enclosed by the contour Γ. This energy change includes the evolution of internal variables of state according eq. (5). Consequently, it is obvious that the J-integral depends on how the contour Γ was chosen. In the case where Γ encloses all dissipative fields, the value of the J-integral is independent of the chosen integration contour.

In elastomeric materials, energy dissipation processes are not restricted. Considering two different contours in a dissipative material, a far field contour Γ_2 and a near crack tip field contour Γ_1 (see fig. 3) the value of the J-integral of the near tip contour

$$J_{\Gamma_1} = J_{\Gamma_2} + \int_{A_{\Gamma_1\Gamma_2}} \sum_i -\frac{\partial w}{\partial \alpha}\frac{\partial \alpha}{\partial x_1} dA \tag{8}$$

can be expressed as the J-integral value of the far field contour and contributions due to the inhomogeneous fields of internal variables of state. In eq. (8), $A_{\Gamma_1\Gamma_2}$ is the area enclosed by the two contours and parts of the crack lines.

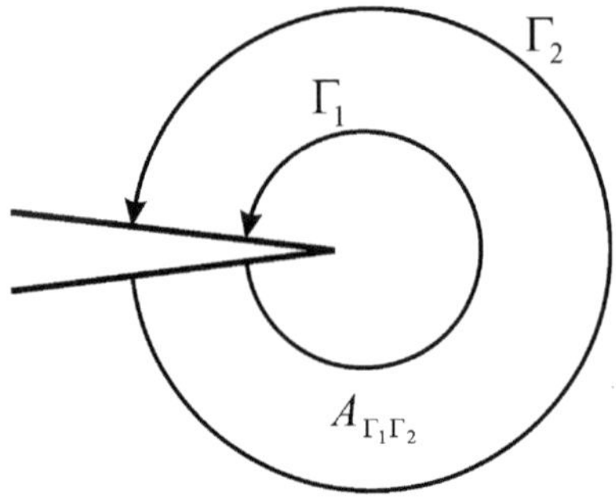

Fig. 3 Definition of two contours encircling the crack tip

It can be easily verified that eq. (8) is equivalent to the material force approach [19-25] when the near tip contour shrinks to the crack tip. In order to reveal the physical significance of material forces in fracture mechanics, eq. (5) should be focused on as an inherent assumption of the J-integral and the material force approach. It prescribes the evolution of internal variables of state that are not controllable. Hence, this assumption must be analyzed with respect to the nature of the dissipative processes. Generally, internal variables of state can be distinguished with and without time scales [17,18]. In the former case, the dissipative process is governed by an explicit time scale, as is found in viscoelastic materials for example. In the latter case, the dissipative process is fast compared to the time scale of modeling, so that it can be regarded as an instantaneous process, e.g. during plastic deformations.

Considering the assumption of steady, i.e. invariant crack tip fields, the evolution of internal variables without a time scale can be described by eq. (5) for every crack tip velocity because the dissipative process occurs instantaneously and is not governed by a time scale. In contrast, if the evolution of an internal variable of state is governed by a time scale, eq. (5) is not fulfilled in every case. Consequently, steady crack tip fields can only be assumed under certain conditions.

It can be therefore concluded, that the J-integral is a reasonable measure of fracture mechanics in the elastoplastic regime for rate-independent dissipative processes, but not always in the viscoelastic regime for rate-dependent dissipative processes. The J-Integral is always related to the energy losses when the all fields within the encircled region are translated in an invariant manner. Otherwise, the general form of energy flux integrals, eq. (3), is recommended to use. Since the tearing energy is related to the total energy loss during crack propagation, energy dissipation outside of the chosen contour due to rate and filler effects must be explicitly considered, as well as the energy flux into the region enclosed by the contour.

The case considered above excludes the possibility of crack deflection. Vectorial generalization

$$J_k = \int_{\Gamma} \left(w n_k - \sigma_{ji} n_j \frac{\partial u_i}{\partial x_k} \right) dL \tag{9}$$

of the J-integral, eq. (7), could therefore be a measure containing information about crack path development. For this reason, the special case of an in-plane mixed mode loaded crack in linear elastic fracture mechanics (see [3]) was considered. However, in this case the angle defined by the J-integral vector, eq. (9), does not coincide with the predicted angles using the accepted crack path criteria of linear elastic fracture mechanics, such as the principle of local symmetry and the maximum energy release rate criterion. It was shown [26] that the interpretation of non-tangential components of the vectorial J-integral (see eq. (9)) in the same way as tangential component J_1, that is equivalent to eq. (7), is not consistent with the accepted crack kinking criteria. Hence, a different interpretation of non-tangential components was proposed; where non-tangential components indicate the sensitivity to crack deflection that has to be compared to the resistance of the material against deflection from the initial crack plane (see also [27,28]). This interpretation reflects the fact that crack direction is likely determined by the interaction between directions preferred by processes on the scale of microstructure and those preferred by the near tip fields. Experimental observations under the quasistatic loading of notched specimens in tensile mode support this point of view. The variation of carbon black types in styrene-butadien rubber under identical test conditions has shown that the material filled with carbon black type N115 tends to a pronounced crack deflection, compared to carbon black types N339 and N550, where no crack deflection could be observed. As a consequence, the interpretation of generalized J-integrals as crack driving forces has to be applied with care when cracks in viscoelastic materials or the deflection of a crack are considered.

2.2 Dissipation Rate Approach

Stable crack propagation in elastomeric materials under quasistatic loading conditions is often characterized by methods based on the crack extension resistance curve of elastic-plastic fracture mechanics [3]. This curve acknowledges the fact that the resistance to fracture increases with growing crack size in elastic-plastic materials. Thus, the stable crack propagation is characterized by the total energy losses required to produce a certain stable, i.e. subcritical crack increment.

In order to analyze the global energy balance, eq. (1), an experiment characterizing the stable crack propagation in elastomeric materials was modified. Highest crack resistance values are determined in highly filled elastomers. Therefore, stable crack propagation in a SBR vulcanizate filled with 40 phr (parts [in weight] per hundred parts rubber) carbon black N234 was analyzed experimentally and by a finite element simulation [29]. However, in such highly filled vulcanizates, the amount of stable crack propagation is very limited, i.e. in the micron range, so that stable crack increment measurement was done by light microscopy. Consequently, a multiple specimen method was applied. Identical notched specimens were loaded in a quasi-static manner up to several points above the crack initiation threshold yet below the point of final specimen fracture. Then, the specimens were unloaded in order to separate stored elastic energy from total energy. Finally, the specimens were cut with a razor blade in order to measure the stable crack increment by distinguishing the morphological differences of the newly created fracture surfaces and the razor-cut surfaces. The analysis of the global energy balance, eq. (1), from the experimentally determined values was compared to the finite element analysis of the experiment. The quasistatic external loading of the samples and the resulting very slow crack propagation in the stable regime of highly filled vulcanizates cause mainly rate-independent dissipative effects. Hence, viscoelastic influences could be neglected and only filler induced dissipation mechanisms must be taken into account in the modeling of material behavior. However, these mechanisms are present in the whole specimen and it is not clear to which extent these mechanisms favor or hinder crack propagation. Therefore, the experiment was simulated by finite element analysis. A rate-independent dissipative behavior due to the change of filler structure with increasing strain was considered. It was modeled with a phenomenological model of endochronic plasticity in consideration of large deformations that capture the effects of such a filler-induced dissipation, especially the hysteretic stress-strain dependency, without a distinctive yield-surface [30].

Comparing both approaches, the results derived from the experiment and simulation, the following conclusions can be drawn. The limited stable crack propagation in such highly filled vulcanizates has only a very small influence on the amount of energy released during crack propagation. The main contributing factor to this effect has its origin in the development and increase of a dissipation zone around the crack tip with increasing load (see fig. 4). Moreover, differences between the J-Integral approach and equivalent derived quantities from the experiment can be attributed to several factors; the fact that for the given experimental setup, the crack propagation had not yet reached a steady state, and that therefore a large part of the energy was still lost in the bulk material due to inelastic processes. The measured energy losses during crack propagation are therefore related to the development and increase of the dissipative zone, and are not related to the invariant translation of such a zone, as assumed for steady-state crack propagation. Indeed, for practical applications it is especially desirable to activate distinct dissipative processes in a region as large as possible in order to increase the material's ability to absorb energy. However, in order to compare the fracture-mechanical properties of various vulcanizates, it is necessary to reach a steady state. Otherwise, reliable comparability is not ensured.

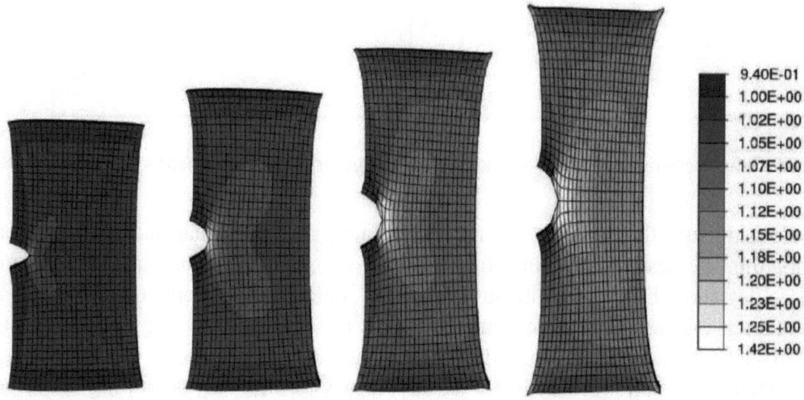

Fig. 4 Development of the dissipation zone during deformation. Values of the equivalent plastic variable as a measure of the amount of filler-induced dissipation are depicted.

In order to reach steady state conditions during crack propagation, similar experiments were conducted avoiding the costly multiple specimen method by using single edge notched constraint tension (pure shear) specimens. Approximation formulas for the determination of tearing energy were derived [2] that are normally used to characterize fracture mechanical behavior. For the case of a single edge notched constraint tension specimen, the approximation formula was compared to the direct evaluation of global energy balance, i.e. to the dissipation rate approach. In deriving the approximation formula for this kind of specimen, it is assumed that the specimen volume can be divided into four differently loaded regions, see fig. 5. In the vicinity of the crack front (region B) quite complicated stress and strain fields can be found. However, the remote fields are assumed to be homogeneous. In front of the crack front there is a region where a homogeneous state of constraint tension (pure shear) can be found (region C) whereas behind the crack front the region is nearly unloaded (region A). Finally, there is a region that is influenced by the outer boundary (region D). Considering infinitesimal crack propagation in such a specimen, it is assumed that the complicated inhomogeneous state in the vicinity is translated with the crack front. Consequently, the size of the region in the pure shear state is reduced increasing the unloaded region. This basic idea leads to the simple formula for tearing energy

$$T = w_{PS} \, h_0 \tag{10}$$

with w_{PS} being the elastic energy density in the constraint tension (pure shear) region and h_0 being the undeformed height of the specimen. Hence, the tearing energy can be determined via the elastic strain energy density in the homogeneous loaded pure shear region, which is measured by the unloading curves from un-notched pure shear specimens at identical boundary displacement where crack propagation occurred in notched specimens.

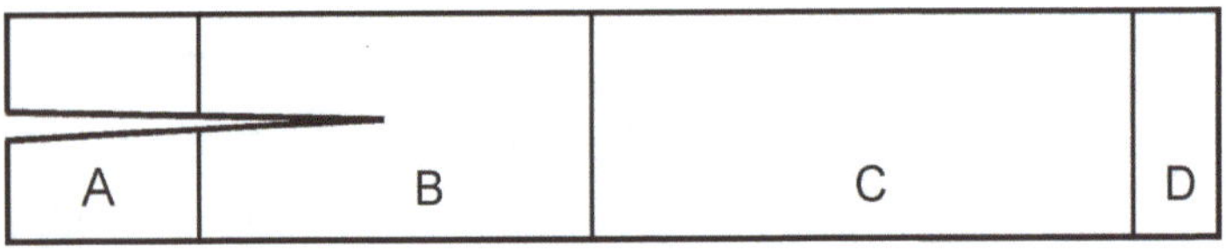

Fig. 5 Characteristic regions of a loaded single edge notched pure shear specimen

Considering the dissipation rate approach, specific quasistatic loading and unloading cycles with a subsequent small increase of maximum displacement were defined. Energy contributions were determined from the loading and unloading curves. Crack increments were measured from digital photographs taken at various instants. In order to compare the quantities to the conventional approach, the same cycles were conducted with the corresponding unnotched specimen. The elastic energy density of eq. (10) was determined from unloading curves.

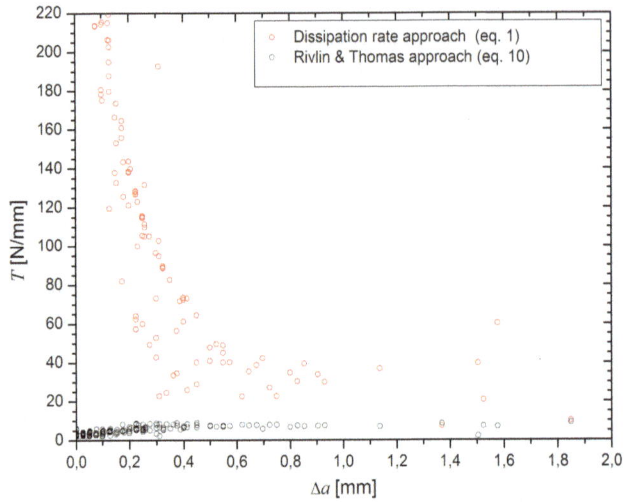

Fig. 6 Dissipation rate approach vs. approximation for an EPDM filled with 60 phr N234

The comparison of both methods of data evaluation for an EPDM filled with 60 phr carbon black N234 can be seen in fig. 6. The dissipation rate decreases with increasing crack increments. Such behavior was also found for metallic materials [8]. The higher energy required at the initial stage of crack propagation with small crack increments is related to the initiation of crack propagation, or to an increase in a dissipation zone in the vicinity of the crack front rather than to a propagation of the crack itself. This is also shown in the preceding analysis of stable crack propagation in a highly filled elastomer. Such higher values could also be required when the crack must be reinitialized after the crack stops. At higher crack increments, lower dissipation rates can be determined, in accordance to the values determined with the approximation (see eq. (10)), for determining the tearing energy. Fitting the experimental determined curve for the dissipation rate approach with an exponential function [8]

$$T = T_{\infty}\left(1+\alpha\exp\left[-m\Delta a\right]\right) \tag{11}$$

allows for determining values to be used for the purpose of comparing various materials. Accordingly, the value T_{∞} characterizes the dissipation rate under steady state conditions. It can be expected that values of the J-integral would be in accordance with T_{∞} if the experiment would be analyzed by means of finite element analysis. Hence, experiments characterizing the stable crack propagation in elastomers should be designed in such a way that steady state conditions can be reached in order to exclude influences on the initial stages of crack propagation that could impact the fracture toughness measurement. Only steady state conditions allow for comparable studies of various elastomeric materials.

3 Viscoelastic Crack Tip Fields and Influence on Crack Propagation Behaviour

Elastomeric materials are viscoelastic, and hence a certain amount of energy can be dissipated within the whole specimen during dynamic loading due to this time-dependent behavior. These materials are usually filled with hard nanoparticles, which results in a filler-induced hysteresis as well. Other main contributions to the tearing energy are the so-called small scale dissipation and the surface free energy that can be related to the different processes of material degradation within the fracture process zone.

Modifying the structure of rubber influences both the energy dissipation far from the crack tip and the fracture process itself. However, most of the energy is dissipated far from the crack tip. Existing models explain well the dissipation mechanism outside the fracture process zone, especially when the material is subjected to small and intermediate deformations. The influence of filler can be understood in the frame of the dynamic flocculation model [13]. The specific microstructure also influences the viscoelastic spectrum. The impact of the viscoelastic spectrum on the tearing under quasistatic loading was emphasized in [31,32]. The enormous increase in tearing energy of non-crystallizing elastomers, from T_0 (resistance to stable crack propagation) at very low crack tip velocities to T_{max} (resistance to unstable crack propagation) at higher crack tip velocities up to fracture instability at $v_c = v_s$ (see fig. 7) can be ascribed to the dissipation of viscoelastic energy in the vicinity of the crack tip, but outside the fracture process zone. In this range stable steady crack propagation with a low crack tip velocity occurs, while unstable crack propagation is hindered due to various dissipative processes. Exceeding the value T_{max} results in unstable crack propagation with a high crack tip velocity. In most applications, unstable crack propagation should be prevented in favor of stable crack propagation. Hence, the understanding of the processes during stable crack propagation and those leading to instability is of practical relevance in the fracture mechanics of elastomers. As the crack velocity at stable crack propagation is very low, inertia contributions can be neglected. The regions of stable and unstable crack propagation on the resulting fracture surfaces can be distinguished in most cases with the naked eye. The surface is much smoother in the case of unstable crack propagation. Differences between stable and unstable crack propagation are supposed to arise due to viscoelastic behavior. The obtained results are related to the specific fracture mechanisms in elastomers, indicating a change that is obviously responsible for the instability (see fig. 7).

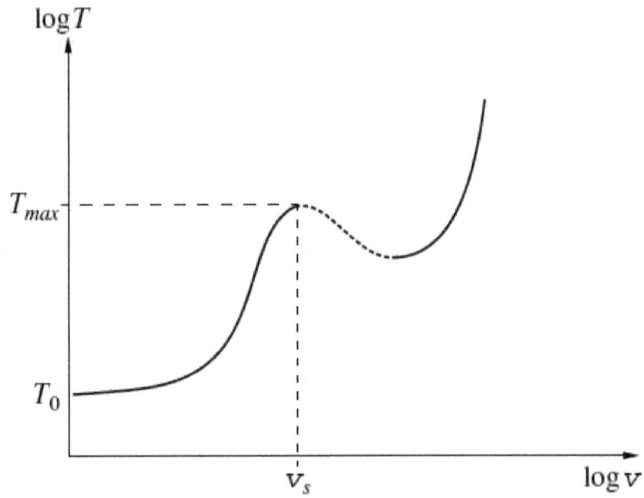

Fig. 7 Schematic relation between tearing energy and crack velocity

Even though most energy dissipation during fracture is amenable by continuum theory, the understanding of the failure mechanism and its modeling is the key to understanding the peculiarities of fracture propagation in rubber materials. The fracture of rubber is generally accompanied by the translation of a process zone where microscopic degradation processes take place. The investigation of fracture surfaces of non-crystallizing SBR compounds indicates that in the low crack tip velocity range the surfaces are rather rough, at least on the macroscopic scale, and the more the compound is filled with carbon black, the smoother the resulting fracture surface is. In contrast, in the high crack tip velocity range, the fracture surfaces are always smooth on the macroscopic scale irrespective of the carbon black content. Surface roughness originates from material degradation processes due to polymer chain rupture in the vicinity of the crack front leading to defects such as microcracks or cavities. As a consequence, the intersection of the growing crack and formed defects results in a rough fracture surface. The ductile process of formation, growth, and coalescence of defects in the low velocity range is very sensitive to microstructural details such as the presence of fillers. In the high velocity range, the number of defects is reduced or suppressed irrespective of the filler content.

It has been shown in [33] that the growth of existing voids is related to an elastic instability, when the hydrostatic stress exceeds a value that is in the order of Young's modulus of rubber elasticity. Extending the considerations of [33], characteristic steady crack tip fields of a propagating crack in a linear viscoelastic material at various finite crack tip velocities under quasistatic loading conditions were calculated separating deviatoric and dilatational contributions.

In the following study, a thin specimen containing a crack is considered. The material behavior can be characterized as homogenous, isotropic, linear viscoelastic. The specimen is loaded in pure Mode I, i.e. an external normal stress field perpendicular to the crack faces is applied. Due to the thin specimen, a plane stress state prevails and the main problem can be reduced to two dimensions. The same assumptions are valid as for linear elastic fracture mechanics [3], i.e. the fracture process zone is small compared to the dimensions of the near tip field. Therefore, the process of material separation will be determined by this field, and thus the stress intensity factor K_I which characterizes the intensity of the square root singularity of the near tip field, acts as a fracture parameter. The dimension of the region of the near tip field is small compared to the crack length and other macroscopic in-plane dimensions of the specimen. It is also assumed that crack faces are free of traction and that the crack propagates without deflection. Kinetic contributions are neglected in the equilibrium equations as well as in the energy balance.

By means of the so-called correspondence principle of linear viscoelasticity, the solution of linear viscoelasticity can be derived from the solution of an equivalent linear elastic problem [34]. Laplace transforms of basic equations of linear viscoelasticity have the same form as the basic equations of linear elasticity. If the solution of the latter problem is known, the solution of the viscoelastic problem is reduced to apply the inverse Laplace transform. In fracture mechanics this method is generally applicable for non-propagating cracks and under certain conditions [35] for propagating cracks. The conditions are as follows: Firstly, crack healing should be prevented. Secondly, the stress field of the corresponding elastic problem should be free of any material parameter. Finally, the elastic displacement field must be written as a product of two functional relationships: a functional relation of the material parameters and a function of other arguments.

To distinguish field quantities of the viscoelastic problem from those of the corresponding elastic problem, the latter are indicated by the upper index *e*. The near tip in-plane stress field in the corresponding case of linear elastic fracture mechanics [3] is dominated by its singular contribution that is given analytically by

$$\sigma_{ij}^{e} = \frac{K_I}{\sqrt{2\pi r}} f_{ij}^{\sigma}(\varphi) \tag{12}$$

where

$$\begin{pmatrix} f^{\sigma}_{11}(\varphi) \\ f^{\sigma}_{22}(\varphi) \\ f^{\sigma}_{12}(\varphi) \end{pmatrix} = \cos\frac{\varphi}{2} \begin{pmatrix} 1-\sin\frac{\varphi}{2}\sin\frac{3}{2}\varphi \\ 1+\sin\frac{\varphi}{2}\sin\frac{3}{2}\varphi \\ \sin\frac{\varphi}{2}\cos\frac{3}{2}\varphi \end{pmatrix}$$

in a Cartesian frame with its origin at the crack tip and moving with it. Due to the stress singularity at the crack tip field, quantities are given as a function of the polar coordinates r and φ. The in-plane displacement field that corresponds to eq. (12) is given by

$$u^e_i = \frac{K_I}{2G}\sqrt{\frac{r}{2\pi}}(\kappa - \cos\varphi) f^u_i(\varphi) \tag{13}$$

where

$$f^u_i(\varphi) = \begin{pmatrix} \cos\frac{\varphi}{2} \\ \sin\frac{\varphi}{2} \end{pmatrix}$$

and

$$\kappa = \frac{5}{3} + \frac{8G}{9K}$$

are considered here.

The viscoelastic crack tip fields can be yielded by an inverse Laplace transform. Then the stress field is the same as in the corresponding elastic case

$$\sigma_{ij} = \sigma^e_{ij}, \tag{14}$$

see eq. (8). The corresponding viscoleastic displacement field is given by

$$u_i = \int_{-\infty}^{t} J(t-\tau)\frac{du^J_i(\tau)}{d\tau}d\tau + \int_{-\infty}^{t} B(t-\tau)\frac{du^B_i(\tau)}{d\tau}d\tau \tag{15}$$

with the functions

$$u^J_i(t) = K_I(t)\sqrt{\frac{r(t)}{2\pi}} f^u_i(\varphi(t))\left(\frac{5}{6} - \frac{1}{2}\cos\varphi(t)\right)$$

and

$$u_i^B(t) = \frac{4}{9} K_I(t) \sqrt{\frac{r(t)}{2\pi}} f_i^u(\varphi(t))$$

describing the time-dependent state of a fixed material point with respect to the crack tip moving with the velocity v_c.

The stress field is the same in both cases; in the linear elastic as well as in the linear viscoelastic. This is because in linear elasticity, the near tip stress field does not depend on the elastic material parameters in plane stress. The constitutive equations only enter the displacement and the strain field. Hence, convolution integrals are only present in the latter fields.

The influence of history is present in the stress intensity factor that reflects external load and crack length as well as in the coordinates of a material point in the moving frame attached to a propagating crack.

Generally, two cases can be distinguished. The first is the case of a loaded but non-propagating crack where the influence of history can only be found in the stress intensity factor due to the influence of external load history. In the second case of a propagating crack, all influences described above must be taken into account, unless additional assumptions are made.

Below, the special case of a propagating crack at different crack tip velocities will be considered [36]. As shown previously, the influence of viscoelasticity is present in the history of the stress intensity factor and the crack tip velocity. The former contains the history of external stresses and the crack length. The latter contains the history of the coordinates of material points with respect to the moving frame attached to the crack tip at finite crack tip velocities.

Assuming that firstly the crack increment is in the order of the dimension of the near tip field and thus at least one order lower than the crack length, and secondly that during the corresponding time increment related to this crack extension the external load does not vary, a nearly constant stress intensity factor then results. Consequently, only the time-dependent coordinates of a material point in the moving frame and hence the crack tip velocity must be taken into account in calculating viscoelastic near tip fields.

For an analytic study of the linear viscoelastic crack tip fields, a simple model was used. Values for the viscoelastic material parameters of the same magnitude as that of a real rubber material were chosen. The shear compliance that describes the shape change behavior, increases at long times to a thousand fold of the short time value. The bulk compliance that describes the dilatational behavior decreases from a finite value at short times to zero at long times. For both transitions the same retardation time was used. Hence, the transition from a glassy and compressible material at short times to a rubbery and uncompressible material was described. Since cavitations due to critical hydrostatic stress are responsible for material degradation on a mesoscale in the vicinity of the crack tip, both the shear and bulk behavior were discussed separately.

The stress field in the vicinity of a propagating crack has the same form for various crack tip velocities, and differs only in the magnitude of the stress intensity factor. Fig. 8 presents the stress field component σ_{22} in the vicinity of the crack tip normalized by the stress intensity factor, as well as hydrostatic and deviatoric stress contributions related to shape and volumetric changes. The corresponding normalized strain field component ε_{22} is shown in figures 9 and 10 at various crack tip velocities, separated by deviatoric and dilatational contributions.

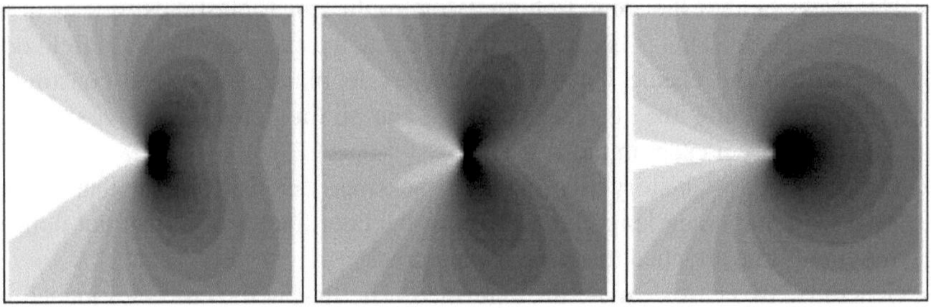

(a) (b) (c)

Fig. 8 Stress field component σ_{22} (a), its deviatoric (b) and hydrostatic (c) stress contribution, each normalized by the stress intensity factor

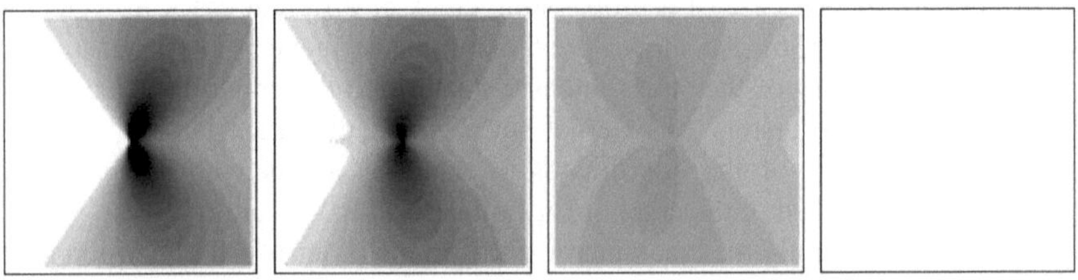

Fig. 9 Normalized deviatoric contribution of the strain field ε_{22} at various crack tip velocities, same scale used, crack tip velocity increases from left to right

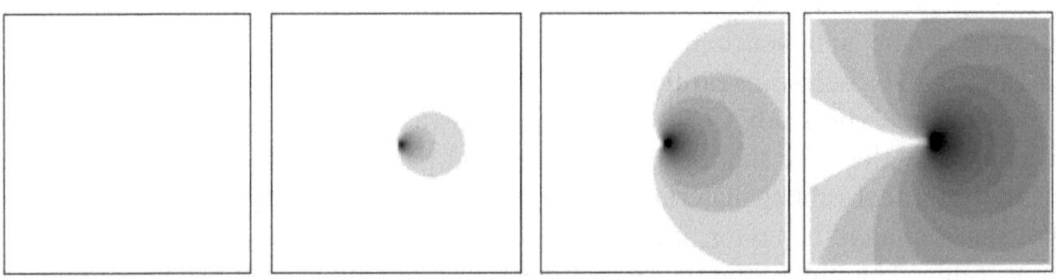

Fig. 10 Normalized hydrostatic contribution of the strain field ε_{22} at various crack tip velocities, same scale used, crack tip velocity increases from left to right

It can be shown that a transition from rubbery to glassy material behavior occurs with increasing crack tip velocity in the vicinity of the crack tip. The increase of tearing energy at higher crack tip velocities can be attributed to the dissipation of viscoelastic energy in the vicinity of the crack tip outside of the fracture process zone [31]. This dissipation can be ascribed to the dramatic decrease of the shear modulus in glass transition. The change in bulk behavior in the vicinity of the crack tip from uncompressible to compressible does not contribute greatly to the viscoelastic dissipation. However, as is shown in fig. 10, at low crack tip velocities the volume of the material cannot increase due to positive hydrostatic stress unless cavities are formed. The threshold value for critical hydrostatic stress is around the magnitude of the long time modulus. At high crack tip velocities, material dilatation is possible to a certain amount and the modulus is dramatically increased. Hence, cavity formation is not favored. These results confirm the conclusions made from fractographic investigations of the fracture surfaces. Generally, the fracture surfaces of rubber specimens created by high velocity crack propagation are smooth on the macroscopic scale. Consequently, it was assumed that the effects of cavitation growth are suppressed. In contrast, the fracture surfaces created by low crack tip velocity propagation are rather rough due to multiple cavitations in the vicinity of the crack tip. Hence, the reason for the instability in the plot of the tearing energy against the crack tip velocity seems to arise due to a transition from a ductile fracture mechanism to a brittle one with increasing crack tip velocity. The cavitation process itself involves energy dissipation. Consequently, more energy is required to shift the fracture process zone when crack advance is due to formation, growth, and coalescence of cavities rather than due to a brittle mechanism.

Under non-quasistatic loading conditions, the dissipation of energy is possible everywhere in the specimen. Consequently, viscoelastic crack tip fields are disturbed due to the rate effects of external loading compared to the calculated case of quasistatic loading. Hence, crack tip velocity as well as the external loading rate influence the viscoelastic near tip fields.

Even though the model is based on linear viscoelasticity and cannot be directly transferred to the case of large deformations, the results show that some kind of multiaxial loading in the vicinity of a crack front combined with incompressible behavior are crucial points in understanding the processes leading to the separation of material.

4 Fracture Surface Analysis

The fracture of a material involves a number of complex processes on multiple length scales. In materials science, increased fracture toughness can be achieved by modifying the structure of the material by inducing dissipative processes on various length scales, which result in increased energy adsorption accompanying the process of crack propagation in this material.

Elastic crack blunting can occur in elastomers due to entropic elasticity, i.e. the crack opens up when the specimen is loaded. In the vicinity of the crack front, the material is extensively stretched and the inhomogeneous structure of the material has a pronounced influence on the way the material separation proceeds within the fracture process zone. As it is visible to the naked eye, both the structural changes and the crack velocity influence the resulting fracture surface morphology of elastomeric materials (see fig. 11).

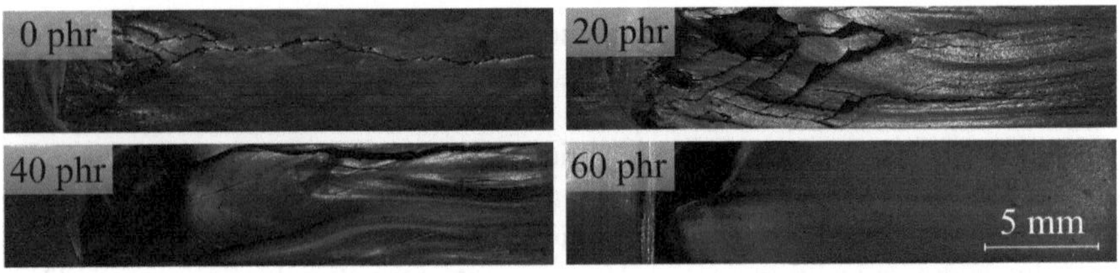

Fig. 11 Photographic top view of various fracture surfaces of elastomeric materials filled with various amounts of carbon black N234. The crack propagated from left to right.

In general, surfaces created by fractures are typically irregular and rough. The morphology of fracture surfaces is a signature of the complex fracture process on the scale of microstructure of the corresponding material. Even though roughness varies with different microstructures, its scaling properties are comparable for many materials [37-40]. Self-affine scaling of a roughness profile is characterized by invariance under a scale transformation

$$(r,h) \rightarrow (\lambda r, \lambda^{\zeta} h) \tag{16}$$

where r is the profile coordinate, h is the height of the profile and ζ is the roughness exponent. Fracture surfaces are found to be self-affine objects that can be described by a universal roughness exponent $\zeta \approx 0.8$ [37,38]. In contrast, recent studies reveal that these fracture surfaces are anisotropic objects that show different roughness exponents across and along the crack propagation direction, denoted in the following by ζ and β. For a wide range of materials including glass, mortar, wood, quasi-crystals and metallic alloys, roughness exponents determined across and along the crack propagation direction are found to be $\zeta \approx 0.8$ and $\beta \approx 0.6$, respectively [39,40]. However, it was found that the fracture surface roughness exponents in glassy ceramics and sandstone are significantly lower: $\zeta \approx 0.4$ and $\beta \approx 0.5$, respectively [39,40], which put the universality of the higher roughness exponents in question. Furthermore, higher roughness exponents describe the scaling behavior on the length scales of the fracture process zone, where a ductile mechanism prevails. On length scales that are larger than the size of the fracture process zone, the morphology proves to be well described by the smaller roughness exponents [39,40].

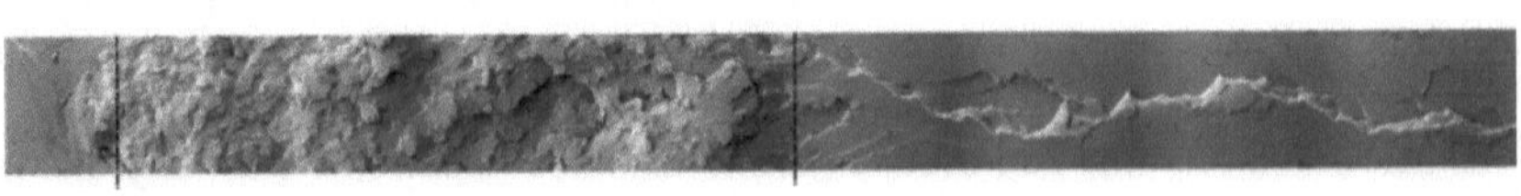

Fig. 12 Top view of a fracture surface of an unfilled SBR vulcanisate. Characteristic morphologies are detectable: notch, stable crack propagation, unstable crack propagation (from left to right).

Fracture surfaces of elastomeric materials without fillers or those with low filler content are quite rough at the macroscopic length scale, especially when they are formed as a result of slow crack propagation with a low crack velocity. In order to analyze the roughness by means of different methods at different length scales, a highly filled vulcanizate with fracture surfaces that are smooth at the macroscopic length scales was used. Surfaces resulting from fast unstable crack propagation (see fig. 12) were analyzed at various length scales by means of atomic force microscopy (Quesant Q-Scope), resolving fine-scale details ranging from nanoscopic to microscopic length scales. The same surfaces were also analyzed by means of confocal light microscopic measurement (Nanofocus μsurf explorer), with lateral resolutions ranging from microscopic to macroscopic length scales.

Fig. 13 Principle of atomic force microscopy (AFM)

Atomic force microscopy (AFM) (see fig. 13) is a type of scanning probe microscopy with very high-resolution. The AFM consists of a cantilever with a sharp tip at its end that is used to scan the specimen surface. When the tip is brought into proximity of a sample surface, forces between the tip and the sample lead to a deflection of the cantilever. Typically, the deflection is measured using a laser spot reflected from the top surface of the cantilever into an array of photodiodes. In order to measure fracture surface topography, the AFM was operated in a dynamic mode in which the cantilever is vibrated. The cantilever is driven to oscillate up and down at near its resonance frequency by a small piezoelectric element mounted in the AFM tip holder. The interaction of forces acting on the cantilever when the tip comes close to the surface cause the amplitude of this oscillation to decrease as the tip gets closer to the sample. An electronic servo uses the piezoelectric actuator to control the height of the cantilever above the sample. The servo adjusts the height to maintain a set cantilever oscillation amplitude as the cantilever is scanned over the sample. Analyzed fracture surfaces were scanned in several $20 \times 20\ \mu m^2$ areas with 1024×1024 data points using dynamic mode at $2\,Hz$ scan frequency.

Fig. 14 Principle of confocal light microscopy (CLM)

The analysis at larger length scales is based on confocal light microscopy (CLM) (see fig. 14), a non-contact surface measurement system. The LED source is focused through a multi-pinhole disc and the objective lens on to the sample surface. The reflected light, reduced by the pinhole to that part which is in focus, is recorded with a CCD camera. The rotation of the multi-pinhole disc enables scanning of the surface which is in focus. Capturing the images at different focal heights allows for reconstruction of surface topography. Fracture surface topography was measured in areas of $260 \times 260\ \mu m^2$ with 512×512 data points.

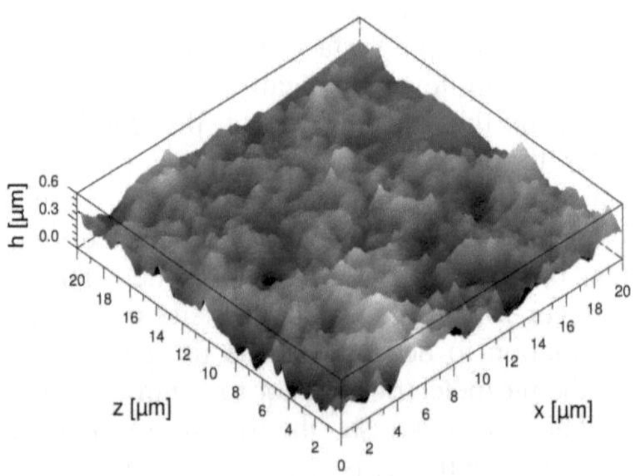

Fig. 15 Example of an atomic force microscopic image of the analyzed fracture surface of a SBR filled with 50 phr N330

For the statistical analysis, the frame of reference is chosen so that the unit vectors $\vec{e}_x$ and $\vec{e}_z$ are parallel to the direction of crack propagation and to the crack front, respectively.

Analyzing the scaling properties of the AFM fracture surface topography of a SBR filled with 50 phr N330 (see fig. 15) [41], profiles along and across the crack propagation direction were extracted from the surface data. For each profile, the 1D height-height correlation functions of the order k [42]

$$C_k(\Delta r) = \left\langle \left| h(r+\Delta r) - h(r) \right|^k \right\rangle_r^{1/k} \tag{17}$$

based on the k-th moment of the height fluctuations

$$\Delta h(\Delta r) = \left| h(r+\Delta r) - h(r) \right|$$

on a scale Δr was calculated· The angular brackets in eq. (17) denote the average of the profile coordinate r. In the case of height fluctuations Δh obeying a Gaussian distribution with a self-affine scaling of the variance with Δr, the height-height correlation functions become

$$C_k^G(\Delta r) = (2\Delta r^{2\zeta})^{1/2} \left(\Gamma((k+1)/2) / \sqrt{\pi} \right)^{1/k}. \tag{18}$$

Consequently the height-height correlation functions (17) normalized by the ratio

$$R_k(\Delta r) = \frac{C_k(\Delta r)}{C_2(\Delta r)}, \tag{19}$$

that yields

$$R_k^G(\Delta r) = \sqrt{2} \left(\frac{\Gamma((k+1)/2)}{\sqrt{\pi}} \right)^{1/k} \tag{20}$$

in this case, collapse for all k [42]. Deviations from the Gaussian distribution of height fluctuations result in a multi-affine scaling regime, i.e. a k-dependent roughness exponent. It was shown that these deviations are due to vertical jump discontinuities in the roughness profile [42], either as a consequence of heterogeneous structure or of poor lateral resolution. Hence, profiles were analyzed using height-height correlation functions normalized by the Gaussian ratio.

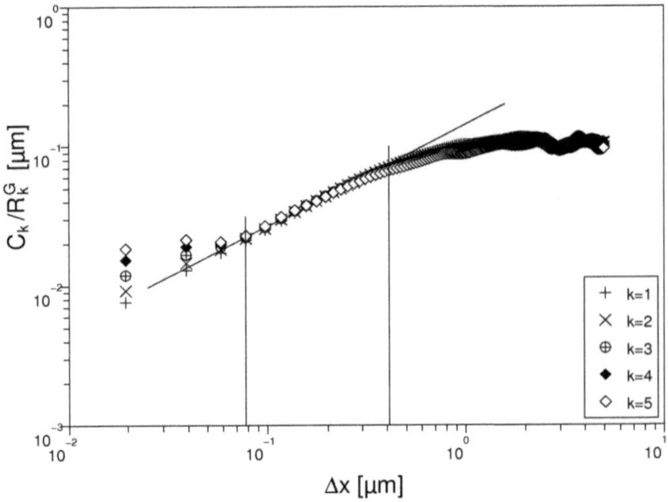

Fig. 16 Typical normalized 1D height-height correlation functions for a profile extracted along the crack propagation direction

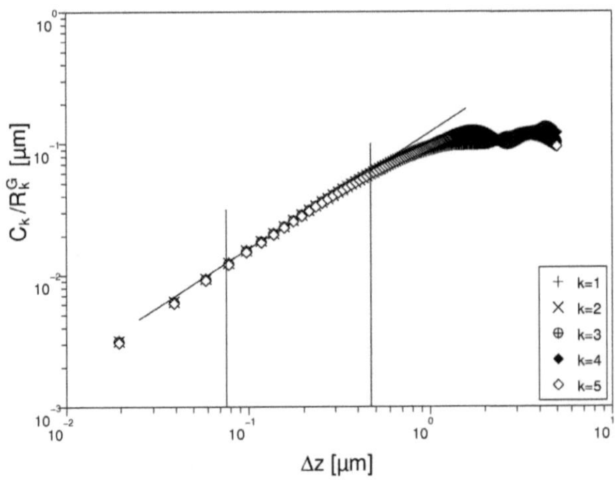

Fig. 17 Typical normalized 1D height-height correlation functions for a profile extracted across the crack propagation direction

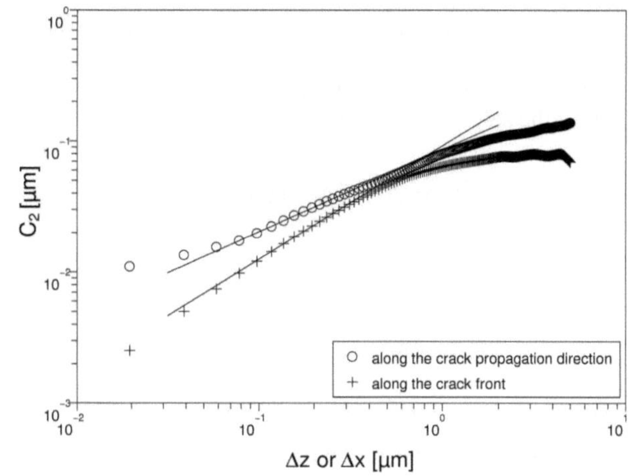

Fig. 18 Typical 1D height-height correlation functions $C_2(\Delta x)$ and $C_2(\Delta z)$ for a profile extracted along and across the crack propagation direction, respectively

Analyzing the fracture surface profiles lead to the results shown in figures 16 and 17. Self-affine scaling exponents were determined within the range of 80 nm to 400 nm. Comparison of both directions and best power law fits can be seen in fig. 18. From the analysis of all possible profiles of various different surface scans, roughness exponents $\zeta = 0.87 \pm 0.03$ along the crack front and $\beta = 0.70 \pm 0.06$ along the crack propagation direction were determined. These exponent values are comparable to those found in non-rubber materials, where the exponents characterize the scaling behavior within the fracture process zone [39,40]. Hence, a ductile fracture mechanism can be conjectured in the analyzed material at unstable crack propagation within the indicated range. The upper cut-off length of 400 nm of this regime is of the order of the fracture process zone [39,40]. Following [41] the lower cut-off length of 80 nm was determined by the fanning of normalized height-height correlation functions as a result of jump discontinuities that can be brought in relation to the microstructure of the material, i.e. the size of carbon black aggregates in the material.

For a complete description the two-dimensional features of the fracture surface, the 2D height-height correlation function

$$C_2^{2d}(\Delta z, \Delta x) = \left\langle \left[h(z+\Delta z, x+\Delta x) - h(z,x) \right]^2 \right\rangle_{z,x}^{1/2} \tag{21}$$

was computed within the range of length scales showing self-affine scaling properties for the 1D analysis. Results can be visualized plotting the corresponding 1D height-height correlation functions for variations Δz at constant values Δx (see inset of fig. 19). However, by normalizing the abscissa and the ordinates by $\Delta x^{1/z}$ and Δx^{β}, a single master curve can be obtained for specific values of z and β that optimize the collapse of 2D height-height correlation functions for different fixed Δx (see fig. 19). Hence, similar to fracture surfaces of non-rubber materials, the elastomeric fracture surface follows a Family-Vicsek scaling [43] within the self-affine regime following

$$C_2^{2d}(\Delta z, \Delta x) \sim \Delta x^{\beta} f(\Delta z / \Delta x^{1/z}) \tag{22}$$

where

$$f(u) = \begin{cases} 1 & \text{if } u << 1, \\ u^{\zeta} & \text{if } u >> 1. \end{cases}$$

The exponents ζ, β and z refer to the roughness exponent, the growth exponent and the dynamic exponent according to the notation of surface growth in which similar scaling properties were observed [43].

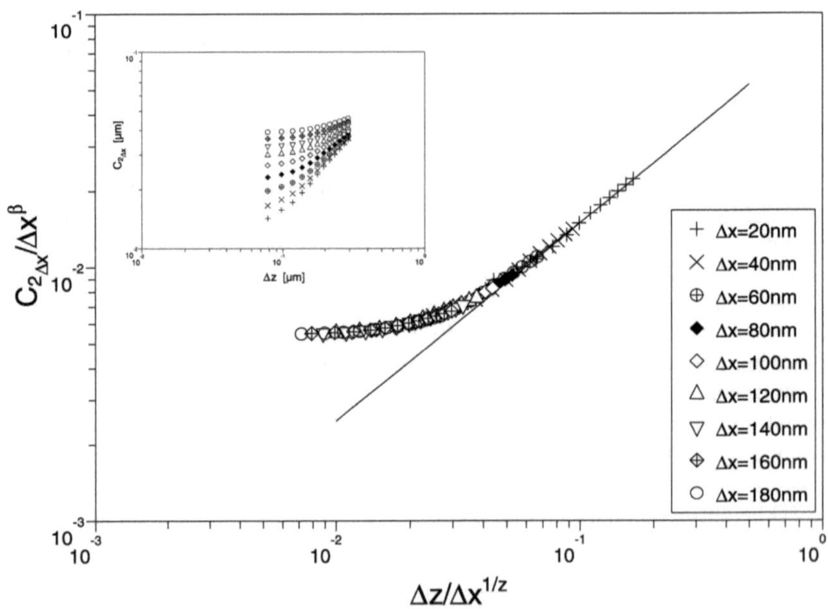

Fig. 19 Typical normalized 2D height-height correlation function. The inset shows the corresponding 1D height-height correlation functions for variations Δz at constant values Δx .

Within the self-affine regime, the three exponents are found to be $\zeta = 0.89 \pm 0.06$, $\beta = 0.76 \pm 0.06$ and $z = 1.21 \pm 0.03$. As $z = \zeta / \beta$, the 2D scaling properties can be characterized by two scaling exponents within this regime that are close to those determined by 1D analysis.

Since heterogeneities can be found over a wide range of length scales in filled elastomers, it is therefore of high practical interest to estimate the relevant length scales of the fracture process. Here, scaling exponents for the fracture surfaces formed in highly filled rubber material as a result of fast crack propagation were determined by means of 1D and 2D height-height correlation functions. The exponents were found to be close to those determined for other, mostly brittle materials, on the length scale where a ductile mechanism prevails. Such a process is related to a microscopic damage mechanism over a limited range of length scales characterized by the formation, growth and coalescence of voids or microcracks [39,40]. Hence, the upper cut-off length of about 400 nm can be used as a rough estimate of the dimension of the fracture process zone within the fracture plane.

The considered material is a weakly cross-linked polymeric system with a high content of carbon black N330. In this material, there is a pronounced polymer-filler interaction that stems from a strong chemical and physical binding due to the roughness of the filler surface [13]. The mean size of primary particles is about 30 nm, but the natural form of the filler consists of primary aggregates with a mean diameter of about 130 nm. Due to the high filler content, a weakly bonded superstructure is formed from the interpenetrating filler clusters as a result of the flocculation of primary aggregates. Mechanical connectivity between the filler particles that belong to different primary aggregates is provided by a flexible, nanoscopic bridge of glassy polymers, formed due to the immobilization of the rubber chains close to the gap between different primary aggregates [13]. Moreover, with increasing strain the stress-induced breakdown of rigid filler clusters occurs, due to destroying nanoscopic bridges connecting them. The large strains in the ligament developed during quasistatic loading of the sample before final fast crack propagation induce a total breakdown of the filler network, implying cluster sizes in the order of the aggregate size, i.e. 130 nm. Consequently, the concentration of destroyed nanoscopic bridges between filler aggregates is quite high due to the high filler content. Therefore, the multi-affine scaling regime on scales smaller than 80 nm for profiles along the crack propagation direction can be attributed to profile overhangs due to voids developed from destroyed nanoscopic bridges between different filler aggregates.

In order to compare fracture surfaces analyzed by AFM and CLM, topographic data from atomic force microscopy and confocal microscopy of the same fracture surface at regions created by unstable crack propagation were analyzed. Considering the fracture surface of a SBR vulcanizate filled with 60 phr carbon black N234 (see fig. 20) the normalized height-height correlation functions are plotted in figs. 21 and 22.

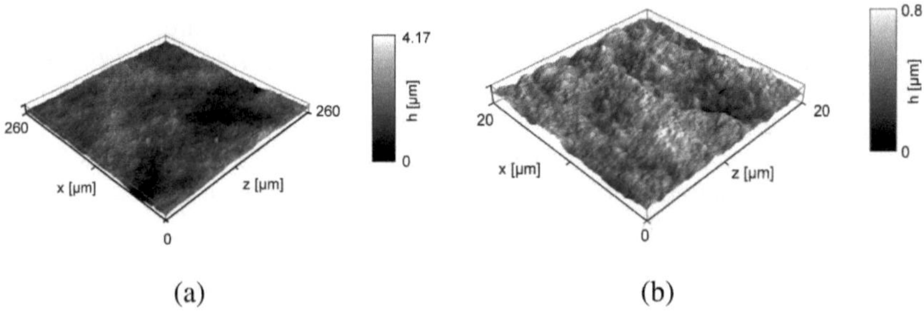

Fig. 20 Topography measured by CLM (a) and by AFM (b) in the unstable region of an SBR vulcanizate filled with 60 phr N234

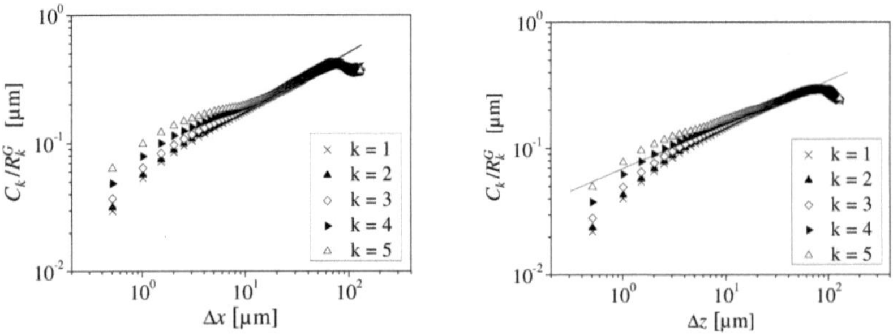

Fig. 21 Normalized height-height correlation functions calculated from CLM data for a fracture surface of an SBR vulcanizate filled with 60 phr carbon black N234

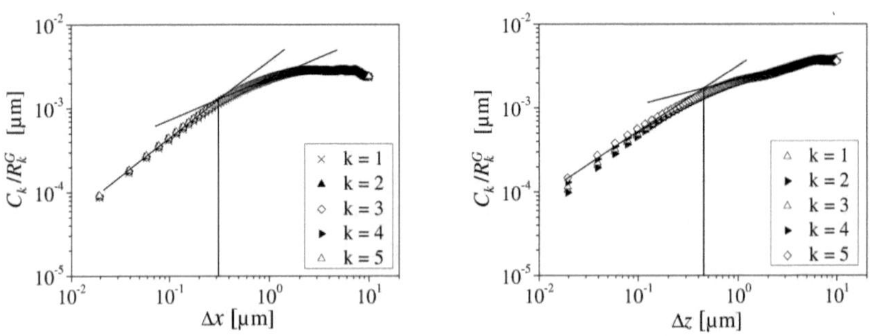

Fig. 22 Normalized height-height correlation functions calculated from AFM data for a fracture surface of an SBR vulcanizate filled with 60 phr carbon black N234

Self-affine scaling behavior can be observed for both directions in any case. However, it can be seen that results obtained from CLM data show only one scaling regime, whereas the two different scaling regimes can be detected from AFM data.

Considering confocal microscopic results, self-affine exponents are found to be $\zeta = 0.44$, and $\beta = 0.36$. It should be noted that multiaffine scaling is due to the limited lateral resolution of the method. Considering atomic force microscopic results, the scaling regime at smaller scales is characterized by $\zeta = 0.89$, and $\beta = 0.78$, whereas the exponents $\zeta = 0.43$, and $\beta = 0.38$ could be found for the scaling regime at larger length scales. It is evident that identical scaling regimes characterized by lower roughness exponents could be identified with both measuring methods. However, the results show that the scaling regime characterized by larger roughness exponents can only be identified with methods measuring topography at small length scales. Taking into account the AFM data, the transition length between both scaling regimes can be taken as a rough estimate of the process zone size. Analyzing fracture surfaces of differently filled SBR vulcanizates by means of atomic force microscopy (see fig. 23) these scaling regimes could be detected in all cases. Differences can be only detected in the value of the transition length (see table 1).

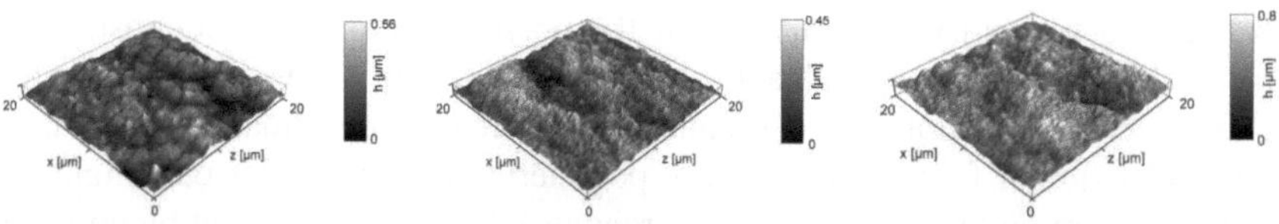

Fig. 23 AFM topographic images taken from fracture surfaces of a SBR vulcanizate filled with various amounts of filler N234: 0phr (a), 40phr (b) and 60phr (c)

Table 1 Roughness exponents and transition lengths of analyzed AFM topography from SBR vulcanizates filled with various amounts of filler N234

Filler amount	ζ_I	ζ_{II}	$l_z[\mu m]$	β_I	β_{II}	$l_x[\mu m]$
0 phr	0.92 ± 0.02	0.43 ± 0.02	0.48 ± 0.07	0.80 ± 0.02	0.37 ± 0.06	0.54 ± 0.07
40 phr	0.90 ± 0.02	0.38 ± 0.04	0.33 ± 0.02	0.69 ± 0.01	0.31 ± 0.02	0.39 ± 0.01
60 phr	0.89 ± 0.04	0.43 ± 0.08	0.36 ± 0.08	0.78 ± 0.03	0.38 ± 0.07	0.40 ± 0.01

Depending on the method applied to analyze the fracture surface topography, different self-affine scaling properties and hence characteristic length scales of the fracture process can be determined. However, the choice of the appropriate method is crucial. Analyzing topography with atomic force microscopy on the smallest length scales is only possible if height differences are not too large, as can be found in fracture surfaces created by unstable crack propagation. Analyzing the topography of the much rougher stable fracture surfaces, an alternative method might be the photogrammetric reconstruction of topography from various inclined scanning electron microscopic images [44]. However, it should be noted that this method is very costly. For poorer lateral resolution, light microscopic methods can be used. However, roughness scaling regimes characterized by higher exponents cannot be detected when larger length scale properties are analyzed.

The analysis of fracture surfaces allows on the one hand the study of the structural influence on the fracture process, on the other hand the determination of a length scale. The damage localizes within the fracture process zone that can be used to model this process.

Even though the investigated part of the fracture surface was created by fast crack propagation, it should be noted that the fracture occurs in the quasi-static regime, meaning that stress wave propagation is irrelevant for the fast crack propagation considered here. However, viscoelasticity is essential for capturing the fracture behavior of elastomeric materials [31]. Crack velocity influences the local loading rate of the material in the vicinity of the crack front, which has an impact on the deformation mechanism in such polymeric materials. Higher crack velocities induce higher local loading rates in the vicinity of the crack front, leading to a glassy-like deformation of the polymeric chains. In contrast, lower crack velocities trigger lower loading rates in the vicinity of the crack tip, leading to a rubbery deformation and, consequently, to extensive deformations even under quite low external forces. In this case, polymeric network chains can align, and finally they are highly stretched and a pronounced elastic crack front blunting can be observed.

The fracture of elastomeric materials is always caused by the loss of integrity of the material in the fracture process zone, thus the basic requirement of this is the rupture of overloaded polymeric network chains. Hence, the deformation mechanism of polymeric chains in the vicinity of the crack front influences the fracture process. For the case considered, polymeric chains in the ligament far from the crack tip are moderately stretched during quasi-static loading of the sample. However, further local loading due to the fast propagation of the crack implies a high local loading rate in the vicinity of the crack front, and thus glassy-like deformation behavior and a corresponding overloading of chains in non-equilibrium. Nonetheless, in increasing the distance from the crack tip, local loading rates decrease, which results in viscoelastic energy losses. Consequently, even though the amount of the tearing energy as a fracture mechanical parameter is governed by viscoelastic losses outside the fracture process zone [31], the fracture process itself seems to be driven by the glassy-like behavior of moderate pre-strained polymeric chains at short times.

Because of the presence of large entropic deformations of the polymer network chains in slow crack propagation, blunting effects become important. This causes a different crack mechanism, resulting in the formation of a discontinuous crack front, which is also observed in other soft materials such as gelatin [45]. Due to discontinuous crack front, characteristic cross-hatched patterns can often be observed in this case. The appearance of a transition from a cross-hatching fracture surface morphology at slow fracture to a smoother fracture surface at fast fracture is therefore related to the molecular dynamics of polymeric chains, which results in a different fracture process for both cases. In order to study the peculiarities of these fracture mechanisms, the change in molecular dynamics due to a decreased mobility of highly stretched polymeric chains and consequently the change of relaxation properties at large deformations must be taken into account [46].

5 Modeling of Material Degradation Due to the Rupture of Polymer Chains

In loaded pre-cracked components, regions in the vicinity of the crack front with localized deformations exist. The formation of these regions is a consequence of softening due to material degradation. The approach of continuum damage mechanics enables the coherent consideration of these effects by incorporating the features of micro-structural damage evolution into the material models. However, numerical analyses of the ductile fracture process based on local damage models are often found to depend on the mesh size used, i.e. the size of grid for the numerical method. Damage tends to localize in only one element layer. As a consequence, increasingly finer discretization grids can lead to earlier crack initiation and faster crack growth. The reason behind this non-physical behavior is the loss of ellipticity of the set of partial differential equations and consequently the non-uniqueness of the solution. Discontinuities may then arise in the displacement solution, which results in a singular damage rate field. Displacement discontinuities and damage rate singularities can be avoided by adding a nonlocal damage term into the constitutive equation of the material model, which carries the material length scale information. With this enhanced continuum description, smooth damage fields can be achieved, in which the localization of damage is limited to a given length, i.e. the characteristic length of the material (see fig. 24). Hence, local material models are not appropriate to model such a process. The localization of material degradation in the fracture process zone must therefore be modeled by means of nonlocal material models in the framework of continuum mechanics. Nonlocal quantities are introduced as weighted averages of the local internal variables of points in the vicinity of the considered point [47]. Moreover, explicit and implicit gradient approximations of the nonlocal approach have been introduced, which are favorable from the computational point of view. However, only the implicit gradient approximation of nonlocal damage is able to describe physically meaningful behavior [48]. Such an approach has been successfully applied to model the damage process in quasi-brittle and elastoplastic materials [49-53] and will be adapted for the modeling of degradation behavior in elastomers considering finite deformations [54].

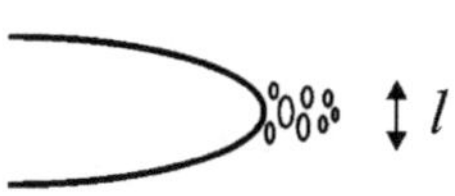

Fig. 24 Localization of damage in the vicinity of the crack tip

Considering the implicit gradient enhanced continuum damage model proposed in [47] by replacing the spatial averaging used in the integral formulation with an inhomogeneous Helmholtz equation, the boundary value problem is described by the following set of governing equations:

$$\frac{\partial P_{ji}}{\partial X_j} = 0 \tag{23}$$

and

$$\bar{z} - l^2 \frac{\partial^2 \bar{z}}{\partial X_i \, \partial X_i} = z \tag{24}$$

with the boundary conditions

$$\begin{aligned} u_i &= g_i, \\ P_{ji} n_j &= t_i \end{aligned} \tag{25}$$

at the corresponding parts of the boundary and

$$\frac{\partial \bar{z}}{\partial X_i} n_i = 0 \tag{26}$$

at the whole boundary. In eqs. (23) and (24), P_{ji} is the first Piola-Kirchhoff stress tensor, u_i is the displacement vector, z is a scalar variable that drives the damage evolution and $\bar{z}$ is the corresponding non-local counterpart. The parameter l is the material-dependent characteristic length that governs the localization of the material degradation process within the fracture process zone. This length scale can be determined from the analysis of fracture surface topography.

The problem in eqs. (23)-(26) is fully described with the relations between the first Piola-Kirchhoff stress tensor and the displacement vector as well as the equations of coupling.

Assuming that the material is loaded in a quasi-static manner and that therefore only hyperelastic material behavior must be taken into account, the rupture of polymer chain segments can only occur in equilibrium, i.e. when chain segments are fully stretched. In this case, the relation between the first Piola-Kirchhoff stress tensor P_{ij} and the displacement vector u_i is given by

$$P_{ij} = F_{ik} S_{kj}, \tag{27}$$

$$F_{ij} = I_{ij} + \frac{\partial u_i}{\partial X_j}, \tag{28}$$

$$S_{ij} = 2(1-\omega)\frac{\partial w}{\partial C_{ij}}, \tag{29}$$

$$C_{ij} = F_{ki}F_{kj} \tag{30}$$

where F_{ij} is the deformation gradient, C_{ij} is the right Cauchy-Green deformation tensor, and S_{ij} is the second Piola-Kirchhoff stress tensor. Hyperelastic behavior is described by the free energy density w, where the extended non-Gaussian tube model with non-affine tube deformation [55] determined by

$$w = \frac{G_c}{2}\left\{\frac{\left(\sum_{i=1}^{3}\lambda_i^2 - 3\right)\left(1-\frac{T_e}{n_e}\right)}{1-\frac{T_e}{n_e}\left(\sum_{i=1}^{3}\lambda_i^2 - 3\right)} + \ln\left[1-\frac{T_e}{n_e}\left(\sum_{i=1}^{3}\lambda_i^2 - 3\right)\right]\right\} + 2G_e\left(\sum_{i=1}^{3}\lambda_i^{-1} - 3\right) \tag{31}$$

dependent on stretch ratios λ_i was chosen. This model has a complete physical origin and the expression w can be traced back to the Hamiltonian of the polymer network using principles of statistical thermodynamics. In eq. (31), G_c and respectively G_e are the elastic moduli, where G_c is proportional to the density of network junctions and G_e is proportional to the entanglement density of the rubber. The ratio n_e / T_e defines the number of statistical chain segments between two permanent trapped entanglements, where n_e is the number of statistical chain segments between two successive entanglements, and the trapping factor T_e ($0 < T_e < 1$) characterizes the fraction of elastically active entanglements. For the limiting case $n_e / T_e = \lambda_1^2 + \lambda_2^2 + \lambda_3^2 - 3$, the free energy density of eq. (31) exhibits a singularity, which appears when the chains between successively trapped entanglements become fully stretched. Incompressibility of the material was accounted for by incorporate a penalty function.

Both partial differential equations (23) and (24) are coupled on the one side by the definition of an appropriate local variable z from the solution of eq. (23) that acts as a driving variable for damage, on the other side by the evolution equation for the damage variable ω (see eq. (29)) which depends on the nonlocal counterpart $\bar{z}$ as solution of eq. (24). This evolution equation was chosen to be similar to the case of plasticity [56], described by

$$f = \bar{z} - \kappa \leq 0, \quad f \leq 0, \dot{\kappa} \geq 0, \quad f\dot{\kappa} = 0 \tag{32}$$

where κ is a threshold value. Below this value, no damage evolution occurs. For values above this threshold, damage evolves and the threshold value κ will be updated. It is assumed that the damage variable ω evolves according to a linear relation to the actual value of κ up to a certain limit, which is set by the nearly fully damaged state. However, it should be noted that damage in elastomeric materials is related to a local reduction of network density, rather than the creation of local voids due to the characteristic chain dynamics above the glass transition temperature. The transition to the formation of defects is continuous, as distance between load carrying chains increases and the reduction of local network density above a certain level leads to the formation and growth of defects. The definition of the local variable z depends on the way that polymer chains break. Considering the case of hyperelasticity, damage can only occur when polymer chains are fully or nearly fully stretched. The damage is then coupled to a deformation measure. Referring to the extended non-Gaussian tube model, eq. (31), the variable z should be related to $\lambda_1^2+\lambda_2^2+\lambda_3^2-3$, i.e. the rupture of polymer chains can only be a consequence of overloading fully or nearly fully stretched polymer chains. Hence, this variable is bounded by the material parameter n_e/T_e. Taking into account that the parameter defines a mean limiting value, it motivates the assumption that degradation due to chain breaking of nearly fully stretched chains starts at levels below the value n_e/T_e, again taking into account the existence of shorter chain length segments.

Considering viscoelastic influences, chain rupture can also occur as a consequence of high viscous overstress at high loading rates due to overloading chains in non-equilibrium, and is therefore not related to reaching a certain level of deformation. The local variable z is therefore related to a stress measure. However, bearing in mind the roughness differences of fracture surfaces between stable and unstable crack propagation, the parameter l controlling the damage localization is not constant. In the case of stable crack propagation, polymer chains can align and chain fracture is the consequence of overloading fully stretched polymer chains, as described above. When unstable crack propagation starts, polymer chains in the ligament are stretched to a much lower degree. For this reason, it is assumed that the length scale of localization l depends on the actual value of stretching. It should be noted that in the case of viscoelasticity, eq. (29) must be modified to

$$S_{ij}=(1-\omega)\left(2\frac{\partial w}{\partial C_{ij}}+S_{ji}^{visc}\right) \tag{33}$$

where S_{ji}^{visc} is the viscous overstress whose evolution can be described by a Maxwell model [34]

$$\dot{S}_{ji}^{visc}+\frac{S_{ji}^{visc}}{\tau}=k_0 2\frac{\partial w}{\partial C_{ij}} \tag{34}$$

with τ and k_0 as the relaxation time and the free energy factor [57] and with the initial condition

$$\left.S_{ji}^{visc}\right|_{t=0}=0. \tag{35}$$

The proposed model was implemented into FEniCS [58,59] allowing to study the difference between stable and unstable crack propagation. A pre-cracked specimen was simulated taking into account viscoelastic effects. The specimen was pre-stretched perpendicular to the notch in a quasistatic manner, and then a rapid loading was applied in the same direction. This leads to a specific loading similar to the experimental case in which stable crack propagation precedes the unstable crack propagation during quasistatic external loading. The damage field resulting from this specific loading case can be seen in fig. 25. As a result of overloading fully stretched polymer chains, the damage localizes in a wider region in the vicinity of the crack tip. Due to the subsequent rapid loading, the rupture of polymer chains less stretched in the ligament lead to the localization of damage in a narrow zone.

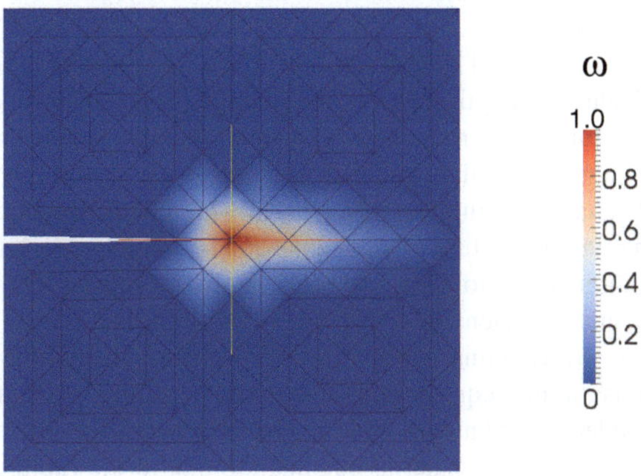

Fig. 25 Localization of damage during rapid loading of a pre-stretched pre-cracked specimen

6 Conclusion

Due to its ability to obtain large deformations even under quite small external forces, the fracture-mechanical characterization of elastomeric material can only be based on a global energy balance. However, rate-dependent and rate-independent dissipative mechanisms are not only located in the vicinity of the crack front, but as well throughout the whole specimen or structural component. Moreover, these dissipative mechanisms are in most cases not uniquely related to crack propagation, they also depend on the specific loading conditions and the specimen geometry. Because of these factors, the fracture-mechanical approach of elastomers, in comparison to other structural materials, is complicated even more.

Therefore, the analysis of global energy balance is crucial for both the experiment and the finite element simulation. Most approaches, such as the J-integral approach and the formulas derived for the experimental determination of tearing energy, assume inherently steady state fields in the vicinity of the crack front. However, it was shown that this assumption is not always fulfilled in the case of stable crack propagation, due to the development of a dissipation zone rather than the shifting of this zone.

Moreover, the viscoelastic behavior of elastomeric materials leads to a situation in which steady state fields depend not only on the external loading but also on the crack velocity and its preceding values. In order to separate the influence of viscoelasticty on crack propagation behavior in elastomers, near tip fields were calculated within plane linear viscoelasticity. It can be shown that a transition from a rubbery to a glassy material behavior occurs with increasing crack tip velocity in the vicinity of the crack tip. The increase in tearing energy at higher crack tip velocities can be attributed to the dissipation of viscoelastic energy in the vicinity of the crack tip outside of the fracture process zone. However, the size of the fracture process zone is much smaller in the unstable regime compared to the stable crack propagation.

The statistical analysis of the fracture surface topography allows for the determination of relevant length scales of the fracture process. Considering height-height correlation functions, roughness exponents can be determined within a certain range of length scales that are close to those found for the fracture surfaces of other materials characterizing length scales, in which a ductile process prevails. Consequently, the size of fracture process zone can be roughly estimated.

Crack propagation in elastomeric materials is always related to the rupture of network chain segments. Generally, rupture is the consequence of overloading chains, either by overloading fully stretched chains in equilibrium or by overloading chains in non-equilibrium due to fast loading. A nonlocal model based on the extended non-Gaussian tube model with non-affine tube deformation was proposed to describe the material degradation due to the rupture of polymer chain segments. It was shown that this model describes the changes of damage localization between stable and unstable crack propagation, as observed in experiments.

References

[1] Griffith, A.A.: The phenomena of rupture and flow in solids. Philosophical Transactions of the Royal Society of London. Series A, pp. 163–198 (1921)
[2] Rivlin, R.S., Thomas, A.G.: Rupture of rubber. I. Characteristic energy for tearing. Journal of Polymer Science 10, 291–318 (2003)
[3] Kanninen, M.F., Popelar, C.L.: Advanced fracture mechanics. Oxford University Press (1985)
[4] Sumpter, J.D.G.: An alternative view of R curve testing. Engineering Fracture Mechanics 64, 161–176 (1999)
[5] Sumpter, J.D.G.: Energy rates and crack stability in small scale yielding. International Journal of Fracture 130, 667–681 (2004)
[6] Sumpter, J.D.G.: The energy dissipation rate approach to tearing instability. Engineering Fracture Mechanics 71, 17–37 (2004)
[7] Turner, C.E.: A re-assessment of ductile tearing resistance, Part I: The geometry dependence of J_R curves in fully plastic bending. Part II: Energy dissipation rate and associated R-curves on normalized axes. In: Firraro, D. (ed.) Fracture Behaviour and Design of Materials and Structures, Turin, pp. 933–949 (1990)

[8] Brocks, W., Anuschewski, P.: Parametrizing ductile tearing resistance by four parameters. Engineering Fracture Mechanics 71, 127–146 (2004)
[9] Memhard, D., Brocks, W., Fricke, S.: Characterization of ductile tearing resistance by energy dissipation rate. Fatigue & Fracture of Engineering Materials & Structures 16, 1109–1124 (1993)
[10] Kolednik, O.: On the physical meaning of the J-Δa-curves. Engineering Fracture Mechanics 38, 403–412 (1991)
[11] Kolednik, O., Turner, C.E.: Application of energy dissipation rate arguments to ductile instability. Fatigue & Fracture of Engineering Materials & Structures 17, 1129–1145 (2007)
[12] Strobl, G.R.: The physics of polymers: concepts for understanding their structures and behavior. Springer (2007)
[13] Klüppel, M.: The role of disorder in filler reinforcement of elastomers on various length scales. Advances in Polymer Science 164, 1–86 (2003)
[14] Rice, J.R.: A path independent integral and the approximate analysis of strain concentration by notches and cracks. Journal of Applied Mechanics 35, 379–386 (1968)
[15] Cherepanov, G.P.: Crack propagation in continuous media. Applied Mathematics and Mechanics (PMM) 31, 503–512 (1967)
[16] Freund, L.B.: Dynamic fracture mechanics. Cambridge University Press (1998)
[17] Maugin, G.A., Muschik, W.: Thermodynamics with internal variables. Part I. General concepts. Journal of Non-Equilibrium Thermodynamics 19, 217–249 (1994)
[18] Maugin, G.A., Muschik, W.: Thermodynamics with internal variables. Part II. Applications. Journal of Non-Equilibrium Thermodynamics 19, 250–289 (1994)
[19] Maugin, G.: Material Inhomogeneities in Elasticity. Chapman & Hall (1993)
[20] Maugin, G.: Material forces: concepts and applications. Applied Mechanics Reviews 48, 213–245 (1995)
[21] Steinmann, P.: Application of material forces to hyperelastostatic fracture mechanics. I. Continuum mechanical setting. International Journal of Solids and Structures 37, 7371–7391 (2000)
[22] Steinmann, P., Ackermann, D., Barth, F.: Application of material forces to hyperelastostatic fracture mechanics. II. Computational setting. International Journal of Solids and Structures 38, 5509–5526 (2001)
[23] Mueller, R., Kolling, S., Gross, D.: On configurational forces in the context of the finite element method. International Journal for Numerical Methods in Engineering 53, 1557–1574 (2001)
[24] Näser, B., Kaliske, M., Müller, R.: Material forces for inelastic models at large strains: application to fracture mechanics. Computational Mechanics 40, 1005–1013 (2007)
[25] Simha, N., Fischer, F., Shan, G., Chen, C., Kolednik, O.: J-integral and crack driving force in elastic–plastic materials. Journal of the Mechanics and Physics of Solids 56, 2876–2895 (2008)
[26] Horst, T., Lauke, B., Heinrich, G.: Application of configurational forces in the context of a kinked crack. Proceedings of Applied Mathematics and Mechanics 6, 173–174 (2007)
[27] Hakim, V., Karma, A.: Crack path prediction in anisotropic brittle materials. Physical Review Letters 95, 235501 (2005)
[28] Hakim, V., Karma, A.: Laws of crack motion and phase-field models of fracture. Journal of the Mechanics and Physics of Solids 57, 342–368 (2009)

[29] Netzker, C., Horst, T., Reincke, K., Behnke, R., Kaliske, M., Heinrich, G., Grellmann, W.: Analysis of stable crack propagation in filled rubber based on a global energy balance. International Journal of Fracture (2013), doi:10.1007/s10704-013-9816-5
[30] Netzker, C., Dal, H., Kaliske, M.: An endochronic plasticity formulation for filled rubber. International Journal of Solids and Structures 47, 2371–2379 (2010)
[31] Gent, A.N., Wang, C.: Fracture mechanics and cavitation in rubber-like solids. Journal of Materials Science 26, 3392–3395 (1991)
[32] Persson, B.N.J., Albohr, O., Heinrich, G., Ueba, H.: Crack propagation in rubber-like materials. Journal of Physics: Condensed Matter 17, R1071–R1142 (2005)
[33] Persson, B.N.J., Brener, E.A.: Crack propagation in viscoelastic solids. Physical Review E 71, 36123 (2005)
[34] Tschoegl, N.W.: The phenomenological theory of linear viscoelastic behavior: An introduction. Springer (1989)
[35] Graham, G.A.C.: The correspondence principle of linear viscoelasticity theory for mixed boundary value problems involving time-dependent boundary regions. Quarterly of Applied Mathematics 26, 167–174 (1968)
[36] Horst, T., Heinrich, G.: Crack propagation behavior in rubber materials. Polymer Science Series A 50, 583–590 (2008)
[37] Bouchaud, E., Lapasset, G., Planes, J.: Fractal dimension of fractured surfaces: a universal value. Europhysics Letters 13, 73–79 (1990)
[38] Bouchaud, E.: The morphology of fracture surfaces: A tool for understanding crack propagation in complex materials. Surface Review Letters 10, 797–814 (2003)
[39] Ponson, L.: Crack propagation in disordered materials: how to decipher fracture surfaces. Annales de Physique 32, 1–120 (2007)
[40] Bonamy, D.: Intermittency and roughening in the failure of brittle heterogeneous materials. Journal of Physics D: Applied Physics 42, 214014 (2009)
[41] Horst, T., Reincke, K., Ilisch, S., Heinrich, G., Grellmann, W.: Fracture surface statistics of filled elastomers. Physical Review E 80, 46120 (2009)
[42] Santucci, S., Maloy, K.J., Delaplace, A., Mathiesen, J., Hansen, A., Haavig Bakke, J., Schmittbuhl, J., Vanel, L., Ray, P.: Statistics of fracture surfaces. Physical Review E 75, 16104 (2007)
[43] Family, F., Vicsek, T.: Dynamics of fractal surfaces. World Scientific (1991)
[44] Marinello, F., Bariani, P., Savio, E., Horsewell, A., De Chiffre, L.: Critical factors in SEM 3D stereo microscopy. Measurement Science and Technology 19, 065705 (2008)
[45] Baumberger, T., Caroli, C., Martina, D., Ronsin, O.: Magic nagles and cross-hatching instability in hydrogel fracture. Physical Review Letters 100, 178303 (2008)
[46] Febbo, M., Milchev, A., Rostiashvili, V., Dimitrov, D., Vilgis, T.A.: Dynamics of a stretched nonlinear polymer chain. Journal of Chemical Physics 129, 154908 (2008)
[47] Peerlings, R.H.J.: Enhanced damage modelling for fracture and fatigue. PhD thesis, Technische Universiteit Eindhoven (1999)
[48] Peerlings, R.H.J., de Borst, R., Brekelmans, W.A.M., de Vree, J.H.P., Spee, I.: Some observations on localisation in non-local and gradient damage models. European Journal of Mechanics, Series A: Solids 15, 937–954 (1996)
[49] Geers, M.G.D., Peerlings, R.H.J., Brekelmans, W.A.M., de Borst, R.: Phenomenological nonlocal approaches based on implicit gradient-enhanced damage. Acta Mechanica 144, 1–15 (2000)

[50] Peerlings, R.H.J., Massart, T.J., Geers, M.G.D.: A thermodynamically motivated implicit gradient damage framework and its application to brick masonry cracking. Computer Methods in Applied Mechanics and Engineering 193, 3403–3417 (2004)
[51] Engelen, R.A.B., Geers, M.G.D., Baaijens, F.: Nonlocal implicit gradient-enhanced elasto-plasticity for the modelling of softening behavior. International Journal of Plasticity 19, 403–433 (2003)
[52] Areias, P.M.A., César de Sá, J.M.A., António, C.A.: A gradient model for finite strain elastoplasticity coupled with damage. Finite Elements in Analysis and Design 39, 1191–1235 (2003)
[53] Bui, Q.V.: Initiation of damage with implicit gradient-enhanced damage models. International Journal of Solids and Structures 47, 2425–2435 (2010)
[54] Holzapfel, G.A.: Nonlinear solid mechanics: a continuum approach for engineering. John Wiley & Sons Ltd. (2000)
[55] Heinrich, G., Straube, E., Helmis, G.: Rubber elasticity of polymer networks: theories. Advances in Polymer Science 85, 33–87 (1988)
[56] Maugin, G.A.: The thermomechanics of plasticity and fracture. Cambridge University Press (1992)
[57] Holzapfel, G.A.: On large strain viscoelasticity: continuum formulation and finite element applications to elastomeric structures. International Journal for Numerical Methods in Engineering 39, 3903–3926 (1998)
[58] `http://www.fenicsproject.org`
[59] Logg, A., Mardal, K.-A., Wells, G.N.: Automated Solution of Differential Equations by the Finite Element Method. Springer (2011)

Macroscopical Modeling and Numerical Simulation for the Characterization of Crack and Durability Properties of Particle-Reinforced Elastomers

Ronny Behnke[1], Hüsnü Dal[2], Gordon Geißler[3], Bastian Näser[4], Christiane Netzker[5], and Michael Kaliske[1]

[1] Institut für Statik und Dynamik der Tragwerke, Technische Universität Dresden, Fakultät Bauingenieurwesen, 01062 Dresden, Germany
michael.kaliske@tu-dresden.de
http://www.tu-dresden.de/isd

[2] Institut für Mechanik (Bauwesen), Universität Stuttgart, Fakultät 2: Bau- und Umweltingenieurwissenschaften, Pfaffenwaldring 7, 70550 Stuttgart, Germany

[3] DYNAmore GmbH Niederlassung, Donaustr. 7, 85049 Ingolstadt, Germany

[4] Dassault Systemes Deutschland GmbH, Joseph-Wild-Str. 20, 81829 München, Germany

[5] DaimlerChrysler AG, EP/SPB, HPC X271, 71059 Sindelfingen, Germany

Abstract. Numerical modeling of particle-reinforced or filled elastomers is a challenging task and includes the adequate representation of finite deformations, nonlinear elasticity, local damage as well as rate-dependent and rate-independent dissipative properties. On the structural scale, the permanent alteration of the material is visible as formation and propagation of discrete cracks, especially in the case of catastrophic crack growth and fatigue crack propagation. In this chapter, macromechanically formulated material models for finite viscoelasticity and endochronic elasto-plasticity of filled elastomers are presented in order to describe the material response of the undamaged continuum. On the FE-discretized structural scale, crack sensitivity of the material is assessed by the material force method. Material forces are used for the computational determination of fracture mechanical parameters of dissipative rubber material. Finally, arbitrary crack growth on the structural level is addressed by an adaptive implementation of cohesive elements. In a first application, crack propagation starting from an initial side notch in a tensile rubber specimen under mixed-mode loading is numerically predicted and compared to experimental observations. In a second example, averaged stress and energy based criteria are studied and compared with respect to their crack path predictability. In a third example, the durability of a tire design is numerically assessed by using the material force method.

Keywords: numerical simulation, macroscopical modeling, elastomeric material, visco-elasticity, endochronic plasticity, fracture, durability, material forces.

W. Grellmann et al. (Eds.): *Fracture Mechanics & Statistical Mech.*, LNACM 70, pp. 167–226.
DOI: 10.1007/978-3-642-37910-9_5

1 Introduction

In industry and our today's life, rubber plays an important role and can be found e.g. in form of mountings, bearings, shafts and various other rubber components. By reinforcing the rubber material with fibers of different type, advanced fiber-reinforced structures like tires, fan belts, drive belts and conveyor belts are obtained which provide a distinctive stiffness in their loading direction. However, the mechanical properties of rubber can also be favorably influenced by adding fillers, e.g. carbon black or silica, in order to increase wear resistance, modify damping characteristics and the degree of cross-links. If fillers are added, rubber is termed particle-reinforced or filled rubber. Fillers also modify the thermal properties of rubber like thermal heat conductivity and heat capacity.

In dependency of its filler content, rubber shows different mechanical behavior with respect to mechanical loading. From experiments on rubber compounds,

- nonlinear finite elasticity,
- stress softening,
- rate-dependent inelasticity, and
- rate-independent inelasticity

can be observed as the main characteristics which are associated with large strains and finite deformations. Today, the design process of rubber components and complex rubber structures includes experimental and numerical studies in order to assess quality and durability of the specimens under investigation. In this context, numerical simulations offer the possibility to decrease prototype costs and carry out virtual experiments by avoiding complex test set-ups. In addition, numerical analyses with respect to shape and material features provide a fast convergence to an optimal design. However, in this case, numerical simulations have to be reliable in terms of the realistic representation of the observed material behavior and characteristics which are of significance. In this context, the short term material response as well as the long term structural resistance to the operating conditions are of importance. Hence, constitutive material descriptions and fracture mechanical models are needed to represent the afore-introduced main characteristics of rubber components as well as their durability and failure properties as illustrated in Fig. 1.

In this chapter, a realistic representation of the material and fracture behavior of filled elastomers within numerical simulations is outlined. A realistic modeling requires micro-, meso- and macroscopical considerations with respect to the material and its alteration mechanisms in form of micro damage and rupture. First in Sect. 2, phenomenological models derived from micromechanical features are introduced and applied to the macroscopical modeling of typical material characteristics associated with filled elastomers. The first model focuses on the visco-elastic behavior. Therefore, the algorithmic implementation of a nonlinear visco-elastic model proposed by Bergström

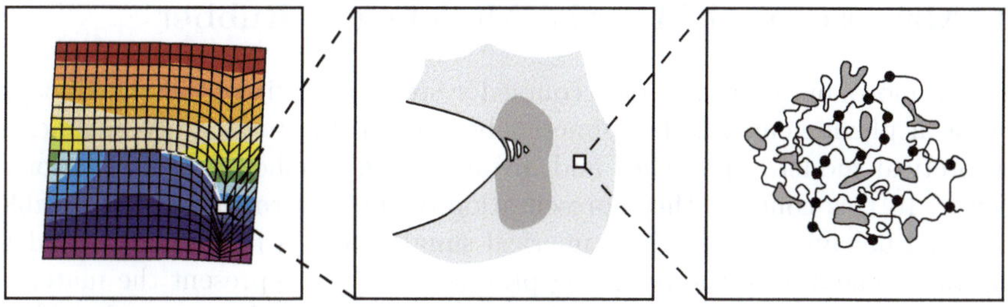

Fig. 1 Micro-meso-macro levels of investigation

and Boyce with strain-type internal variables and an implicit update scheme are demonstrated. The second model deals with a suitable form of elasto-plasticity. In this context, an endochronic formulation for yield-surface-free plasticity of rate-independent elastomeric material is proposed in analogy to the visco-elastic setting in order to facilitate a combination of both material models in one generalized numerical approach. As a starting point for further structural investigations including crack propagation, Sect. 3 presents the characterization of fracture sensitivity, crack propagation criteria, the prediction of the direction of crack propagation and a numerical model for representing finite cracks within numerical simulations. First, a micro-macro damage approach is shown which constitutes a link between micro damage and macroscopically observable material alteration. Second, the adaption and extension of the material force method to inelasticity will permit the computation of fracture mechanical quantities in inelastic materials at finite strains and deformations. As a result, a relation between numerically and experimentally determined fracture mechanical parameters in case of a distinctive inelastic process zone at the crack tip becomes possible. In order to model discrete cracks in the structure, cohesive elements are described in Sect. 4 in combination with adaptive meshing strategies. Several examples on the structural scale are shown in Sect. 5 and illustrate the application of the derived models and methods in the framework of the finite element method (FEM). In the first example, crack propagation in a tensile rubber specimen under mixed-mode loading is numerically simulated. The second structural example illustrates crack propagation in rubber plates with arbitrary crack paths by using the developed adaptive meshing strategy and different crack propagation criteria. In the third example, the material force method is used to assess the durability of a given tire design under consideration of the dwell-effect due to visco-elastic rubber material. At the end, a brief conclusion and outlook on related topics and further research are given in Sect. 6.

2 Macroscopical Material Modeling of Rubber

Today, more and more complex coupled systems including different kinds of physical phenomena can be numerically studied due to steadily advancing theoretical model approaches and an increasing number of computational power. In this context, the representation of the experimentally observable material behavior within the numerical simulation is a fundamental modeling task. The aim of the modeling process is to best represent the material behavior for further arbitrary loading conditions within the range of verification and validation of the model. The latter requires an intensive testing of the correct implementation (verification) and of the right model assumptions (validation).

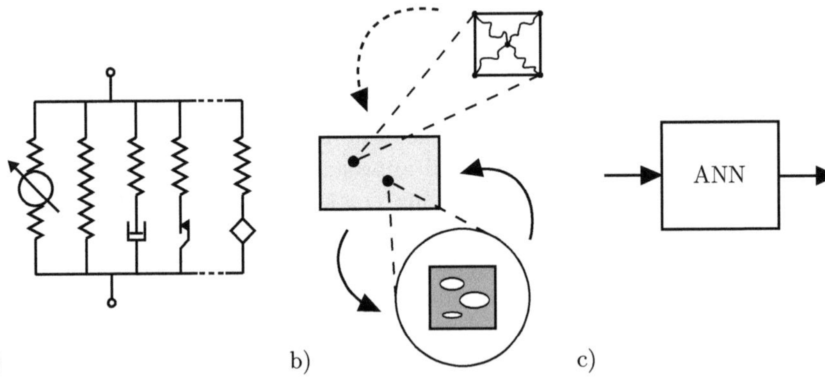

a) b) c)

Fig. 2 Different approaches and methods to represent the material response in numerical simulations: a) rheological elements; b) micro-macro transition; c) artificial neural networks

Fig. 2 illustrates different approaches to represent the material behavior in numerical simulations on the macroscopical scale. Different approaches exist which base on purely phenomenological relations, physical models and model-free methods which use artificial neural networks (ANN) with and without memory effects.

Phenomenological relations can be easily derived by a pure mapping of the input variables (e.g. strain state) to the experimentally observed results (e.g. stress state). The mapping can be realized by simple pairs of input and output quantities recorded during experiment or a mathematical function which has been determined by regression from the experimental data pairs. Therefore, a good agreement between experimental and simulation results can be found for similar test modes. If other loading scenarios are considered, the described approach can suffer from a false representation of the material response since it is not physically based. In addition, this procedure offers no possibility to derive physical quantities like the energy dissipation directly. The latter can be overcome by allocating to each macroscopically observed material property its corresponding rheological element. In this case, the material properties observed during experiments are represented by rheological elements with evolution laws based on phenomenological considerations on the macroscale. Due to the introduction of internal variables and evolution laws, it is possible to derive other physical quantities like material forces, energy dissipation, local damage gradients, etc. directly.

If observations of the microscale enter the model process, the material modeling is already started at the micro level. In this case, an approach consists of an analytical derivation of micromechanically inspired macroscopical models. For rubber, the molecular-statistical theory of polymer networks can be applied by considering one single molecular chain with respect to possible positions and deformation characteristics via statistical thermodynamics. In this framework, Gauß statistics, Langevin statistics and tube concepts are used which describe free motion of single chains without or with constraints due to entanglements with surrounding polymer chains. Another approach in the framework of micro-macro bridging consists of a permanently applied homogenization of the material response of a microstructure to obtain the material response at the macroscale. Due to the difference in scale, these simulations are also termed two-scale simulations. In this case, the microstructure is also discretized and included in the simulation process. First, the macroscopical deformation state is projected to the microstructure. Second, the boundary value problem is solved on the microstructure. Third, the obtained stress response from the microstructure is projected again to the macro structure to yield the stress response on the macroscale. In comparison to the analytical approach, two-scale simulations are computationally expensive and require detailed information of the material characteristics and its structure on the microscale.

Similar to the phenomenologically based approaches, ANN can be used as a model-free description to represent a material or a structural behavior within numerical simulations with or without taking history effects into account.

In this section, macroscopical models for rubber are introduced which are based on physical, micromechanical effects. Therefore, a brief characterization of the chemical and inner structure of rubber is presented in order to derive the significant properties of rubber and possible model approaches to represent them in a numerical model. In a consecutive step, the continuum mechanical basis for further derivations is introduced. For the main rubber properties of interest, which are finite visco-elasticity and finite elasto-plasticity, two macroscopical and micromechanically inspired models are derived. A note on parameter identification, which is an important task to conclude the modeling phase of a material and to make it available within numerical simulations, is presented at the end of this section.

2.1 Inner Structure of Elastomers

Unfilled Elastomers. Elastomers are polymeric materials with macromolecules which consist of chain forming monomers. Chemical cross-links between the molecules form a wide-meshed network of polymer chains. Beside entangled molecular chains, free chain segments can rotate freely in combination with Brownian motions. These motions are possible due to the free volume. The entropy becomes maximal for a disordered structure. The motion of the free chain segments can be constrained by

- chemical links in the network,
- entangled chains, and
- crystallized domains.

Crystallized domains can be formed by aligned polymer chains in a disordered, amorphous environment and depend highly on temperature. In addition to temperature, also an externally applied loading can provoke strain-induced crystallization (SIC) due to the imposed orientation of the polymer chains. In all cases, the crystallization leads to an increase of the material's stiffness.

Filled Elastomers. By adding fillers, one or more properties of the elastomer can be optimized by usually affecting other properties as well. The fillers consist of primary particles which already form aggregates when they are added. Beside the filler type (active, inactive), also the size and the form of the aggregates influence the resulting properties. Since the free volume is partially substituted by nearly rigid fillers, the polymer chains are more extended during external loading in comparison to unfilled rubber which can be observed as a stiffening of the material. The stiffening is also related to the filler-elastomer links and the filler-filler linkage. In the first case, filler-elastomer links are formed by molecular chain bonding, chemical bonding and physical bonding due to chain entanglements with aggregates. The result is a high cross-linking level of the elastomer in combination with free chain segments. The high cross-linking density decreases also the initial length of free chain segments compared to the unfilled elastomer. The filler-filler linkage stems from the interaction of filler particles with filler particles which form together filler aggregates. By increasing the filler amount, filler clusters become visible. Filler clusters can form an own filler network by a further increase of the filler content. An illustration of the inner structure of filled elastomers with different types of bondings and cross-links is depicted in Fig. 3.

2.2 Material Properties

From the knowledge of the general inner structure of filled elastomers, resulting material properties can be derived. These properties originate from the molecular structure and their related micromechanical processes.

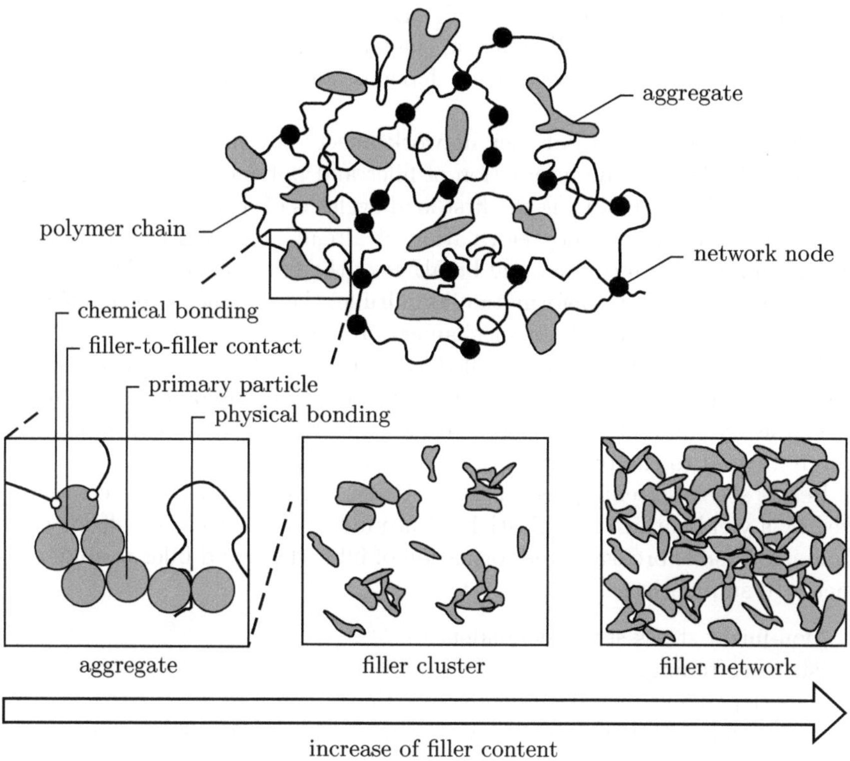

Fig. 3 Representation of the inner structure of filled elastomers

Micromechanical processes are e.g. rearrangement of cross-links between polymer chains, chain reptations and irreversible sliding of filler-to-filler and filler-to-matrix contacts due to an externally applied loading.

Elastomers show nonlinear elasticity at large strains. Their entropic elasticity can be explained by the motion of single polymer chains which are more or less constrained by linkages and entanglements of the network. During external loading, the entangled polymer chains are ordered which decreases mainly the entropy by nearly unchanged inner molecular energy. With increasing temperature above the glass transition temperature, the entropic elasticity becomes even more pronounced due to the enforced Brownian molecular motions. In this case, the increasing degree of order yields a stiffer material behavior for unfilled elastomers. For filled elastomers, elasticity can change due to filler-chain interactions. In consequence, even a stiffness loss with increasing temperature can be observed for filled elastomers. Below the glass transition temperature, elastomers show energetic elasticity with significantly reduced molecular Brownian motions and chain deformations. In this case, elasticity stems mainly from deformation at the atomic scale.

Since the relative motion of the polymer chains is related to internal friction and a consecutive energy dissipation, elastomers show nonlinear visco-elastic material behavior, i.e. the relaxation depends nonlinearly on the deformation state.

Adding fillers to the polymer network affects the mechanical properties significantly. Due to the interaction of filler and polymer chains, stress softening (Mullins effect) during the first load cycles is observed. Stress softening is caused by breakage and reformation of already highly stretched chains, change in filler-chain contact and breakage of the filler network. After several load cycles at constant deformation amplitude, the so-called preconditioning process is terminated and the mechanical properties stabilize.

The visco-elastic properties of filled elastomers are dominated by the Payne effect which is known as an amplitude dependency of visco-elasticity. In this case, the dynamic stiffness decreases with increasing amplitudes which is purely related to the presence of fillers in elastomers.

Beside stress softening and visco-elasticity, filled elastomers are characterized by a yield-surface-free plasticity. Rate-independent inelastic effects stem from irreversible breakage and formation of filler-chain and filler-to-filler contacts. To summarize,

- non-linear stress-strain dependencies,
- stress softening,
- rate-dependent inelasticity (visco-elasticity), and
- rate-independent inelasticity (yield-surface-free elasto-plasticity)

are characteristic properties of filled elastomers (see also Grambow [1]). In the following, rheological elements are associated with the characteristic material properties, which have been identified from mechanical tests, chemical analysis and microscopical observations, in order to propose a suitable macroscopical model for the mechanical behavior of filled elastomers.

2.3 Configurations

For further theoretical derivations in this chapter, the continuum mechanical background is introduced. Therefore, a continuous body $\mathfrak{B} \subset \mathbb{R}^3$, which is formed by material particles $\mathbf{X} \in \mathfrak{B}$ of the reference or material configuration at time $t = 0$, is considered. In the current or spatial configuration at time $t > 0$, these material particles occupy the position $\mathbf{x}$ of the deformed body $\mathfrak{b} \subset \mathbb{R}^3$. The deformation map $\varphi(\mathbf{X}, t) : \mathfrak{B} \times \mathbb{R}^3 \to \mathbb{R}^3$ constitutes the link between the reference and the current configuration by the relations $\mathbf{x} = \varphi(\mathbf{X}, t)$ and $\mathfrak{b} = \varphi(\mathfrak{B}, t)$, respectively. The gradient of the mapping is termed deformation gradient $\mathbf{F} := \nabla_X \varphi(\mathbf{X}, t)$ and will be used in the following as a fundamental measure for the constitutive material modeling under consideration of finite deformations and strains. Its determinant $J := \det \mathbf{F} > 0$ can be interpreted as the ratio of volume change of one infinitesimal volume element $\mathrm{d}v$ in the current configuration compared to its infinitesimal volume element $\mathrm{d}V$ in the reference configuration,

$$J = \frac{\mathrm{d}v}{\mathrm{d}V} \quad \text{or} \quad J\,\mathrm{d}V = \mathrm{d}v\,. \tag{1}$$

The operator symbol $\nabla_X\Box$ stands for the gradient operator $\mathrm{Grad}\,\Box$ with respect to the reference configuration while $\nabla_x\Box$ is used for the gradient operator $\mathrm{grad}\,\Box$ in the current configuration. The absolute temperature of a material particle $\mathbf{x}$ is introduced with $\Theta(\mathbf{x}) = \Theta(\varphi(\mathbf{X})) > 0$. $\mathbf{u}$ denotes the displacement of a material point at time $t \geq 0$ with $\mathbf{u} := \mathbf{x} - \mathbf{X}$.

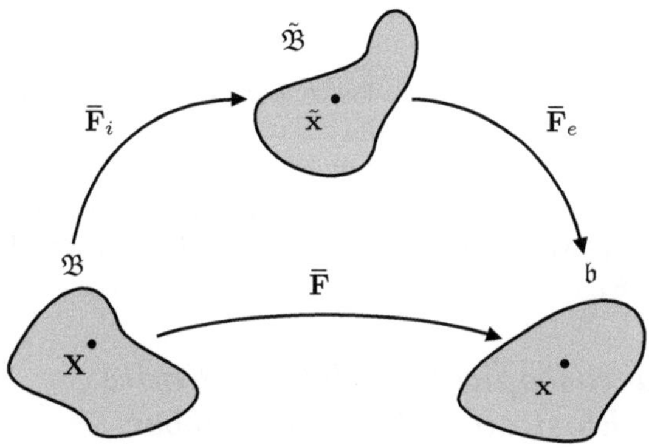

Fig. 4 Configurational setting for isochoric finite visco-elasticity and isochoric finite elasto-plasticity

The right Cauchy-Green strain tensor $\mathbf{C} := \mathbf{F}^{\mathrm{T}}\mathbf{F}$ and the left Cauchy-Green strain tensor $\mathbf{b} := \mathbf{F}\,\mathbf{F}^{\mathrm{T}}$ are defined as a measure of strain in the Euclidean space. In this case, a Cartesian coordinate system is used which possesses an orthonormal coordinate basis vector field for which the metric tensor $\mathbf{g}$ yields $\mathbf{g} = \mathbf{1}$. The material behavior will be represented by different constitutive models for the material's isochoric and volumetric response. Therefore, the deformation gradient is multiplicatively decomposed,

$$\mathbf{F} = \bar{\mathbf{F}}\,\mathbf{F}_{\mathrm{vol}}, \quad \mathbf{F}_{\mathrm{vol}} = J^{\frac{1}{3}}\mathbf{1}, \quad \bar{\mathbf{F}} = J^{-\frac{1}{3}}\mathbf{F}\,. \tag{2}$$

$\bar{\mathbf{F}}$ represents the volume-preserving (isochoric) part and $\mathbf{F}_{\mathrm{vol}}$ the volumetric part of the deformation. In analogy to the right and left Cauchy-Green strain tensor, the isochoric right and left Cauchy-Green strain tensor are defined as

$$\bar{\mathbf{C}} = \bar{\mathbf{F}}^{\mathrm{T}}\bar{\mathbf{F}} = J^{-\frac{2}{3}}\mathbf{C} \quad \text{and} \quad \bar{\mathbf{b}} = \bar{\mathbf{F}}\,\bar{\mathbf{F}}^{\mathrm{T}} = J^{-\frac{2}{3}}\mathbf{b}\,, \tag{3}$$

respectively. To capture finite inelastic deformations of viscous or plastic type, a multiplicative split of the isochoric deformation gradient $\bar{\mathbf{F}}$ is applied,

$$\bar{\mathbf{F}} = \bar{\mathbf{F}}_e\,\bar{\mathbf{F}}_i\,. \tag{4}$$

In general, the same multiplicative split into elastic parts $\overline{\mathbf{F}}_e$ and inelastic parts $\overline{\mathbf{F}}_i$ as carried out for $\overline{\mathbf{F}}$ can be analogously applied to the volumetric part of the deformation $\mathbf{F}_{\text{vol}}$. For sake of simplicity, the inelastic deformations are assumed to be of purely isochoric nature so that they can be exclusively associated with $\overline{\mathbf{F}}_i$ in the following. This assumption is a good approximation for the nearly incompressible material behavior of filled elastomers. For further derivations, it is worth mentioning that

$$\det \overline{\mathbf{F}} \equiv 1, \quad \det \overline{\mathbf{F}}_e \equiv 1 \quad \text{and} \quad \det \overline{\mathbf{F}}_i \equiv 1 \tag{5}$$

hold. The conditions specified in Eq. (5) imply that the evolution law for the inelastic deformations has to preserve the isochoric nature of the inelastic deformations expressed via $\overline{\mathbf{F}}_i$.

The multiplicative split into elastic parts $\overline{\mathbf{F}}_e$ and inelastic parts $\overline{\mathbf{F}}_i$ of the isochoric deformation can be illustrated by introducing a fictitious intermediate configuration $\tilde{\mathfrak{B}}$ as depicted in Fig. 4. As already mentioned, this general concept will be used in the following to model finite visco-elasticity and finite elasto-plasticity where in both cases the inelastic deformations are represented by $\overline{\mathbf{F}}_i$.

2.4 A Macroscopical Model for Finite Visco-Elasticity of Filled Rubber

In the following, the focus is on a material model for finite nonlinear visco-elasticity which is formulated on the macroscale but inspired by micromechanics. Therefore, rheological elements are used on the macroscale in order to build up a rheological material model. The properties and functional relations of each rheological element are derived from micromechanical considerations. A continuum mechanical quantity is then used to constitute a micro-macro transition.

In order to represent nonlinear elastic material properties, which are attributed to the so-called equilibrium (EQ) or ground state response of the material, a nonlinear hyperelastic material description is required for which a rheological spring element can be employed. The nonlinear inelastic properties – in the present case the viscosity of the material – are formulated as the non-equilibrium (NEQ) response of the material by a nonlinear spring-dashpot combination. The complete rheological material model is depicted in Fig. 5.

For the hyperelastic properties, several phenomenological strain energy functions are available, see e.g. Mooney [2] or Yeoh [3]. A physically inspired model is the 8-chain model proposed by Arruda and Boyce [4]. The description idealizes the polymeric network structure by eight chains in a cubic reference volume where each chain is connected to the center of the cube and to one of its corners. Other physically based models for rubber are proposed e.g. by Miehe et al. [5] and a comparison of different models to represent experimental data has been studied by Marckmann and Verron [6]. The extended

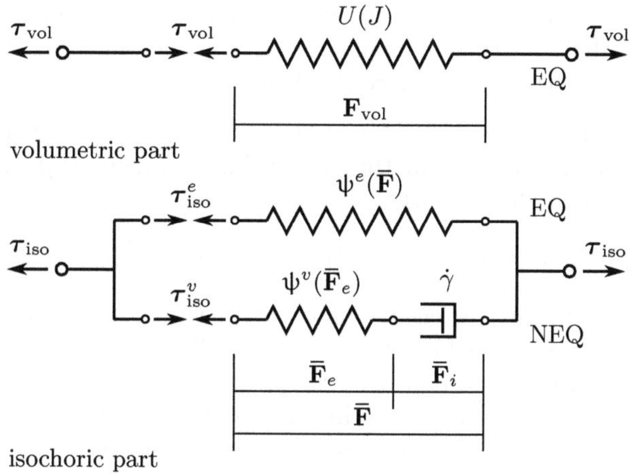

Fig. 5 Rheology of the visco-elastic material model for the volumetric and isochoric part in 1D representation

tube model introduced by Kaliske and Heinrich [7] considers the interaction of polymer chains including topological constraints and limited chain extensibility of network chains. The strain energy function is derived from physics on the molecular scale with parameters representing the stiffness of the cross-links in the rubber network and the stiffness due to topological constraints. The model is based on the assumption that the polymer chains of rubber are contained in a tube-like geometry with constant diameter. The topological constraints result from neighboring links. Therefore, parameters describing the form and extensibility of network chains are included in the strain energy function. For the numerical implementation and parameter studies, the 8-chain model has been used in Dal and Kaliske [8] while the extended tube model has been employed in the same visco-elastic framework by Behnke et al. [9, 10]. Beside the presented model for vulcanized rubber, research has also been done on material formulations representing the behavior of unvulcanized rubber as documented in Kaliske et al. [11] and Dal et al. [12].

Creep and relaxation are phenomena of the rubber's rate-dependent inelastic behavior. Both lead to a frequency-dependent strain-stress response, e.g in form of different hystereses under cyclic loading conditions. A physical interpretation can be found in network rearrangements consisting of cross-link breakage between polymer chains and their reformation (see e.g. Green and Tobolsky [13]). Another explication is given by the reptation theory proposed by Doi and Edwards [14] which considers the motion of a single chain in a polymer gel in form of Brownian and reptational motions. Bergström and Boyce [15] formulated their approach to finite visco-elasticity based on the reptation theory. Other viscoelastic models have been presented by Reese and Govindjee [16] or Miehe and Keck [17]. Nonlinear evolution laws are used by

Miehe and Göktepe [18] to describe micromechanically inspired finite visco-elasticity. For the numerical implementation in the present case, the approach proposed by Bergström and Boyce [15] is followed.

Free Energy Functions. As depicted in Fig. 5, the free energy function ψ contains contributions form the equilibrium branch (EQ) and the non-equilibrium branch (NEQ) with

$$\psi(J, \bar{\mathbf{F}}, \bar{\mathbf{F}}_e) = U(J) + \psi^e(\bar{\mathbf{F}}) + \psi^v(\bar{\mathbf{F}}_e)\,. \tag{6}$$

$U(J)$ is the volumetric free energy function for which various types of strain energy potentials are possible. However, the bulk modulus κ of the material should be chosen large enough in order to act as a penalty parameter and, in consequence, enforce the nearly incompressibility of the rubber material. For the isochoric equilibrium and non-equilibrium free energy,

$$\psi^e(\lambda_r) = \mu\, N \left(\lambda_r\, \mathcal{L}^{-1}(\lambda_r) + \ln \frac{\mathcal{L}^{-1}(\lambda_r)}{\sinh \mathcal{L}^{-1}(\lambda_r)} \right) \tag{7}$$

and

$$\psi^v(\lambda_r^e) = \mu_v\, N_v \left(\lambda_r^e\, \mathcal{L}^{-1}(\lambda_r^e) + \ln \frac{\mathcal{L}^{-1}(\lambda_r^e)}{\sinh \mathcal{L}^{-1}(\lambda_r^e)} \right) \tag{8}$$

have been used, respectively.

$$\lambda_r = \sqrt{\frac{I_1}{3\,N}} \quad \text{and} \quad \lambda_r^e = \sqrt{\frac{I_1^e}{3\,N_v}} \tag{9}$$

are the relative network stretches which are functions of the first invariants $I_1 = \operatorname{tr} \bar{\mathbf{C}}$ and $I_1^e = \operatorname{tr} \bar{\mathbf{C}}_e$. μ and μ_v are the shear moduli of the chain networks while N and N_v denote the segment numbers of the networks. $\mathcal{L}(\lambda_r)$ denotes the Langevin function with the argument λ_r. The inverse of the Langevin function can be computed by a Padé approximation via

$$\mathcal{L}^{-1}(x) \approx x\, \frac{3 - x^2}{1 - x^2} \tag{10}$$

as proposed by Cohen [19]. An illustration of the 8-chain model is given in Fig. 6.

Stress Response. Form the free energy function ψ, the Kirchhoff stress tensor

$$\boldsymbol{\tau} = \boldsymbol{\tau}_{\text{vol}} + \boldsymbol{\tau}_{\text{iso}}^{e} + \boldsymbol{\tau}_{\text{iso}}^{v} \tag{11}$$

can be derived as a sum of the equilibrium ($\boldsymbol{\tau}_{\text{vol}} + \boldsymbol{\tau}_{\text{iso}}^{e}$) and non-equilibrium ($\boldsymbol{\tau}_{\text{iso}}^{v}$) stress response of the material.

Evolution Law. Special attention is paid now to the evolution law for the inelastic network deformations. Therefore, the micromechanical relaxation process of a single chain in a polymer network is considered as illustrated in Fig. 7.

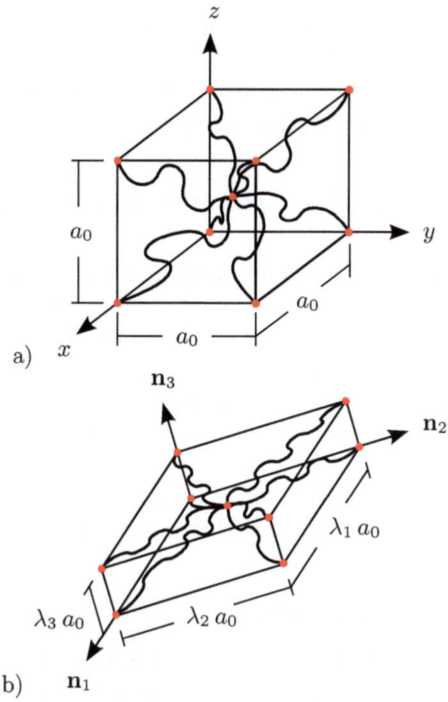

Fig. 6 8-chain model of a rubber network: a) undeformed reference cube; b) deformed configuration in the eigenspace with principal stretches λ_1, λ_2 and λ_3

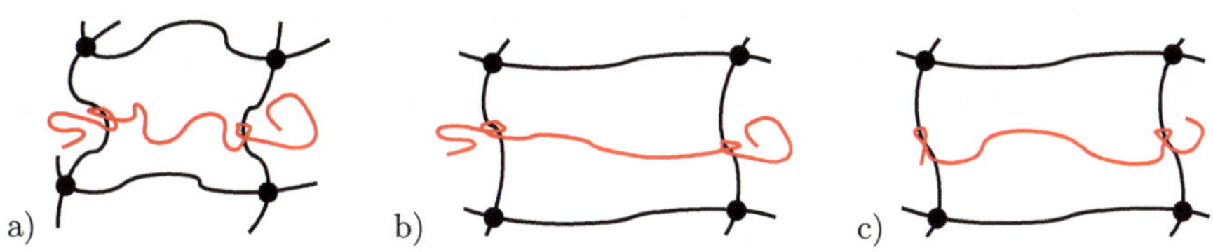

Fig. 7 Superimposed free chain in an elastic underlaying network: a) undeformed state; b) instantaneous loading; c) relaxed state

Bergström and Boyce [15] defined the evolution of the inelastic rate of deformation tensor

$$\tilde{\mathbf{d}}_i = \dot{\gamma}\,\mathbf{N} \tag{12}$$

in the current configuration as the creep process in the viscous stress directions with the stress projection tensor

$$\mathbf{N} = \frac{\boldsymbol{\tau}^v_{\text{iso}}}{\|\boldsymbol{\tau}^v_{\text{iso}}\|}, \quad \|\boldsymbol{\tau}^v_{\text{iso}}\| = \sqrt{\boldsymbol{\tau}^v_{\text{iso}} : \boldsymbol{\tau}^v_{\text{iso}}}\,, \tag{13}$$

where $\boldsymbol{\tau}^v_{\text{iso}}$ are the isochoric viscous Kirchhoff stresses and $\dot{\gamma} \geq 0$ denotes the effective creep rate. The motion of a free chain in an elastic network – as depicted in Fig. 7 – can be described with Brownian and reptational motions (see Doi and Edwards [14]). Bergström and Boyce [15] formulated in a similar way the evolution of the inelastic network stretch as a function of time where after several operations the effective creep rate

$$\dot{\gamma} = \dot{\gamma}_0 \left[\lambda^i_{\text{chain}} - 1\right]^c \left(\frac{\tau_v}{\hat{\tau}}\right)^m \quad \text{with} \quad \tau_v = \frac{\|\boldsymbol{\tau}^v_{\text{iso}}\|}{\sqrt{2}}, \quad \lambda^i_{\text{chain}} = \sqrt{\frac{I^i_1}{3}} \tag{14}$$

is obtained. The link between microscale and macroscale is established by relating the stretch of the relaxing single polymer chain (see Fig. 7) to the continuum mechanical quantity $\operatorname{tr} \overline{\mathbf{C}}_i$ which enters the effective creep rate expression by the first invariant $I^i_1 = \operatorname{tr} \overline{\mathbf{C}}_i$ with $\overline{\mathbf{C}}_i = \overline{\mathbf{F}}^{\mathrm{T}}_i \overline{\mathbf{F}}_i$. In Eq. (14), $\dot{\gamma}_0$ denotes the reference creep rate, $\hat{\tau}$ is a parameter for conversion of dimensions with the same units of τ_v, c is a parameter to define the kinetics of chain relaxation and m controls the energy activation of the flow process.

In order to know the evolution of the elastic deformation part $\overline{\mathbf{F}}_e$, Eq. (12) has to be solved for any time step within a numerical simulation. Therefore, the identity

$$-\frac{1}{2}\overline{\mathcal{L}}_v(\overline{\mathbf{b}}_e)\,\overline{\mathbf{b}}^{-1}_e = \tilde{\mathbf{d}}_i = \dot{\gamma}\,\mathbf{N} \tag{15}$$

is used where the evolution law can be advantageously integrated by employing an operator split of the material time derivative of $\overline{\mathbf{b}}_e = \overline{\mathbf{F}}_e\,\overline{\mathbf{F}}^{\mathrm{T}}_e$ with

$$\dot{\overline{\mathbf{b}}}_e = \mathbf{l}_{\text{iso}}\,\overline{\mathbf{b}}_e + \overline{\mathbf{b}}_e\,\mathbf{l}^{\mathrm{T}}_{\text{iso}} + \overline{\mathcal{L}}_v(\overline{\mathbf{b}}_e)\ . \tag{16}$$

$\mathbf{l}_{\text{iso}}$ denotes the isochoric spatial velocity gradient and $\overline{\mathcal{L}}_v(\overline{\Box})$ is a unimodular operator similar to the Lie time derivative of the unimodular quantity $\overline{\Box}$. After an initial elastic trial step, an inelastic corrector step is applied in which the evolution law results in

$$\dot{\overline{\mathbf{b}}}_e = -\,[2\,\dot{\gamma}\,\mathbf{N}]\,\overline{\mathbf{b}}^{\text{tr}}_e\ . \tag{17}$$

Eq. (17) can be solved by an exponential mapping which leads to

$$\overline{\mathbf{b}}_e = \exp\left[-2\int_{t_n}^{t_n+\Delta t} \dot{\gamma}\,\mathbf{N}\,\mathrm{d}t\right]\overline{\mathbf{b}}^{\text{tr}}_e \tag{18}$$

and yields after the approximation of the integral, the notation in the principal stretch directions of $\boldsymbol{\tau}^v_{\text{iso}}$, $\mathbf{N}$ and $\overline{\mathbf{b}}^{\text{tr}}_e$ as well as the logarithm

$$\ln\lambda^e_a \approx -\frac{\Delta t}{\sqrt{2}}\,\frac{\dot{\gamma}}{\tau_v}\,[\boldsymbol{\tau}^v_{\text{iso}}]_a + \ln\lambda^{e\,\text{tr}}_a\ . \tag{19}$$

Eq. (19) is a nonlinear relation of the elastic logarithmic stretches $\varepsilon_a = \ln\lambda^e_a$ and $\varepsilon^{e\,\text{tr}}_a = \ln\lambda^{e\,\text{tr}}_a$ which can be iteratively solved by formulating a residual expression

$$r_a = \varepsilon_a^e + \frac{\Delta t}{\sqrt{2}} \frac{\dot{\gamma}}{\tau_v} \left[\boldsymbol{\tau}_{\text{iso}}^v \right]_a - \varepsilon_a^{e\,\text{tr}} \rightarrow 0 \tag{20}$$

and a local Newton iteration at the integration point of each finite element. Further details on the complete derivation of the stress and consistent material tangent for a quadratic convergence within a finite element code are given in Dal and Kaliske [8].

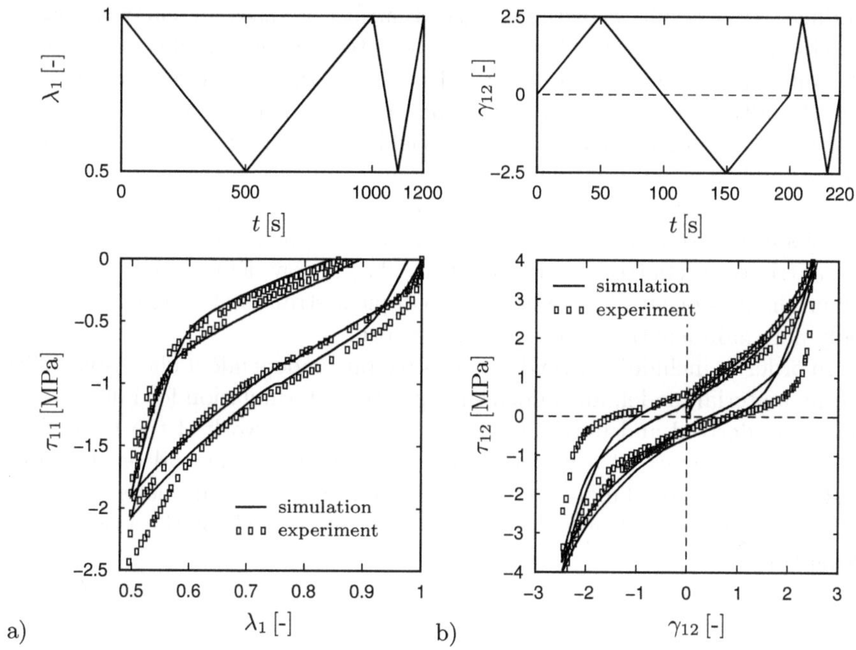

Fig. 8 Cyclic tests on HDR with different strain rates and comparison of simulation results with experimental results (Amin et al. [20]): a) uniaxial compression; b) simple shear

The implemented material model has been compared to experimental data published by Amin et al. [20] for high damping rubbers (HDR). The cyclic loading at different strain rates and for different deformation modes is depicted together with the experimentally and numerically obtained results in Fig. 8. For the simulation of the uniaxial compression and the simple shear test, the same model parameters $\kappa = 500\,\text{MPa}$, $\mu = 0.70\,\text{MPa}$, $\mu_v = 2.0\,\text{MPa}$, $N = 5.5$, $N_v = 6$, $\dot{\gamma}_0/\hat{\tau}^m = 0.2\,\text{s}^{-1}\text{MPa}^{-m}$, $c = -0.8$ and $m = 4$ are used. A further comparison with experimental data and a detailed parameter study on the influence of each model parameter can be found together with a note on the algorithmic setting in Dal and Kaliske [8] as well as in Dal [21].

2.5 A Macroscopical Model for Finite Endochronic Plasticity of Filled Rubber

As already pointed out in Sect. 2.2, experimental observations have revealed that the material behavior of filled rubber exhibits also rate-independent irreversible plastic deformations. During monotonic loading, carbon black filled rubber shows irreversible deformations in combination with a stress increase instead of a constant yield stress normally associated with a distinctive yield surface. As a result, a distinctive yield surface as used by classical elasto-plastic models does not well represent the rate-independent effects of rubber. A multi-yield surface approach resulting from the use of several rheological Prandtl elements in parallel as proposed by Kaliske and Rothert [22] overcomes the limitations of a single yield surface approach but requires a large number of material parameters. The endochronic theory (see Valanis [23]) seems a promising approach to represent the continuous slope of the quasi-static hystereses resulting from changes in the filler-to-filler and filler-to-matrix contacts (see e.g. Kilian et al. [24]) at low deformation rates. The inner changes occur at different macroscopical strain states due to different polymer chain lengths in the rubber network.

In order to include also yield-surface-free rate-independent elasto-plasticity in the material model, an endochronic plasticity formulation for filled rubber has been derived and implemented within the framework of the FEM. The current stress state of a material point is assumed to depend on its strain history. The material model requires only eight material parameters and uses the same configurational setting as already introduced for the visco-elastic behavior.

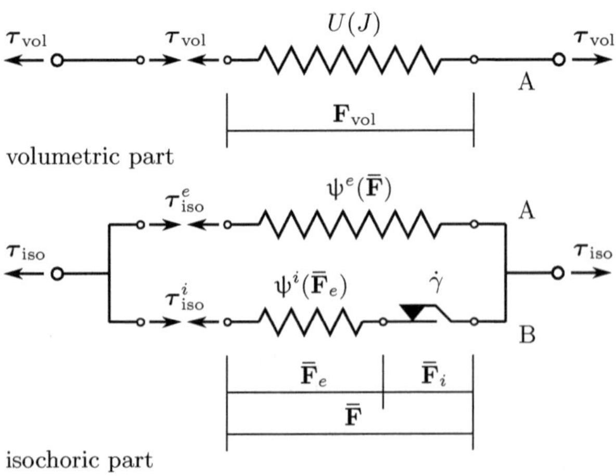

Fig. 9 Rheology of the endochronic elasto-plastic material model for the volumetric and isochoric part in 1D representation

Free Energy Functions. Fig. 9 illustrates the rheology of the elasto-plastic material model. The volumetric stress response is assumed to be independent of inelastic effects and is exclusively modeled by a hyperelastic spring. The isochoric stress response includes an elastic branch (A) and an endochronic branch (B). The endochronic branch is assembled from a hyperelastic spring and an endochronic friction element in series. As a result, the free energy function ψ for the volumetric and isochoric part reads

$$\psi(J, \bar{\mathbf{F}}, \bar{\mathbf{F}}_e) = U(J) + \psi^e(\bar{\mathbf{F}}) + \psi^i(\bar{\mathbf{F}}_e) \,. \tag{21}$$

Again, $U(J)$ stands for the volumetric free energy function containing the bulk modulus κ of the material. For the isochoric elastic and endochronic free energy functions,

$$\psi^e(\lambda_r) = \mu\, N \left(\lambda_r\, \mathcal{L}^{-1}(\lambda_r) + \ln \frac{\mathcal{L}^{-1}(\lambda_r)}{\sinh \mathcal{L}^{-1}(\lambda_r)} \right) \tag{22}$$

and

$$\psi^i(\lambda_r^e) = \mu_i\, N_i \left(\lambda_r^e\, \mathcal{L}^{-1}(\lambda_r^e) + \ln \frac{\mathcal{L}^{-1}(\lambda_r^e)}{\sinh \mathcal{L}^{-1}(\lambda_r^e)} \right) \tag{23}$$

have been used, respectively. Compared to the visco-elastic formulation described before, the same underlaying elastic material model (8-chain model) is chosen.

$$\lambda_r = \sqrt{\frac{I_1}{3\,N}} \quad \text{and} \quad \lambda_r^e = \sqrt{\frac{I_1^e}{3\,N_i}} \tag{24}$$

are the relative network stretches which can be computed form the first invariants $I_1 = \operatorname{tr} \bar{\mathbf{C}}$ and $I_1^e = \operatorname{tr} \bar{\mathbf{C}}_e$ as well as the material specific segment numbers N and N_i. μ and μ_i stand for the shear moduli of the chain networks.

Stress Response. Similar to the visco-elastic material model, the Kirchhoff stress tensor

$$\boldsymbol{\tau} = \boldsymbol{\tau}_{\text{vol}} + \boldsymbol{\tau}_{\text{iso}}^e + \boldsymbol{\tau}_{\text{iso}}^v \tag{25}$$

can be derived as a sum of the ground state elastic stress response ($\boldsymbol{\tau}_{\text{vol}} + \boldsymbol{\tau}_{\text{iso}}^e$) and the inelastic contribution stemming from the endochronic branch ($\boldsymbol{\tau}_{\text{iso}}^i$).

Evolution Law. The evolution of the inelastic deformations in the endochronic branch is governed by a constitutive evolution law which represents the flow rule for the internal variable of plastic strain,

$$-\frac{1}{2}\bar{\mathcal{L}}_v\left(\bar{\mathbf{b}}_e\right)\, \bar{\mathbf{b}}_e^{-1} = \tilde{\mathbf{d}}_i = \dot{\gamma}\,\mathbf{N} \quad \text{with} \quad \mathbf{N} = \frac{\boldsymbol{\tau}_{\text{iso}}^i}{\left\|\boldsymbol{\tau}_{\text{iso}}^i\right\|} \,. \tag{26}$$

$\mathbf{N}$ represents the direction of inelastic flow which coincides with the directions of the isochoric inelastic Kirchhoff stress tensor $\boldsymbol{\tau}_{\text{iso}}^i$. Since no rate-dependent effects are considered, the endochronic plastic behavior is formulated without any physical time measure. Therefore, the Lagrange multiplier $\dot{\gamma}$ is a function of an intrinsic time measure $\dot{z}$ and the plastic network stretch λ^p (see Miehe and Keck [17]) representing the absolute value of the inelastic strain rate tensor $\tilde{\mathbf{d}}_i$,

$$\dot{\gamma} = \dot{z}\,\lambda^p \quad \text{with} \quad \dot{z} = \|\dot{\boldsymbol{\varepsilon}}\|\,, \quad \lambda^p = \sqrt{\frac{2}{3}}\frac{1}{\eta_i}\left(\frac{\tau_i}{\hat{\tau}}\right)^n, \quad \tau_i = \frac{\|\boldsymbol{\tau}^i_{\text{iso}}\|}{\sqrt{2}}\,. \tag{27}$$

$\dot{z}$ represents the strain history by considering the change of arc length in total logarithmic strains $\boldsymbol{\varepsilon}$. The material parameters η_i and n control the evolution of the inelastic deformations and $\hat{\tau}$ renders the evolution law non-dimensional. By combining the afore-introduced steps, the evolution law for the inelastic rate of deformation tensor yields

$$-\frac{1}{2}\mathcal{L}_v(\bar{\mathbf{b}}_e)\,\bar{\mathbf{b}}_e^{-1} = \sqrt{\frac{2}{3}}\frac{\dot{z}}{\eta_i}\left(\frac{\tau_i}{\hat{\tau}}\right)^n \mathbf{N}\,. \tag{28}$$

For the numerical implementation of the evolution law and its solution, the strain history is discretized in time intervals t_n and t_{n+1} with a time step Δt,

$$\dot{z} = \|\dot{\boldsymbol{\varepsilon}}\| \approx \frac{\sqrt{\Delta\boldsymbol{\varepsilon} : \Delta\boldsymbol{\varepsilon}}}{\Delta t}, \quad \text{with} \quad \boldsymbol{\varepsilon} = \frac{1}{2}\ln\bar{\mathbf{C}}\,. \tag{29}$$

The intrinsic time can now be written as

$$\dot{z} = \frac{1}{2}\frac{1}{\Delta t}\sqrt{\left(\ln\bar{\mathbf{C}}_{t_{n+1}} - \ln\bar{\mathbf{C}}_{t_n}\right) : \left(\ln\bar{\mathbf{C}}_{t_{n+1}} - \ln\bar{\mathbf{C}}_{t_n}\right)}\,. \tag{30}$$

As already applied in the framework of the visco-elastic formulation, the evolution law is integrated by using an operator split of the material time derivative of $\bar{\mathbf{b}}_e$ as stated in Eq. (16). After the initial elastic trial step, the inelastic corrector step yields

$$\dot{\bar{\mathbf{b}}}_e = -\sqrt{\frac{8}{3}}\frac{\dot{z}}{\eta_i}\left(\frac{\tau_i}{\hat{\tau}}\right)^n \mathbf{N}\,\bar{\mathbf{b}}_e^{\text{tr}}\,. \tag{31}$$

Using again the solution method of the exponential mapping results in

$$\bar{\mathbf{b}}_e = \exp\left[-\sqrt{\frac{8}{3}}\int_{t_n}^{t_n+\Delta t}\frac{\dot{z}}{\eta_i}\left(\frac{\tau_i}{\hat{\tau}}\right)^n \mathbf{N}\,\mathrm{d}t\right]\bar{\mathbf{b}}_e^{\text{tr}} \tag{32}$$

and yields after an approximation of the integral expression, the notation in principal stretch directions of $\boldsymbol{\tau}^i_{\text{iso}}$, $\mathbf{N}$ and $\bar{\mathbf{b}}_e^{\text{tr}}$ as well as the logarithm

$$\ln\lambda^e_a \approx -\sqrt{\frac{2}{3}}\,\Delta t\,\frac{\dot{z}}{\eta_i}\left(\frac{\tau_i}{\hat{\tau}}\right)^n \frac{\left[\boldsymbol{\tau}^i_{\text{iso}}\right]_a}{\left\|\left[\boldsymbol{\tau}^i_{\text{iso}}\right]_a\right\|} + \ln\lambda_a^{e\,\text{tr}}\,. \tag{33}$$

With principal logarithmic stretches $\varepsilon_a = \ln\lambda_a$ and $\varepsilon_a^{e\,\text{tr}} = \ln\lambda_a^{e\,\text{tr}}$, a residual expression of the form

$$r_a := \varepsilon_a^e + \sqrt{\frac{2}{3}}\,\Delta t\,\frac{\dot{z}}{\eta_i}\left(\frac{\tau_i}{\hat{\tau}}\right)^n \frac{\left[\boldsymbol{\tau}_{\mathrm{iso}}^i\right]_a}{\left\|\left[\boldsymbol{\tau}_{\mathrm{iso}}^i\right]_a\right\|} - \varepsilon_a^{e\,\mathrm{tr}} \to 0 \tag{34}$$

can be found. For a numerical solution, the residual expression given in Eq. (34) is iteratively solved by a local Newton iteration at the integration point within a finite element analysis. More details on the derivation and numerical implementation in the framework of the FEM are given in Netzker et al. [25].

The proposed material model for endochronic plasticity of filled rubber has been used for the simulation of cyclic tension and compression tests where the comparison of the numerical results with the experimental results reveals a good agreement as reported in Netzker et al. [25]. For a simple shear test on a filled rubber specimen with five strain levels, a numerical simulation has been carried out with the model parameters $\kappa = 1000\,\mathrm{MPa}$, $\mu = 0.05\,\mathrm{MPa}$, $\mu_i = 0.37\,\mathrm{MPa}$, $N = 4.4$, $N_i = 2.2$, $\eta_i = 5.85$, $\hat{\tau} = 1.00\,\mathrm{MPa}$ and $n = 0.00$. The numerical results are plotted in Fig. 10 together with the experimental data which have been taken from Besdo and Ihlemann [26].

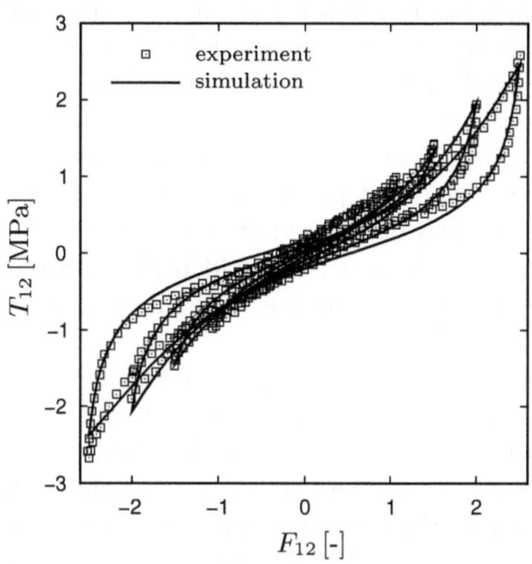

Fig. 10 Comparison of numerical and experimental results (Besdo and Ihlemann [26]) for a cyclic simple shear test on a filled rubber specimen loaded at five strain levels

2.6 Generalized Rheological Models

As already discussed in the context of material modeling (see Sect. 2), rheological elements allow the representation of different material properties which are observed on the macroscale. A combination of several rheological branches in series or in parallel yields finally a more or less complex material model for the total material response.

Regarding filled elastomers, the ground state elastic response together with visco-elastic and elasto-plastic phenomena can be modeled by adding one or more visco-elastic and elasto-plastic rheological branches to the equilibrium branch of the material model. Since both of the afore-presented models are formulated in the same framework, a combination to a visco-elastic-plastic material model is feasible. Such a generalized model with several additional rheological branches takes the real material behavior better into account which shows in general a wide and not clearly distinguishable spectrum of all possible material properties where some are more pronounced than others.

As a result, parameter identification becomes a challenging task due to the high number of material parameters and different material behavior which occur only in a combined form during experimental testing.

2.7 Parameter Identification

The parameter identification for material models is an important task to adapt the chosen material formulation to the experimentally observable material behavior. Only by knowing the model parameters, a derived material model can be employed to represent the material behavior within numerical simulations. Beside the correct identification of the underlaying constitutive description including the significant material characteristics, also adapted experimental test set-ups and test procedures have to be developed in order to already separate best each of the significant material characteristics in the experiment. Decoupling allows a unique identification of the corresponding model parameters and, in consequence, will simplify the parameter identification process significantly.

The model parameters for the derived material models have been identified using an inhouse parameter identification procedure. The parameter identification algorithm is often a more or less iterative approach to obtain the desired model parameters. Therefore, an analytically founded strategy is of high interest to render the parameter identification an efficient, automated task. One possible strategy consists in formulating a mathematical optimization problem in which an optimum of the objective function $Z(\mathbf{X})$ is desired,

$$Z(\mathbf{X}) \rightarrow \text{opt.} \tag{35}$$

In this case, a set of n a priori unknown input parameters $\mathbf{X}_0$ are the material model parameters to be identified. The least mean square error between experimental and numerical data of force-displacement or stress-strain relations can be used as objective function $Z(\mathbf{X})$. With the help of a parameter modification procedure, e.g. based on a genetic evolution algorithm with inheritance and mutation, a new improved set of input parameters $\mathbf{X}_k$ can be obtained in the k-th step.

In the context of material parameter identification, the advantage of micromechanically inspired or physical material models becomes also obvious. For these models, at least some parameters have a physical meaning and can be directly derived from molecular physics or experimental measurements in order to consider them as fixed parameters during the iterative identification procedure.

3 Material Durability, Failure and Crack Propagation Criteria

Durability and failure investigations for materials or components are of high interest to assess risk and safety aspects in industrial applications. In addition, increasing the life time of a component in service decreases waste and saves resources. Any material is subjected to alterations over time. As a result of chemical and physical processes, the material is altered on the microscale, e.g. by network rearrangements or rupture of polymer chains. In consequence, cavity growth and void coalescence lead to an onset of crack initiation and crack growth in the crack tip region on the macroscale.

Therefore, it is important to clearly define the term durability of a material or a structural component. In industry, a standard criterion is crack initiation to define that the observed component has failed during service. The latter shows that reliable predictions with respect to durability under ultimate and fatigue loading have to take local damage formation, crack initiation and crack propagation into account. To summarize, information about

- crack initiation,
- crack propagation (fatigue, catastrophic failure), and
- crack direction

are required from experimental as well as numerical analyses in order to enable durability prediction and, in a consecutive step, failure prediction.

Conventional fracture mechanical concepts rely on purely elastic material behavior in combination with only small deformations. They are classically applied to analyze brittle materials or metals. Since elastomers show large deformations before rupture, these classical concepts are not well suited to predict the failure characteristics of rubberlike materials. The FEM is a widely used numerical approximation method to solve initial boundary value problems for the governing equations in a continuous body. The incorporation of crack sensitivity measures in the FEM is a still ongoing task and different methods and approaches have been developed to account for the characterization and representation of crack propagation within this numerical methodology.

The challenge for numerical simulation of material durability and failure lies in the incorporation of failure processes taking place at the microscale, mesoscale and macroscale as already pointed out in Sect. 1 (see e.g. Fig. 1). In this section, a micro-continuum-mechanical model for failure of elastomers, to describe the material failure on the microscale, is introduced. Next, crack sensitivity measures for crack propagation and crack direction are discussed. To assess crack sensitivity, the J-integral is commonly used in finite element codes together with purely elastic material behavior. The material force method is an alternative method (Maugin [27], Gurtin [28], Kienzler and Herrmann [29]) to efficiently compute fracture mechanical quantities, especially the crack driving force, for elastic and also inelastic material. Hence, the extension of the material force method to inelasticity and its application to elastomeric material is presented at the end of this section.

3.1 A Micro-continuum-Mechanical Model for Failure of Rubber

Before macroscopically observable cracks are formed under monotonic loading, irreversible alterations of the polymer network occur. These alterations can originate from an ultimate stretch of single polymer chains (bond rupture) or failure of network nodes (chain-network rupture). In standard hyperelastic constitutive approaches, the elastic stored energy tends to infinity for steadily increasing strain. Therefore, a locking stretch (significant upturn of the stress response at a certain stretch level) is represented but not failure of the polymer chain and, in consequence, failure of the material. In addition, the free energy functions for hyperelastic material are derived by considering mainly the entropic elasticity of elastomeric material. This assumption is valid for small up to moderate stretches. But with increasing deformation, the single polymer chains attain their ultimate stretch in a non-Gaussian region for which also energetic contributions stemming from the interatomic displacements become significant.

From these reflections and to overcome limitations of ideal entropic hyperelasticity models, a micromolecular representation of material failure of elastomeric material is discussed in the following. In order to concentrate on the main idea of the proposed failure model, purely hyperelastic material behavior is considered. Starting at the microscale, the rheology of the micro-continuum model consists of two springs in series as depicted in Fig. 11. The first rheological element is a Langevin-type spring which models the energy storage stemming from conformational changes of a polymer chain as a result of deformation. The second spring accounts for the bond potential and models the stored energy in the polymer chain as a result of the interatomic displacements activated especially at large strains. For this second spring, the Morse potential models the stored energy of the interatomic bonds and governs the failure of the chain caused by bond rupture. The micro-macro transition is accomplished via the non-affine kinematics of the micro-sphere model (see Miehe et al. [5]).

Free Energy Function. The macroscopical free energy function

$$\psi(\mathbf{F}) = n\,\psi_\eta(\lambda_\eta) + n\,\psi_e(\gamma) \tag{36}$$

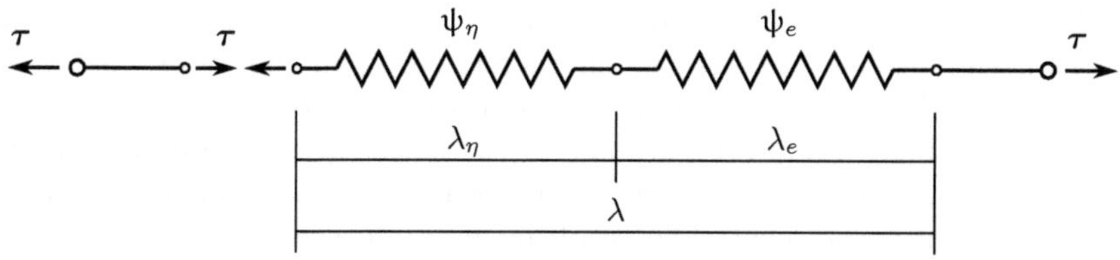

Fig. 11 Rheology of the micro-continuum-mechanical model in 1D representation

is composed of the Langevin part $\psi_\eta(\lambda_\eta)$ and the energetic part $\psi_e(\gamma)$ for a single chain where n denotes the chain density. For the average network stretch

$$\lambda = \langle\lambda\rangle_p = \lambda_\eta\,\lambda_e\;, \tag{37}$$

the kinematical framework of the micro-sphere model is used. λ_η and λ_e are the entropic and the energetic part of the p-root averaged network stretch

$$\langle\lambda\rangle_p := \left\{\frac{1}{|\mathcal{S}|}\int_{\mathcal{S}} \left(\bar{\lambda}\right)^p \,\mathrm{d}S\right\}^{1/p}\;, \tag{38}$$

which relates the non-affine micro-stretch λ to the affine macro-stretch $\bar{\lambda}$ on the unit-sphere of the micro-sphere model. For more details about the micro-sphere approach, the reader is referred to Miehe et al. [5].

$$\psi_\eta(\lambda_\eta) = N\,k_{\mathrm{B}}\,\Theta\left(\lambda_\eta\,\mathcal{L}^{-1}(\lambda_\eta) + \ln\frac{\mathcal{L}^{-1}(\lambda_\eta)}{\sinh\mathcal{L}^{-1}(\lambda_\eta)}\right) \tag{39}$$

takes the changes in the free energy resulting from isochoric entropic effects into account. N denotes the number of chain segments which gives the locking stretch $\lambda_{\mathrm{lock}} = \sqrt{N}$. k_{B} is the Boltzmann constant and Θ is the absolute temperature of the network. The inverse Langevin function can be computed in analogy to Eq. (10). The energetic contribution to the free energy function

$$\psi_e(\gamma) = \hat{e}\left[\exp\left(-2\,\gamma\right) - 2\,\exp\left(-\gamma\right) + 1\right] \tag{40}$$

with

$$\gamma := \alpha_1\left(\lambda_e - 1\right) + \alpha_2\left(J-1\right)^2 \tag{41}$$

is a modified version of the Morse potential (see Morse [30])

$$\Phi(r) = \hat{e}\left[\exp\left(-2\,\alpha\left(r-r_0\right)\right) - 2\,\exp\left(-\alpha\left(r-r_0\right)\right)\right]\;, \tag{42}$$

where its partial derivative with respect to the separation r of a pair of atoms yields the force-displacement function

$$f = 2\,\hat{e}\,\alpha\left[-\exp\left(-2\alpha\left(r-r_0\right)\right) + \exp\left(-\alpha\left(r-r_0\right)\right)\right]\;. \tag{43}$$

In this case, $\hat{e}$ stands for the dissociation energy for a pair of atoms and $r = r_0$ is the interatomic distance characterized by an energy minimum. As illustrated in Fig. 12, a stable bond can be observed up to the maximum force at $r_m = r_0 + \ln 2/\alpha$. Via the parameter $\gamma = \gamma(\lambda_e, J)$, higher order energetic interactions of molecules are supposed to be possible mechanisms of volume change at continuum level. The resulting micro-forces of the entropic and energetic part with the modified Morse potential given in Eq. (40) are

$$f_\eta := \frac{\partial \psi_\eta}{\partial \lambda_\eta} = k_\mathrm{B}\, \Theta\, \frac{3\,N - \lambda_\eta^2}{N - \lambda_\eta^2} \tag{44}$$

and

$$f_e := \frac{\partial \psi_e}{\partial \lambda_e} = 2\,\alpha_1\, e_0 \left[-\exp\left(-2\,\gamma\right) + \exp\left(-\gamma\right)\right] \;, \tag{45}$$

respectively.

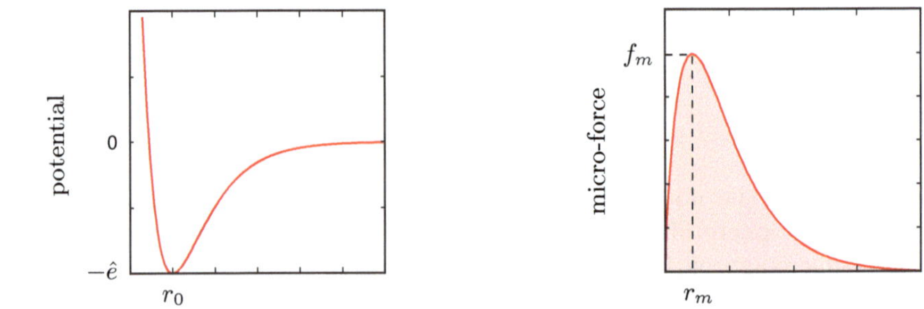

Fig. 12 Morse potential: a) potential in function of separation; b) associated micro-force in function of separation

The equilibrium constraint for a given deformation state characterized by λ is

$$f_\eta\, \lambda_\eta = f_e\, \lambda_e \tag{46}$$

and results from the serial construction of the two springs in the rheology of the model under finite deformations. The equilibrium constraint has to be solved at the micro-level for the non-affine micro-stretch λ since the network is represented by only one free energy function given in Eq. (36). From this equilibrium constraint, an iterative solution strategy can be formulated by using the residual expression

$$r := f_\eta\, \lambda_\eta - f_e\, \lambda_e \rightarrow 0 \quad \text{with} \quad \lambda_\eta\, \lambda_e = \lambda\,. \tag{47}$$

In this case, λ is the given micro-stretch and λ_η as well as λ_e have to be found. More information on the configurational setting and the numerical implementation are given in Dal and Kaliske [31].

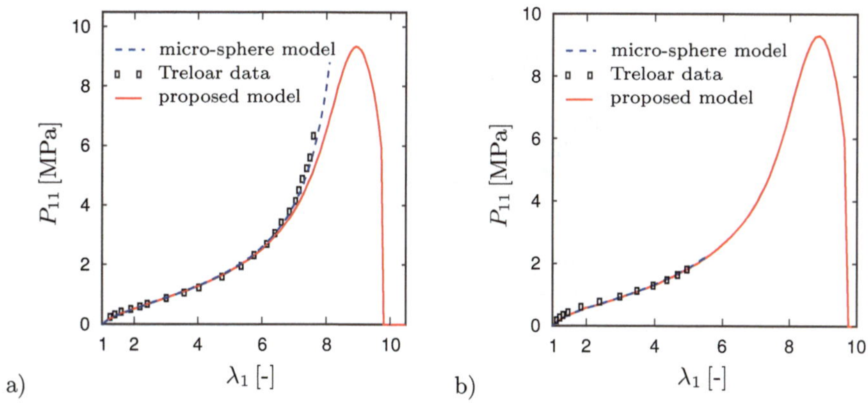

Fig. 13 Comparison of results obtained by original micro-sphere model non-affine stretch part, the proposed model and experimental data (Treloar [32]): a) uniaxial tension; b) pure shear

The model for the micromolecular representation of material failure of elastomeric material has been compared to the original micro-sphere model non-affine stretch part. The parameters $\mu = n\, k_{\mathrm{B}}\, \Theta = 0.325\,\mathrm{MPa}$, $N = 22.263$, $p = 1.39$ are the same as used by Miehe et al. [5] to simulate the uniaxial tension and pure shear test reported by Treloar [32] as depicted in Fig. 13. The remaining parameters of the novel failure model have been determined to $\alpha_1 = 7$, $\alpha_2 = 100$, $n\,\hat{e} = 22\,\mathrm{MPa}$ where a high value for α_2 enforces the material's incompressibility until a moderate stretch. The original entropic micro-sphere model and the proposed failure model show a nearly identical behavior up to moderate stretches. Due to the combination of a soft and a hard spring, the energetic contribution is only activated at large stretches leading to bond rupture.

With the proposed model, several failure modes can be represented (uniaxial tension, equibiaxial tension, pure shear, pure dilatation failure modes) due to the fact that the volumetric and isochoric material response are not decoupled. The coupling between the volumetric and isochoric material response leads to a unified failure criterion for all the deformation modes. To simplify the model, volumetric deformations are completely attributed to the energetic contributions via $\gamma(\lambda_e, J)$.

In order to investigate the failure prediction under various loading conditions, the proposed failure model has been fitted to stress-strain curves for natural rubber (NR) and styrene-butadiene rubber (SBR). The parameters for NR are $\mu = 0.60\,\mathrm{MPa}$, $N = 9.02$, $p = 1.6$, $\alpha_1 = 2.12$, $\alpha_2 = 20$, $n\,\hat{e} = 144\,\mathrm{MPa}$ and for SBR $\mu = 1.3\,\mathrm{MPa}$, $N = 6.68$, $p = 1.3$, $\alpha_1 = 1.54$, $\alpha_2 = 20$, $n\,\hat{e} = 171\,\mathrm{MPa}$. In Fig. 14, the experimentally investigated normalized failure elongations taken from Hamdi et al. [33] are plotted together with the numerically obtained results of the proposed model. Failure

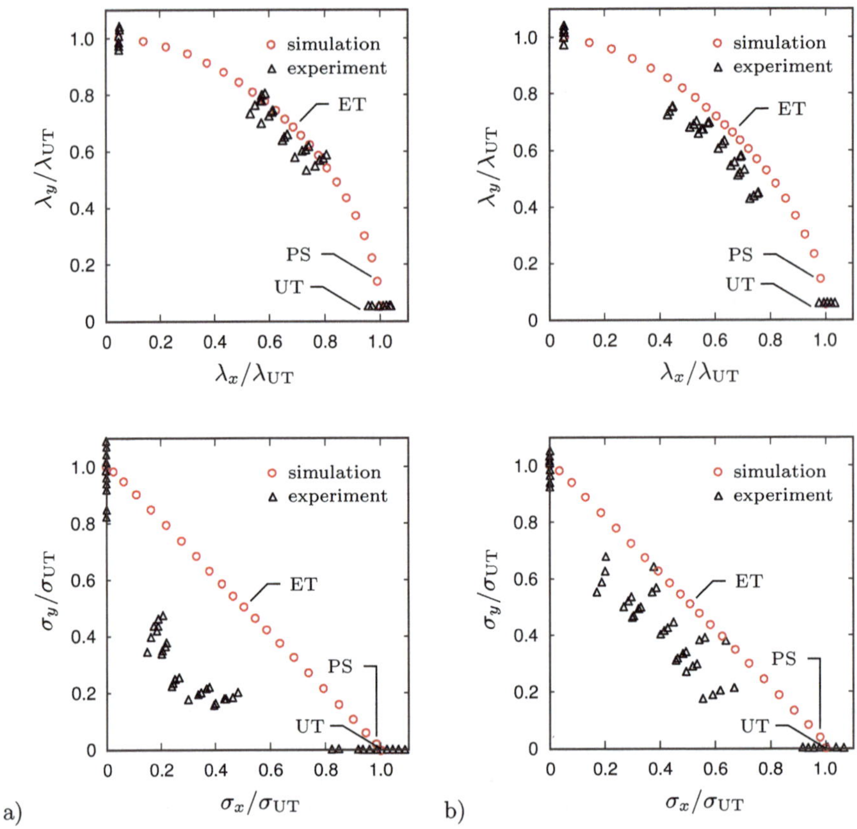

Fig. 14 Comparison of ultimate stretches normalized by uniaxial ultimate stretch λ_{UT} and ultimate Cauchy stresses normalized by uniaxial ultimate Cauchy stress σ_{UT} obtained from experiment (Hamdi et al. [33]) as well as simulation with proposed model: a) natural rubber; b) styrene-butadiene rubber

elongations corresponding to uniaxial tension (UT), pure shear (PS) and equibiaxial tension (ET) tests are used for comparison. A good agreement between experiment and simulation can be recognized for NR and SBR while the failure of SBR is slightly overestimated by the simulation.

In addition to the afore-presented two examples, the proposed failure model is also applied to represent the ageing-induced surface cracking of a rubber component as reported in Dal and Kaliske [31].

3.2 Crack Criteria

Beside modeling of crack initiation, which can be accomplished by e.g. considering processes on the microscale as discussed before, especially the experimental and numerical characterization of crack growth is an important indicator to assess fracture resistance of a structural component.

Once the threshold measure of crack resistance is exceeded in the current loading state, crack growth, as an on the macroscale observable combination of micromechanical processes, takes place. In this case, the threshold value of the characteristic crack resistance and the crack propagation direction is of interest.

The energy balance for an infinitesimal time increment $\mathrm{d}t$ separating two configurations of an elastic cracked body can be stated as

$$0 = \mathrm{d}\Pi_a + \mathrm{d}\Pi_e + \mathrm{d}\Gamma_{\mathrm{cr}} \; . \tag{48}$$

$\mathrm{d}\Pi_a$ denotes the infinitesimal change of external work, $\mathrm{d}\Pi_e$ is the infinitesimal change of the internal stored elastic strain energy and $\mathrm{d}\Gamma_{\mathrm{cr}}$ is the infinitesimal energy which leads to an infinitesimal crack growth of newly created crack face area $\mathrm{d}A$. In experimental fracture investigations, often thin plate-shaped specimens of thickness t are used for which Eq. (48) can be reformulated with $\mathrm{d}A = t\,\mathrm{d}a$ to

$$-\frac{\partial\,\Pi_e}{t\,\partial a} = \frac{\partial \Gamma_{\mathrm{cr}}}{t\,\partial a} \; , \tag{49}$$

where the change in the external potential has been assumed to be zero ($\mathrm{d}\Pi_a = 0$) and the sensitivity with respect to straight crack growth with a length of $\mathrm{d}a$ has been derived. A possible criterion for crack growth could be

$$-\frac{\partial\,\Pi_e}{t\,\partial a} = \frac{\partial \Gamma_{\mathrm{cr}}}{t\,\partial a} \leq T_c \; , \tag{50}$$

where T_c is a characteristic material parameter termed tearing energy (energy per unit crack area to initiate crack extension). The tearing energy in elastic material can be interpreted as the amount of work $W_{\mathrm{sep}} = T_c\,t\,\mathrm{d}a$ which is needed to increase an initial crack in a plate of thickness t by the length $\mathrm{d}a$. Eq. (50) is known as the Griffith criterion introduced by Griffith [34] for the fracture of brittle materials. The Griffith criterion is a widely used testing methodology which can be also applied to rubber. However, for rubber, the tearing energy T_c is a function of temperature and crack tip velocity. In Fig. 15 a), an example of an initially notched specimen with the initial notch length a_0 and the crack increment $\mathrm{d}a$ due to crack propagation is depicted.

In fracture mechanics, the J-integral (see Rice [35])

$$J := \int_{\Gamma} \mathbf{n} \left(\psi\, \mathbf{1} - \mathbf{F}^{\mathrm{T}}\, \mathbf{P} \right) \mathbf{q} \,\mathrm{d}\Gamma \tag{51}$$

is a means to compute the energy release rate at a crack tip with stress-free crack faces in elastic material. $\mathbf{n}$ is the normal vector on the contour path Γ and $\mathbf{q}$ stands for the possible crack propagation direction for which the energy release rate is computed as depicted in Fig. 15 b). The term in brackets in Eq. (51) is the so-called energy momentum tensor or Eshelby stress

a) 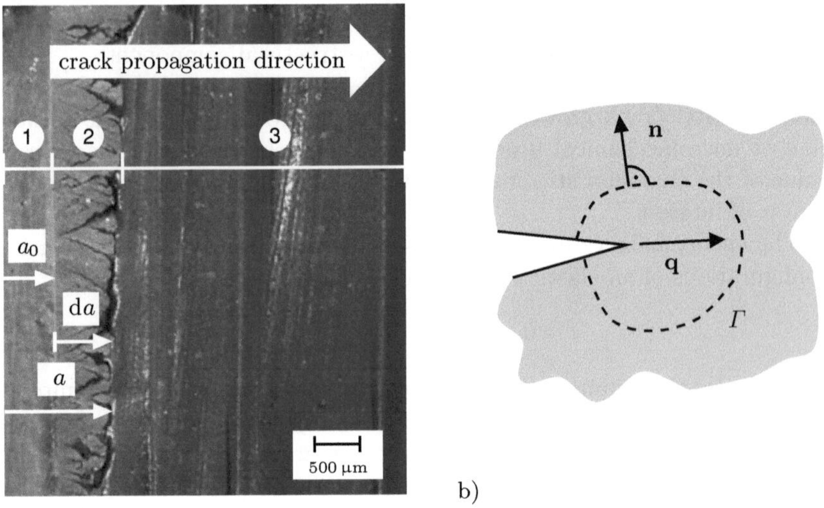

b)

Fig. 15 Definition of crack length and contour of crack tip: a) microscopical image of cut through fracture-plane of an initially notched rubber specimen, 1 – initial notch, 2 – crack increment due to crack growth, 3 – post experimental cut through rest of specimen; b) illustration of contour path Γ around the crack tip

tensor $\boldsymbol{\Sigma} = \psi\,\mathbf{1} - \mathbf{F}^{\mathrm{T}}\,\mathbf{P}$, see Eshelby [36]. The path-independent contour integral is a standard approach to evaluate crack sensitivity of elastic structures within numerical simulations and, therefore, available in several commercial FE codes. However, the computation of the J-integral requires the definition of a contour around the crack tip on which the energy momentum tensor is considered and – as already pointed out – its interpretation as energy release rate holds only for the elastic case.

For a straight crack with stress-free crack faces in elastic material, the integral in Eq. (51) is path-independent. However, its numerical evaluation within the FEM requires a discretization and shows a path dependency. Since artificial stress singularities can directly appear at the crack tip due to discretization reasons, the contour path should not fall within a critical radius around the crack tip.

It is worth mentioning that up to now, purely elastic material behavior has been assumed. Since rubber material shows also dissipative effects which are due to inelastic material behavior, a more general fracture mechanical measure is the energy release rate $\mathcal{G}$. From now on, dissipative material behavior is assumed and, in consequence, Eq. (48) yields

$$0 = \mathrm{d}\Pi_a + \mathrm{d}\Pi_e + \mathrm{d}\Delta + \mathrm{d}\Delta_{\mathrm{cr}} + \mathrm{d}\Gamma_{\mathrm{cr}} \tag{52}$$

as the incremental energy balance for an inelastic cracked body (see e.g. Netzker et al. [37]). Due to the material dissipation, the new energy term $\mathrm{d}\Delta_{\mathrm{cr}}$ representing the energy dissipated in a dissipative process zone around the crack tip and $d\Delta$ representing the energy dissipated in the rest of the considered body appear. All the terms given in Eq. (52) are illustrated in Fig. 16 which depicts an initially notched specimen with inelastic material behavior and an inelastic process zone subjected to uniaxial tension.

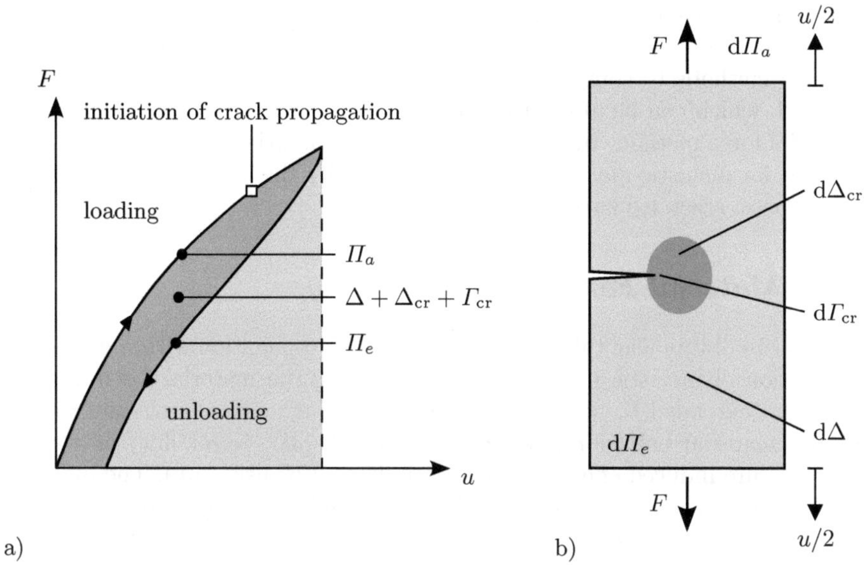

Fig. 16 Energy contributions for the loading and unloading of a specimen showing crack propagation: a) force-displacement relation; b) infinitesimal changes of energy terms due to crack propagation and change of external loading

The already mentioned energy release rate can be written as

$$\mathcal{G} = \frac{\partial \Delta + \partial \Delta_{\text{cr}} + \partial \Gamma_{\text{cr}}}{\partial a} = -\frac{\partial \Pi_a + \partial \Pi_e}{\partial a} = -\frac{\partial \Pi}{\partial a} \tag{53}$$

with $\Pi = \Pi_a + \Pi_e$. The energy release rate can be interpreted as the amount of energy which is released in an inelastic body due to an increase of the crack length by $\mathrm{d}a$. By comparing Eq. (48) and Eq. (52), it can be seen that the energy dissipation of the inelastic material (process zone and rest of the body) contributes to the crack growth resistance. Especially for filled elastomers, it can be observed that the energy dissipation in a plastic process zone at the crack tip and the energy dissipation in the rest of the body are significant before the crack starts to propagate.

However, within numerical simulations, a material specific fracture mechanical measure is required which governs crack growth. By considering Eq. (52), the so-called crack driving force

$$\frac{\partial \varGamma_{\mathrm{cr}}}{\partial a} = -\frac{\partial \varPi_a + \partial \varPi_e + \partial \Delta + \partial \Delta_{\mathrm{cr}}}{\partial a} \tag{54}$$

can be identified. The crack driving force considers in analogy to the tearing energy in elastic materials only the crack tip and includes the amount of energy, which is exclusively needed to create new free surfaces at the crack tip. Since it is hard to measure the crack driving force experimentally by separating the energy dissipation of the material response (see Fig. 16), numerical methods to compute the crack driving force are advantageous. One method, which can be also formulated for inelastic material behavior, is the material force method. In the next section, the material force method will be derived for inelastic material behavior to identify the crack driving forces at an inelastic crack tip zone.

3.3 Material Forces in Inelasticity

In order to obtain a suitable criterion for crack propagation within numerical simulation of inelastic material at large strains, the material force method has been extended to account for stress softening, visco-elastic and elasto-plastic material behavior (see e.g. Näser et al. [38], Näser [39]) as well as temperature induced effects (see e.g. Behnke and Kaliske [40]). The material force approach is a more general and easy to handle method to assess fracture mechanical sensitivities within standard FE simulations in a post-processing step. Material forces can be consistently derived for inelastic materials and, in consequence, are not restricted to the purely elastic case. For applications of the material force method in numerical simulations and fracture mechanics, it is referred to e.g. Mueller et al. [41], Steinmann et al. [42] and Liebe et al. [43], respectively. The basic concept is briefly outlined in this section.

As already introduced, the fracture mechanical sensitivity of a component can be expressed by the energy release rate where the energy release is defined as the change of the potential energy due to an advancing of the crack by an infinitesimal length da. For elastic material, the energy release rate and the crack driving force coincide. For inelastic material, the crack driving force is different to the energy release rate since the crack driving force is defined as the energy which is needed to advance the crack tip. In this context, material forces allow for the computation of fracture mechanical quantities in case of linear and nonlinear elastic as well as inelastic material behavior. In addition, monotonic loading is not required for the use of material forces as a fracture mechanical parameter.

Material forces are the thermodynamic driving forces on any kind of inhomogeneities and discontinuities in a material. At the crack tip, the material force could be understood as the force preventing the crack to change its position under the current loading conditions in order to reach a neighboring state which is energetically more attractive. Material forces occur in a body due to

- inhomogeneities, e.g.
  + material inhomogeneities,
  + process zone at the crack tip, and
- discontinuities, e.g.
  + inclusions,
  + voids,
  + dislocations,
  + wave fronts, and
  + cracks.

Material forces can be applied in various fields of computational mechanics, e.g. in the optimization process of a finite element mesh in order to get a load case specific, energetically optimized FE mesh. In this case, the nodes of the FE mesh in the material configuration are displaced in the direction of the residual material force vector components which have to be zero for equilibrium in the material space.

Since material forces are vectorial quantities, the material force vector represents the direction of maximum energy release and points therefore in the opposite direction of possible crack propagation. Note that material forces are defined in the material space, therefore, their directional character is valid for the material space.

At the crack tip, material forces correspond to the energy release at the crack tip zone where for inelastic materials the crack driving force and the energy release in the dissipative zone are different fracture mechanical quantities. Therefore, the total energy release rate is the energy change of the system for a displacement of the crack tip and the dissipative zone. Only the crack driving force is independent from dissipative effects and a fracture mechanical measure to characterize the fracture process in a material. It can be numerically computed from material forces and volume forces representing material body forces in the material space. In general, the dissipative process zone leads to a decrease of the crack driving force in comparison to purely elastic material. An inelastic crack tip zone is illustrated in Fig. 17 where the energy release rate is equal to the energy flow in the dissipative zone and the crack driving force is the energy flow to the crack tip (see also Näser et al. [38]).

The energy release due to dissipative effects can be seen as an additional fracture resistance property of the material against crack propagation. This is especially the case for highly filled elastomers which show large dissipative effects before crack propagation. If the initial crack starts to propagate, the range of stable crack propagation is very small and is followed by a direct transition to unstable crack growth and failure. As a result, the crack driving force is an essential fracture mechanical parameter which has to be correctly assessed in numerical simulations. If the material volume forces are neglected in case of dissipative material, an over- or underestimation of the crack driving force is possible.

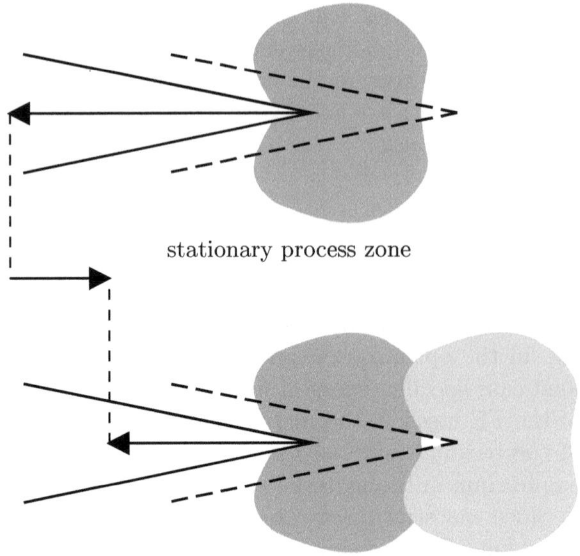

Fig. 17 Material force contributions at a crack tip with inelastic process zone

In case of inelastic material, the J-integral introduced in Eq. (51) is no longer path-independent since an inelastic process zone is observable in front of the crack tip. Therefore, the J-integral will include the driving force on the crack tip and the driving forces on the dissipative zone as a mixed quantity. However, for the fracture process of the material, only the pure crack driving force is of interest. Therefore, the material force method is applied to compute the crack driving force by a summation of the material forces along a circumference of the crack tip which is equal to the J-integral with the identical contour path from which also the material volume forces are subtracted. The J-integral would only yield for the limit case

$$J = \lim_{\Gamma \to 0} \int_{\Gamma} \mathbf{n} \left(\psi \, \mathbf{1} - \mathbf{F}^{\mathrm{T}} \, \mathbf{P} \right) \mathbf{q} \, \mathrm{d}\Gamma \tag{55}$$

the crack driving force for inelastic material. However, the limit consideration is numerically not reliable due to stress singularities resulting from the finite element discretization at the crack tip. If inelastic effects like viscosity are assumed to develop in the whole specimen, it will be even more complicate to clearly distinguish between the inelastic crack tip zone and the rest of the specimen.

For the following derivation of material forces in inelasticity, the multiplicative split of the deformation gradient into volumetric and isochoric as well as the further split of the isochoric part into elastic and inelastic contributions is applied as introduced in Sect. 2.3. In this case, the free energy function

$$\psi(\mathbf{F}, \bar{\mathbf{F}}_i, d, \Theta, \mathbf{X}) = \psi(\mathbf{F}, \bar{\mathbf{F}}_e, d, \Theta, \mathbf{X}) \tag{56}$$

depends on the current deformation state represented by $\mathbf{F}$, the internal variables computed from $\bar{\mathbf{F}}_i$ to express finite visco-elastic or elasto-plastic material behavior, the scalar damage parameter d representing stress softening phenomena, the absolute temperature Θ and the material position $\mathbf{X}$. As already introduced by $\det \bar{\mathbf{F}}_i \equiv 1$, the inelastic deformations are assumed to be volume preserving. As a result, the derivations will simplify in the following. For a more general approach including compressible inelastic deformations, the reader is referred to e.g. Näser [39]. Material forces can be derived starting from the balance of the spatial motion problem

$$\operatorname{Div} \mathbf{P} + \mathbf{b}_0 = \mathbf{0}\,. \tag{57}$$

In this case, $\mathbf{P}$ denotes the first Piola-Kirchhoff stress tensor

$$\mathbf{P} = \frac{\partial \psi}{\partial \mathbf{F}} \tag{58}$$

and $\mathbf{b}_0$ stands for the physical body force vector due to gravity. A pull back operation of Eq. (57) by $\mathbf{F}^{\mathrm{T}}$ yields the local material balance

$$\mathbf{F}^{\mathrm{T}} \operatorname{Div} \mathbf{P} + \mathbf{F}^{\mathrm{T}} \mathbf{b}_0 = \mathbf{0}\,. \tag{59}$$

The complete derivation of the material gradient of the free energy function ψ with respect to all arguments yields

$$\begin{aligned} \nabla_X \psi(\mathbf{F}, \bar{\mathbf{F}}_e, d, \Theta, \mathbf{X}) = {} & \frac{\partial \psi}{\partial \mathbf{F}} : \frac{\partial \mathbf{F}}{\partial \mathbf{X}} + \frac{\partial \psi}{\partial \bar{\mathbf{F}}_e} : \frac{\partial \bar{\mathbf{F}}_e}{\partial \mathbf{X}} \\ & + \frac{\partial \psi}{\partial d} : \frac{\partial d}{\partial \mathbf{X}} + \frac{\partial \psi}{\partial \Theta} : \frac{\partial \Theta}{\partial \mathbf{X}} + \left.\frac{\partial \psi}{\partial \mathbf{X}}\right|_{\text{expl.}} . \end{aligned} \tag{60}$$

By introducing the definition of the fictitious first Piola-Kirchhoff stress tensor

$$\tilde{\mathbf{P}}^i := \frac{\partial \psi}{\partial \bar{\mathbf{F}}_e} \quad \text{with} \quad \mathbf{P}^i = \tilde{\mathbf{P}}^i \, \bar{\mathbf{F}}_i^{-\mathrm{T}} \tag{61}$$

and by using the identity $\nabla_X \psi = \operatorname{Div}(\psi \, \mathbf{1})$ as well as the relation

$$\mathbf{F}^{\mathrm{T}} \operatorname{Div} \mathbf{P} = \operatorname{Div}\left(\mathbf{F}^{\mathrm{T}} \mathbf{P}\right) - \mathbf{P} : \nabla_X \mathbf{F} \tag{62}$$

together with Eq. (59), Eq. (60) can be rewritten as

$$\operatorname{Div} \boldsymbol{\Sigma} + \mathbf{g}_0 + \mathbf{g}_i + \mathbf{g}_d + \mathbf{g}_\Theta + \mathbf{g}_X = \mathbf{0} \tag{63}$$

representing the strong form. As already introduced, $\boldsymbol{\Sigma} = \psi \, \mathbf{1} - \mathbf{F}^{\mathrm{T}} \mathbf{P}$ denotes the Eshelby stress tensor in Eq. (63) and the material volume force contributions stemming from physical body forces

$$\mathbf{g}_0 = -\mathbf{F}^{\mathrm{T}}\mathbf{b}_0 \,, \tag{64}$$

inelastic effects like visco-elasticity or elasto-plasticity

$$\mathbf{g}_i = + \left(\bar{\mathbf{F}}_e^{\mathrm{T}} \, \mathbf{P}^i \right) : \nabla_X \, \bar{\mathbf{F}}_i \,, \tag{65}$$

local continuum damage

$$\mathbf{g}_d = - \frac{\partial \psi}{\partial d} \nabla_X \, d \,, \tag{66}$$

temperature gradient

$$\mathbf{g}_\Theta = - \frac{\partial \psi}{\partial \Theta} \nabla_X \, \Theta \tag{67}$$

and inhomogeneous material properties

$$\mathbf{g}_X = - \left. \frac{\partial \psi}{\partial \mathbf{X}} \right|_{\text{expl.}} \tag{68}$$

have been summarized. From the local material volume forces $\mathbf{g}_\Box$ and the Eshelby stress tensor $\boldsymbol{\Sigma}$, the material nodal forces $\mathbf{G}$ and the material volume nodal forces $\mathbf{G}^{\text{int}}$ can be computed in a post-processing step of a finite element analysis by integration over the discretized finite element volumes. In general, after multiplication of the strong form with a test function and application of the Gauss theorem, material nodal forces are computed in their weak form as

$$\mathbf{G} = - \bigcup_{n=1}^{n_{\text{el}}} \int_{\mathcal{B}_0} \mathbf{B}_I^{\mathrm{T}} \, [\boldsymbol{\Sigma}] \, \mathrm{d}V \tag{69}$$

and material nodal volume forces in analogy

$$\mathbf{G}^{\text{int}} = \bigcup_{n=1}^{n_{\text{el}}} \int_{\mathcal{B}_0} \mathbf{N}_I^{\mathrm{T}} \, [\mathbf{g}_0 + \mathbf{g}_i + \mathbf{g}_d + \mathbf{g}_\Theta + \mathbf{g}_X] \, \mathrm{d}V \,, \tag{70}$$

where the material equilibrium in the body for a static configuration without acceleration

$$\mathbf{G} + \mathbf{G}^{\text{int}} = \mathbf{0} \tag{71}$$

is fulfilled. For each integration point I of a finite element, $\mathbf{N}_I$ denotes the matrix of shape functions and $\mathbf{B}_I$ stands for the matrix of derivatives of the shape functions contained in $\mathbf{N}_I$. For inelastic material behavior represented by internal variables, a projection of the internal variables from the integration points to the nodes of the FE mesh including an averaging at the nodes to account for the contributions stemming from neighboring elements is required. Other approaches take the internal variables as unknowns at the nodes into account. For more details, the reader is referred to e.g. Menzel et al. [44] and Zimmermann [45]. In addition, an adaptive discretization with refinement around the crack tip is advantageous since a coarse discretization

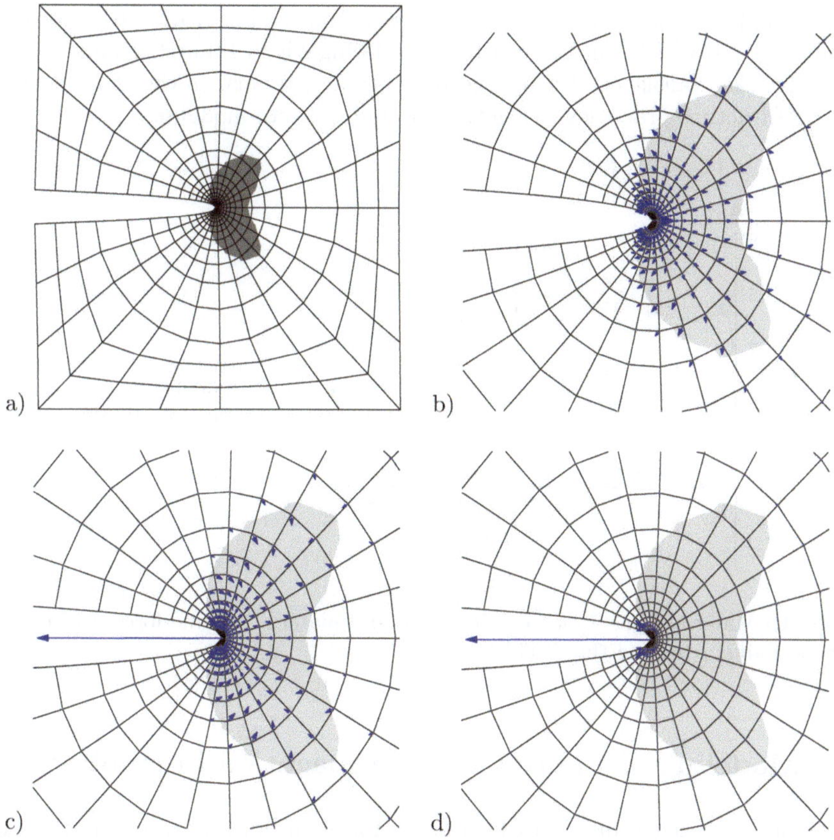

Fig. 18 Plate made of elasto-plastic material with initial notch: a) deformed geometry with opened crack and plastic process zone; b) material volume nodal forces $\mathbf{G}_i^{\text{int}}$ resulting from inelastic deformation; c) material nodal forces $\mathbf{G}$ as driving forces on the crack tip and the process zone; d) crack driving forces $\mathbf{G} + \mathbf{G}_i^{\text{int}}$ from summation of material nodal forces and material volume nodal forces

including inter- and extrapolation errors during the projection algorithm of the internal variables will lead to non-physical residual mesh-dependent material forces due to the discretization.

As numerical example, an initially notched specimen of dimensions 150 mm $\times$ 150 mm $\times$ 1 mm made of an elasto-plastic material (one Hooke and one Prandtl element based on von Mises plasticity) is investigated. The elastic parameters for the Hooke and Prandtl element are $\kappa = 10$ N/mm^2 and $\mu = 5$ N/mm^2, the inelastic deformations are assumed as incompressible and governed by a yield stress of $\sigma_y = 0.9$ N/mm^2. The plate's deformed geometry resulting from a vertical displacement controlled loading of 10 mm is depicted together with the different material force contributions in Fig. 18. Note that the crack driving forces are path-independent while the material forces and the material volume forces depend on the integration path around the crack tip which can be seen by considering Fig. 19. Further information are given in Kaliske et al. [46] together with a summary on the characterization of fracture processes by continuum and discrete modeling.

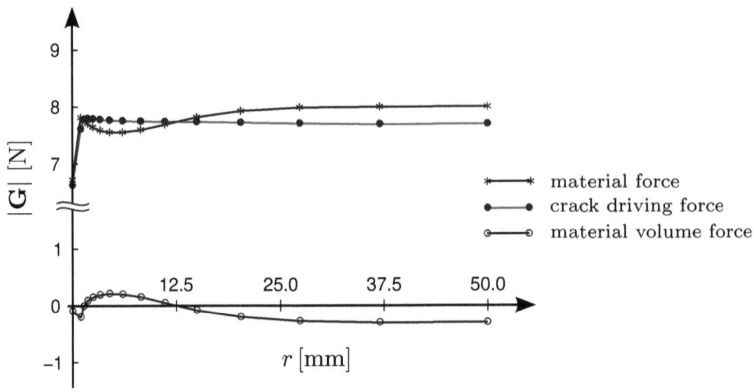

Fig. 19 Material force, crack driving force and material volume force as function of the radius r around the crack tip

4 Modeling of Finite Crack Propagation in Rubber

On the macroscale, finite cracks are observable in form of discontinuities in a previously continuous body. Cracks are characterized by a separation of previously connected material points. This is especially true for a widely opened crack. However, at the crack tip, a transition zone exists in which the material points are still linked by not fully damaged material bridging the emerging crack.

As a result, the modeling process of finite crack propagation has to account for a discrete separation of previously connected points which concerns especially the topology change due to crack propagation, the energy release and the solution field (e.g. displacement field, temperature field, moisture field etc.) which shows a discontinuity at the crack. In addition, for modeling an advancing crack, crack growth and crack direction criteria are required. Another challenge arises from the creation of new surfaces at the crack faces. In comparison to the external boundaries of the body, the internal surfaces of the crack faces can be subjected to boundary constraints like mechanical loading, exclusion of penetration during crack closure and other flux conditions regarding e.g. a thermal solution field.

In consequence, different numerical solution methods have been developed over the last decades. An elegant way to overcome the problem of a fixed finite element mesh within numerical simulations is the use of mesh free methods

(MFM) without fixed relations between neighboring points. In form of so-called hydrocodes, which have been derived from the field of hydrodynamics, it is possible to capture the nonstationarity of the material and structural response under dynamic loading which can cause high fragmentation. Hence, smooth particle hydrodynamics methods (SPH), which base on the solution of the balance equations for mass, momentum and energy (Hiermaier [47]), are mainly employed in the field of shock and impact on structures characterized by fragmentation processes. In the framework of the FEM, other techniques have been developed to overcome the limitations of an initially fixed finite element mesh. In Fig. 20, different approaches to represent emerging finite cracks within the finite element simulation process are depicted.

a) b) c)

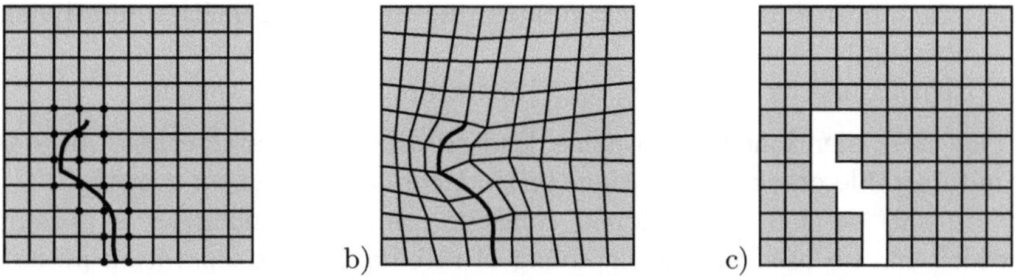

Fig. 20 Modeling approaches for the numerical simulation of arbitrary crack paths within the FEM: a) enrichment functions in the crack domain; b) cohesive elements and mesh adaption; c) deleting of finite elements after exceeding of a failure criterion

Approaches to overcome the fixed element dependency of mesh based solution strategies are e.g.

- elements with embedded discontinuities,
- interface elements (cohesive elements) in combination with mesh adaptivity,
- smeared crack representation or deletion of failed finite elements.

In the first case, elements with embedded discontinuities for the description of discontinues in the considered solution fields (e.g. displacement, temperature, moisture etc.) are used in the framework of the extended finite element method (X-FEM). The X-FEM is based on the introduction of additional degrees of freedom in an assumed region and direction of crack propagation (see e.g. Belytschko and Black [48]) in order to model the jump in the solution field by additional enrichment functions. In the second case, the finite element mesh is adapted to account for the changing topology due to crack propagation by remeshing (see e.g. Neumann [49]). Principally, it is possible to model the forming crack with or without using interface elements representing the crack faces. In this context, interface elements offer the advantage to assign specific characteristics and boundary conditions (e.g. contact conditions, friction, heat flux, moisture transport) to the created surfaces and processes of residual force transmission over the opened crack in the crack tip zone can be captured. Therefore, interface elements are also termed cohesive elements. In the third case, a scale bridging strategy is the representation of finite cracks in a smeared form via damaged continuum elements or their deletion after their total failure. The latter method still suffers from a mesh dependency and the violation of the balance equations of energy and mass if continuum elements are simply deleted. A comparison of the afore-discussed methods for brittle materials can be found e.g. in Song et al. [50]. The challenges for modeling crack propagation in elastomers are, among others,

- large strains and large deformations at the crack tip,
- inelasticity in combination with process zones,
- strain induced crystallization,
- crack branching and crack kinking due to inhomogeneities and dynamic loading,
- tearing energy depending on temperature and crack velocity.

The afore-mentioned phenomena lead to different characteristics of stable and unstable crack propagation (fatigue, catastrophic crack growth).

In this section, a developed cohesive element formulation is presented which can be used for the modeling of rate-dependent separation processes of an initially defined failure layer (e.g. glued film). Together with a mesh adaption algorithm and the adaptive implementation of cohesive elements in a finite element mesh according to a crack growth criterion, it becomes possible to represent arbitrary crack patterns. Finally, the cohesive element formulation and the implemented adaptive meshing strategy are combined with stress and energy based crack propagation criteria formulated on the macroscale to model crack propagation in rubber components.

4.1 Cohesive Element Formulation

With the help of cohesive elements, it is possible to model finite, discrete cracks within the FEM. The advantage of cohesive elements is based on the incorporation of new surfaces with defined boundary conditions, e.g. the transmission of residual stresses at the crack tip for the opening crack. Different crack tip models are shown in Fig. 21 together with the model assumptions of the cohesive element formulation.

Griffith [34] assumed an inelastic process zone with small dimensions and material separation at the crack tip of brittle material to derive the Griffith criterion as discussed before in Sect. 3.2. Other model assumptions base on a process zone with general dimensions. Dugdale [51] associated the stresses in the process zone with the yield stress of the material. In the context of molecular decohesion, Barenblatt [52] assumed the even more general case of nonlinearly distributed cohesive stresses along the crack tip which depend on the opening of the crack.

Applications within the FEM have been initiated by Needleman [53] who introduced and derived the cohesive element formulation in the FEM and

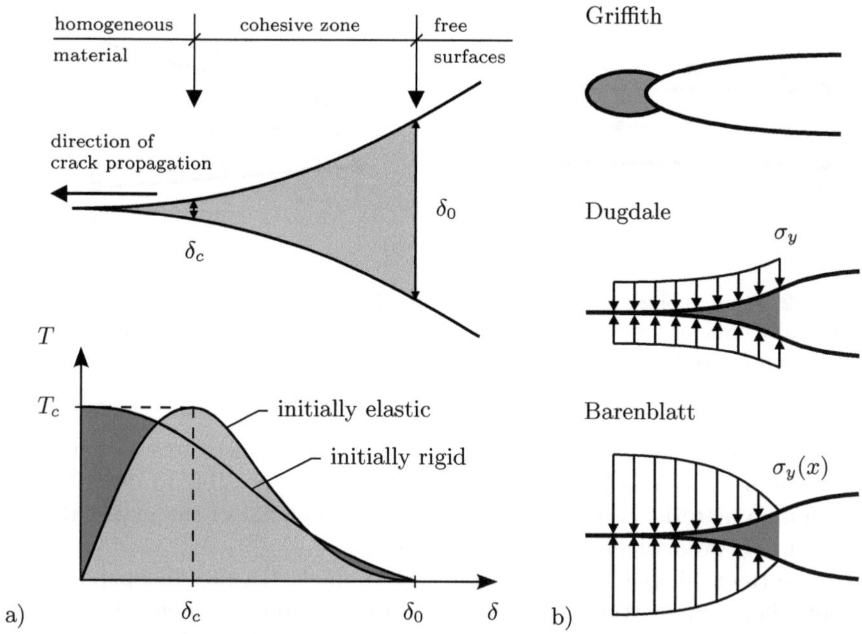

Fig. 21 Assumptions of the cohesive zone model: a) notations; b) crack tip models

formulated a constitutive relation (traction separation law) for the cohesive tractions at the crack tip. The traction separation law (TSL) describes the tractions over the crack as a function of crack face separation as illustrated in Fig. 21, where different functions are possible to represent the material specific characteristics related to fracture. A more detailed overview on cohesive elements in the framework of the FEM can be found in Geißler [54].

In Fig. 22, a 4-node cohesive element is exemplarily depicted in its undeformed initial configuration with separation $\delta = 0$ and the deformed configuration with normal and tangential separation δ_N and δ_T, respectively. The initially connected and then separated crack faces are termed Ω^+ and Ω^-. The number of element nodes per surface Ω^+ or Ω^- has to fit with the number of element nodes per side of the continuum elements. In the present case, a middle surface has been used to describe the opening of the cohesive element via the local reference system $\bar{x}$ and $\bar{y}$. For more details on the element formulation, it is referred to Geißler and Kaliske [55].

With the help of the TSL, it is possible to describe the micromechanical phenomena of void nucleation and micro damage in the process zone which contribute to the mechanical response on the macroscale. In this context, the tractions at the crack faces can be interpreted as the residual mechanical response of material bonds bridging the crack. If the separation increases, the material bonds break and the tractions decrease irreversibly. In addition, the area under the traction-separation curve represents the release of energy

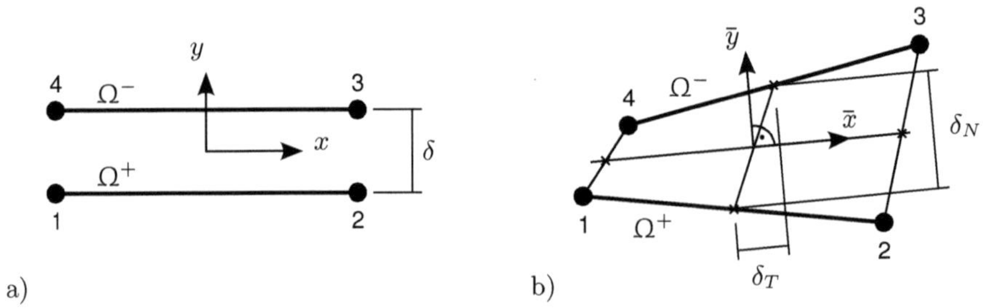

Fig. 22 4-node cohesive element with initially connected surfaces Ω^+ and Ω^-: a) initially undeformed state with separation $\delta = 0$; b) deformed state with normal separation δ_N and tangential separation δ_T

due to the creation of new surfaces and, therefore, constitutes a characteristic fracture mechanical quantity. As a result, it is possible to describe the continuum material response and the fracture behavior of the material with the help of two distinctive model formulations.

Since the available TSL are mainly based on elastic, rate-independent relations between traction and separation rate, a cohesive element formulation with rate-dependent, visco-elastic tractions has been derived and proposed in Geißler and Kaliske [55]. The constitutive relations are based on the visco-elastic material description of elastomers published by Kaliske and Rothert [56]. The rheology of the model governing the visco-elastic traction response is depicted in Fig. 23 and consists of a generalized Maxwell model. The equilibrium branch (EQ) consists of a Hooke element and the N non-equilibrium branches (NEQ) formed by Maxwell elements yield the rate-dependent response.

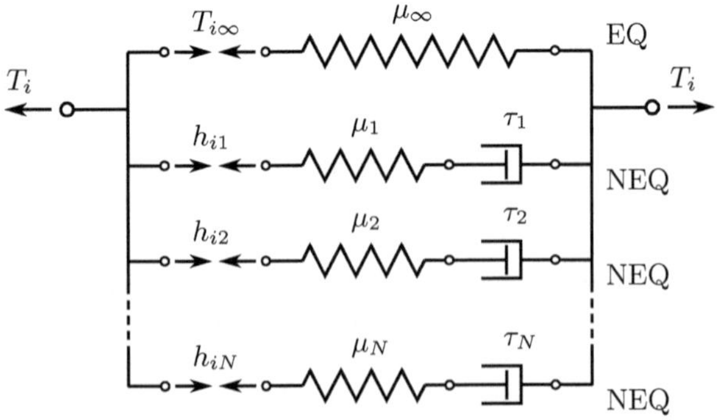

Fig. 23 Rheology of the generalized Maxwell model for the rate-dependent, visco-elastic traction separation law

By considering the equilibrium condition in each non-equilibrium branch between spring and linear viscous dashpot element with relaxation time τ_j, the evolution law

$$\mu_j \dot{T}_{i0} = \dot{h}_{ij} + \frac{1}{\tau_j} h_{ij} \quad \text{with} \quad \mu_\infty + \sum_{j=1}^{N} \mu_j = \mu_0 \tag{72}$$

governing the evolution of the tractions in the j-th branch can be derived. In Eq. (72), μ_j denotes a dimensionless parameter, T_{i0} the instantaneous traction response of the generalized model and h_{ij} is the viscous traction in the j-th branch which becomes the internal variable. With the help of a convolution integral, Eq. (72) can be solved to give after several steps the tractions

$$h_{ij}(t_{n+1}) = \mu_j \int_0^{t_{n+1}} \exp\left(-\frac{t_{n+1} - s}{\tau_j}\right) \frac{\partial T_{i0}(s)}{\partial s} \, \mathrm{d}s \tag{73}$$

at the discrete time t_{n+1}. The integral expression with the local time measure s can be approximated by a linear interpolation from time t_n to t_{n+1} separated by the time step Δt to give

$$h_{ij}^{n+1} = \exp\left(-\frac{\Delta t}{\tau_j}\right) h_{ij}^{n} + \mu_j \frac{1 - \exp\left(-\frac{\Delta t}{\tau_j}\right)}{\frac{\Delta t}{\tau_j}} \left[T_{i0}^{n+1} - T_{i0}^{n}\right] \, . \tag{74}$$

The total traction response T_i^{n+1} at t_{n+1} due to a separation of the crack faces is obtained by a summation over each branch,

$$T_i^{n+1} = T_{i\infty}^{n+1} + \sum_{j=1}^{N} h_{ij}^{n+1} \, . \tag{75}$$

The developed rate-dependent, visco-elastic cohesive element formulation has been applied to several numerical and industrial examples reported in Geißler and Kaliske [55]. In the following, the results of a peel test simulation are shown. The peel system consists of polyethylene/polybutene-1-blend films for sealing and is characterized among other effects by a distinctive rate-dependency. The peel test focuses on the delamination of films and measures the reaction force which is necessary to advance the peel process. In this case, the failure plane can be directly modeled via cohesive elements since the failure plane is known from the beginning. The experimental work is described in detail in the publications by Geißler et al. [57] and Nase et al. [58]. Fig. 24 depicts the numerical simulation results for different peel experiments with different test velocities. From the comparison between simulation results and the experimentally recorded peel force, it can be concluded that the rate sensitivity of the loading is well captured by the visco-elastic cohesive element formulation.

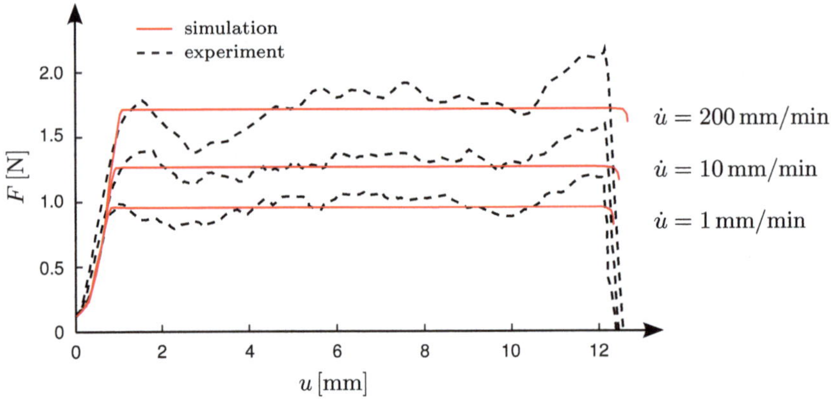

Fig. 24 Comparison of experimentally and numerically obtained peel force-separation relations for delamination of a peel system at different test speeds $\dot{u}$

4.2 Mesh Adaptivity Strategies

As already discussed in the introduction of this section (see Sect. 4), a solution strategy is required which is able to represent not only finite cracks in form of separation in the body, but also an arbitrary path of the crack. The aforementioned smeared approach, where the crack is represented by damaged continuum elements modeling stress softening of the material, is characterized by a mesh dependence due to the representation of the crack in smeared form, i.e. via the displacement field of a continuum element and its volume. With the help of regularization methods taking the element volume into account, it is possible to overcome the drawback of mesh dependency. However, the energy release due to the creation of new surfaces is still not included.

Cohesive elements allow a discrete representation of macroscopically observable material failure and take the energy release during crack propagation into account. In the previous section, cohesive elements have been used to model an initially known failure plane within a numerical simulation. However, the failure plane and its position within an arbitrary component of one material subjected to arbitrary loading conditions is unknown in general. In addition, if not a glued film or a composite material with cohesive interfacial properties is described, the presence of initially elastic cohesive elements will decrease the stiffness of the system in an artificial way. In this context, a TSL with zero traction response for the initially closed case $\delta = 0$ is termed initially elastic since a minimum crack opening is needed to generate and finally transmit tractions over the distinctive cohesive element layer and inner forces in the body.

In order to capture arbitrary crack patterns, the cohesive element approach has been modified in a first step by using an initially rigid TSL for the element formulation and, in a second step, it has been combined with an adaptive meshing strategy. Initially rigid TSL have already been considered by Hillerborg et al. [59]. Camacho and Ortiz [60] as well as Pandolfi and Ortiz [61, 62] applied the initially TSL together with an adaptive implementation of cohesive elements in the structure especially for dynamic fragmentation processes.

The adaptive mesh manipulation and remeshing strategy is briefly outlined in the following and has been implemented in a quasi-static implicit framework of the FEM to capture arbitrary crack paths by the numerical model. Arbitrary crack propagation is characterized by topology changes of the current as well as the reference configuration, the creation of new surfaces and fragmentation of the considered body. As a consequence, the adaptive meshing strategy effects the global program structure. The adaptive cohesive element formulation has been implemented for 8-node hexahedral continuum elements in combination with 8-node cohesive elements. In each simulation step, the failure criterion is evaluated at each element node. If a critical value is exceeded at a node, a mesh update algorithm is invoked. According to the crack propagation direction criterion, the affected element surfaces are identified and a mesh modification for these surfaces is carried out as illustrated by the simple example depicted in Fig. 25. The failure nodes, where the critical value is exceeded, are copied and cohesive elements are inserted in the crack surfaces. At this state, also the element node connections have to be modified for all the neighboring continuum elements.

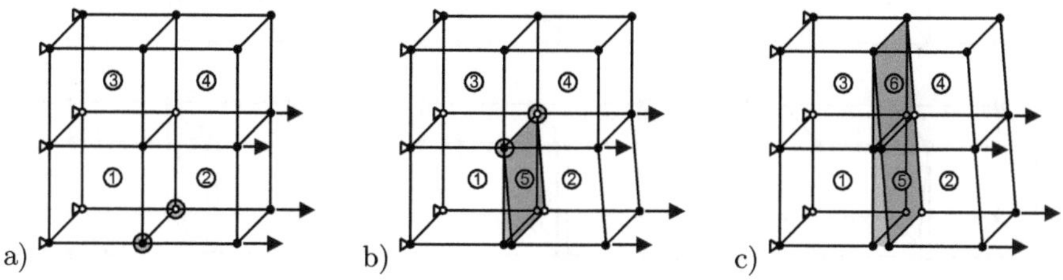

Fig. 25 Adaptive implementation of cohesive elements: a) initial mesh with identified nodes to be doubled; b) insertion of cohesive element; c) further advancing crack modeled by consecutive cohesive elements

In general, one distinguishes between intrinsic failure criteria, which are incorporated in the initially elastic cohesive phase as a critical traction, and extrinsic failure criteria, which governs the topology changes of the structure by conditions formulated on the global finite element level. The mesh modifications should not affect the equilibrium state reached so far, i.e. the modification process should conserve that the load and stress state of the previous system is identical to the load and stress state of the modified system. This condition has been termed as time continuity task (see e.g. Papoulia et al. [63]). As a result, the internal forces at the failure nodes, where cohesive elements are inserted, become the initial traction vector of each initially rigid cohesive element and, as a result, are parameters of the rigid TSL. The initial tractions can be computed from the released residual forces stemming from the cohesive node and the node from the neighboring continuum element during the duplication event (see Fig. 26).

It is worth mentioning that the mesh modification requires also a projection of the internal variables in case of inelastic material behavior. Since the internal variables describing inelastic material properties are normally only available at the integration points of each continuum element, the internal variables have to be projected to the nodes of the old mesh which includes an averaging procedure at each node to obtain a smooth field. Next, the internal variables have to be reprojected again to the new position of the integration points of the modified mesh.

The implemented mesh adaptivity approach (see e.g. Fig. 26) overcomes the limitations of the dependence on the initial finite element discretization and allows the modeling of arbitrary crack patterns during the numerical simulation. In order to predict the crack path, criteria for crack growth and crack direction are still required and have to be incorporated into the simulation approach as discussed in the next subsection.

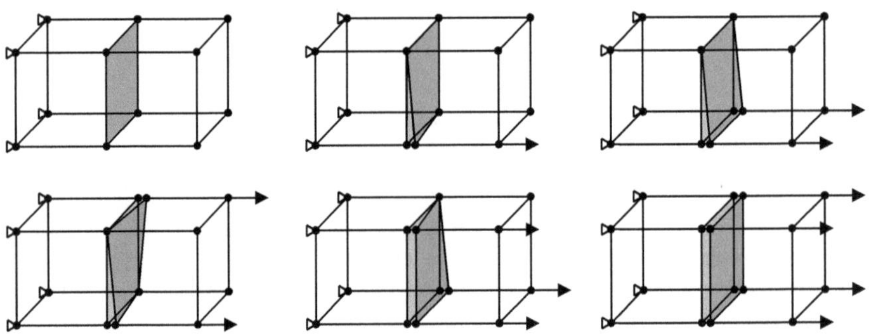

Fig. 26 Adaptive implementation of cohesive elements according to loading conditions

4.3 Combination with Crack Propagation Criteria

Catastrophic crack growth and fatigue crack growth in combination with the afore-mentioned hyperelastic and inelastic material characteristics of elastomers render the crack sensitivity analysis a challenging task. In the presented investigations, the focus is on the assessment of the fracture sensitivity of a structure possessing an initial crack or the tracing of a propagating crack starting from an initial failure location. Therefore, initiation of crack development and propagation are not considered.

As a result, energy based fracture criteria and stress based fracture criteria for crack growth and crack direction have been employed in the context of the afore-described adaptive meshing approach in combination with cohesive elements. The material force is an energy based fracture criterion and offers in contrast to the standard J-integral also vectorial information concerning a possible crack propagation direction. Their application and limitations as fracture mechanical measure are illustrated by structural examples in Sect. 5.

Beside the material forces, an averaged stress criterion (see also e.g. Wells and Sluys [64]) and an averaged principal stress direction criterion have been employed for the crack simulation. In this case, the averaged stress criterion considers not only the crack tip, but a certain field around the crack tip by taking the stress contributions with a weighting function as an averaged stress into account,

$$\bar{\boldsymbol{\sigma}} = \int w(r,l)\, \boldsymbol{\sigma}\, \mathrm{d}V \tag{76}$$

where $w(r,l)$ is a linear weighting function or a weighting function of Gaussian type

$$w(r,l) = \begin{cases} -\frac{r}{l} + 1 & \text{for } r < l \\ 0 & \text{for } r \geq l \end{cases} \quad \text{and} \quad w(r,l) = \frac{1}{(2\,\pi)^{3/2}\, l^3} \exp\left(-\frac{r^2}{2\,l^2}\right), \tag{77}$$

respectively. r is the distance from the crack tip and l denotes a form parameter.

In Fig. 27, a stress based and a material force based criterion are illustrated which lead to a directed crack growth within a finite element discretized structure.

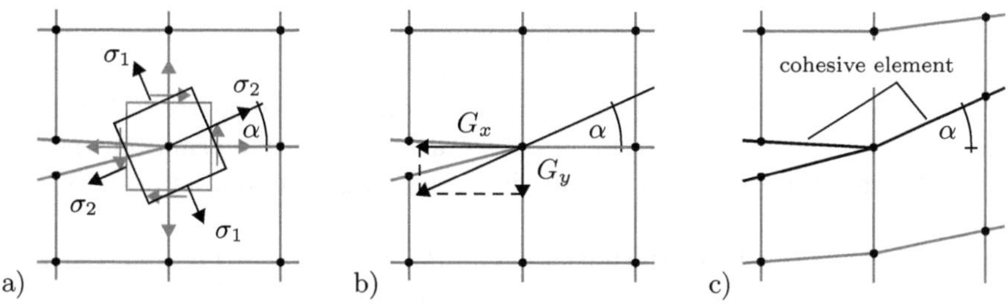

Fig. 27 Crack growth and crack direction criteria at the crack tip: a) averaged principal stress criterion; b) material force criterion; c) adapted mesh after implementation of cohesive element in the crack propagation direction

5 Structural Examples of Durability Investigations and Finite Crack Propagation

In this section, the developed simulation approaches are combined to analyze durability and fracture mechanical behavior of elastomeric components on the structural scale. The first example focuses on a tensile rubber specimen under mixed-mode loading. Crack propagation starting from an initial side notch is numerically predicted and compared to experimental observations. In the second example, the averaged stress and energy based criteria introduced in Sect. 4.3 are studied and compared with respect to their crack path predictability for different fracture modes. The third example deals with the numerical durability assessment of a tire design for which a large scale finite element model of a truck tire is considered. Durability properties, which are characteristic for the inelastic elastomeric material, are discussed in the context of fracture mechanics and numerical methods are proposed to compute the energy release rate by applying the material force method.

5.1 Tensile Rubber Specimen under Mixed-Mode Loading

In this example, crack propagation under mixed mode loading in a tensile rubber specimen is investigated. Therefore, an inclined side notch has been placed in a distance at $h = 3.8\,\text{cm}$ from the lower edge of the specimen. The rubber specimen with dimensions $l = 10.16\,\text{cm}$, $b = 2.54\,\text{cm}$ and $t = 0.21\,\text{cm}$ has been discretized with 3D-continuum elements and subjected to a monotonic displacement controlled tensile loading $u(t)$. The rubber material is assumed to be purely elastic and, in consequence, is characterized by a Young's modulus of $E = 5.5\,\text{MPa}$ and a Poisson's ratio of $\nu = 0.49$ representing nearly incompressibility of the material. The finite element mesh together with the specimen's geometry and boundary conditions are depicted in Fig. 28 a). Plane strain is assumed and modeled by constraining the degree of freedoms in z-direction for all the x-y-planes. As crack propagation criterion, the averaged stress criterion with a Gaussian weighting function, as described in Sect. 4.3, is used.

During the simulation, the initial notch ($a_0 = 0.25\,\text{cm}$, inclination $\alpha = 45°$) starts to grow and tends to a horizontal fracture plane in y-direction as depicted in Fig. 28 b) for different times t. The numerically obtained crack path is in good agreement with the experimentally obtained path reported by Pidaparti et al. [65].

5.2 Study of Stress and Energy Based Criteria: Square Rubber Plate

The influence of the averaged stress criterion and the material force criterion introduced in Sect. 4.3 on the predicted crack path are studied in this

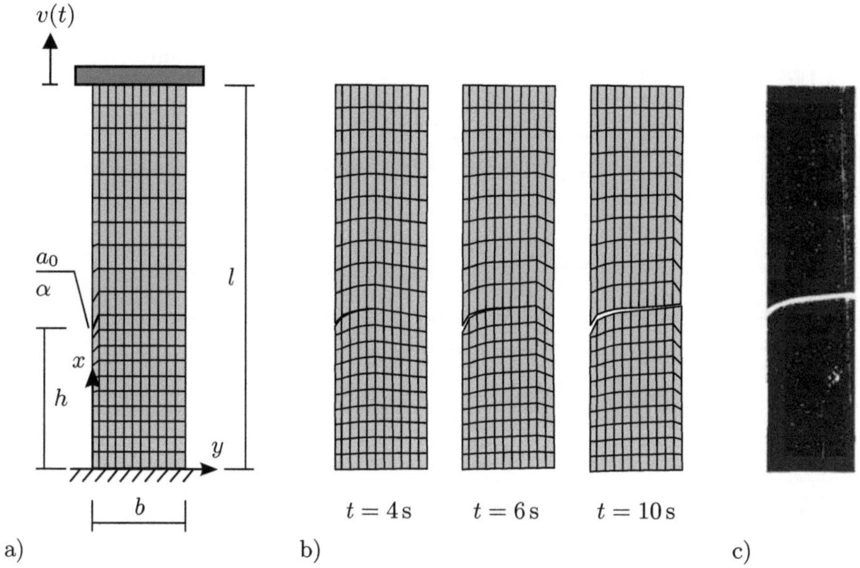

Fig. 28 Rubber specimen with an initially inclined side notch subjected to monotonic, displacement controlled tensile loading: a) finite element model with boundary conditions; b) simulated crack propagation for different times t; c) experimentally observed crack in the specimen (Pidaparti et al. [65])

example. A square plate with a side dimension of $a = 1\,\text{cm}$ and a thickness of $t = 0.1\,\text{cm}$ is subjected in case (A) to tension loading with mode I fracture and in case (B) to shear loading with mode II fracture. An initial side notch of $a_0 = a/2$ is placed in the center of the specimen's height. The finite element model has been assembled using 3D-continuum elements. The rubber material is assumed as elastic and is described by a Young's modulus of $E = 2\,\text{MPa}$ and a Poisson's ratio of $\nu = 0.49$. The applied loading and the boundary conditions for case (A) are illustrated in Fig. 29. The degrees of freedom in z-direction of all the x-y-planes are constrained to represent plane strain conditions.

A parameter study for the value of l of the Gaussian weighting function has been carried out to reveal its influence on the predicted crack path. As depicted in Fig. 29 for case (A), the variation of l has no influence on the mode I crack direction for tensile loading. For case (B) with mode II loading, the variation of l causes only small deviations of the crack path from the anticipated crack path at 45° for pure shear loading. In this context, it has been observed that for decreasing values of l, the deviations from the crack path at 45° increase. As a result, if a sufficient large area of the stress field around the crack tip is considered by the averaged stress criterion, a good correlation can be obtained as shown in Fig. 30.

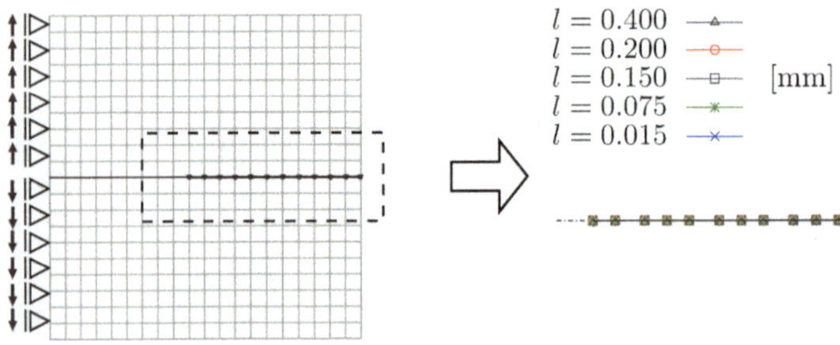

Fig. 29 Square rubber plate: loading, boundary conditions and simulated crack path for mode I fracture using averaged stress criterion

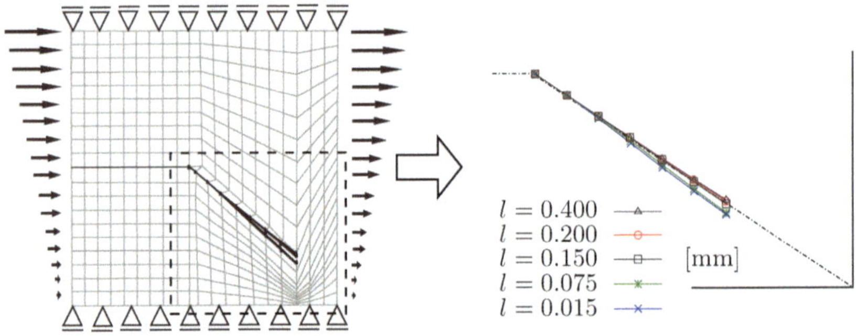

Fig. 30 Square rubber plate: loading, boundary conditions and simulated crack path for mode II fracture using averaged stress criterion

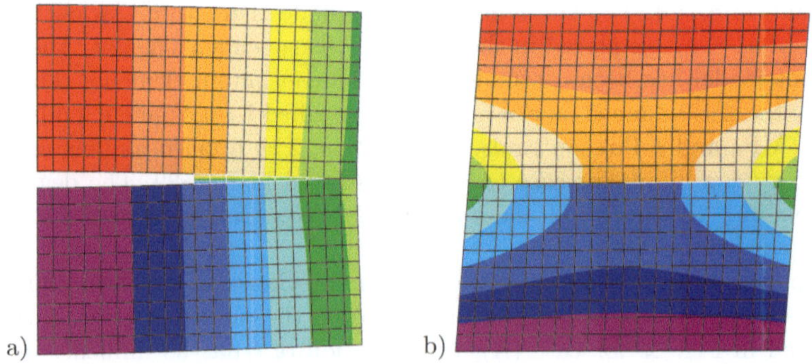

Fig. 31 Square rubber plate with simulated crack path using the material force criterion: a) mode I tension loading; b) mode II shear loading

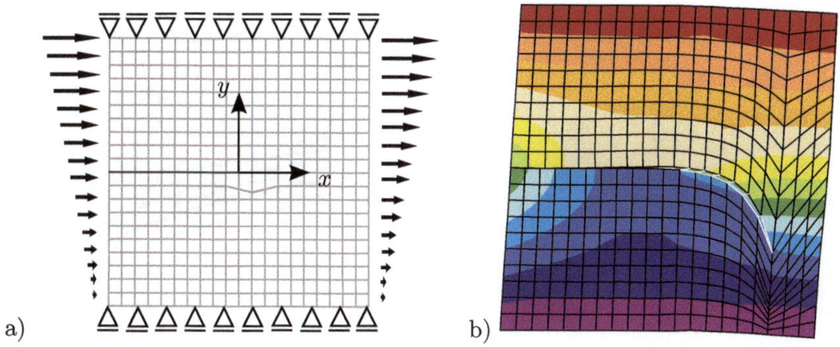

Fig. 32 Square rubber plate: a) modified, distorted mesh; b) crack path for distorted mesh using the material force criterion

The results of the computed crack paths using the material force criterion are depicted in Fig. 31 and Fig. 32. For case (A) with mode I loading, the anticipated horizontal and straight crack path is obtained. However in case (B), the mode II loading shows also a straight crack path for the symmetric mesh which is in contradiction with the expected result. A study on the influence of the mesh distortion revealed that the crack path is influence by the type of the applied mesh distortion. As exemplarily depicted in Fig. 32 a), a vertical displacement of the node in the crack tip region leads to the anticipated downturn of the crack path. However, a vertical shift of the node to the upper side leads to an upturn of the crack path which is in conflict with the anticipated crack direction. As a result, a combination of different crack propagation criteria, based on different approaches, will therefore improve the realistic prediction capabilities of the simulation approach in case of an arbitrary mesh discretization (see also Gürses [66]). Further examples of crack propagation in visco-elastic rubber are reported in Kaliske et al. [67].

5.3 Tire Durability Investigations

As an example for the application of the material force method to evaluate the durability of a tire design (see also Kaliske et al. [68], Nasdala et al. [69], Previati and Kaliske [70]), a pneumatic standard truck tire is considered. Pneumatic tires are subjected to the already described effects of physical and chemical alteration during their service. In addition, tires are composite structures of fibers embedded in a rubber matrix showing hyperelastic, visco-elastic and plastic material behavior. The composite structure privilege tire failure in form of critical delamination at the interfaces of steel fibers and the rubber matrix as depicted in Fig. 33.

The delamination is triggered by a weak bonding between rubber matrix and fibers as well as initial micro cracks at the belt edges. The cyclic loading of the tire under service conditions leads to crack growth. For a U-shaped

a) 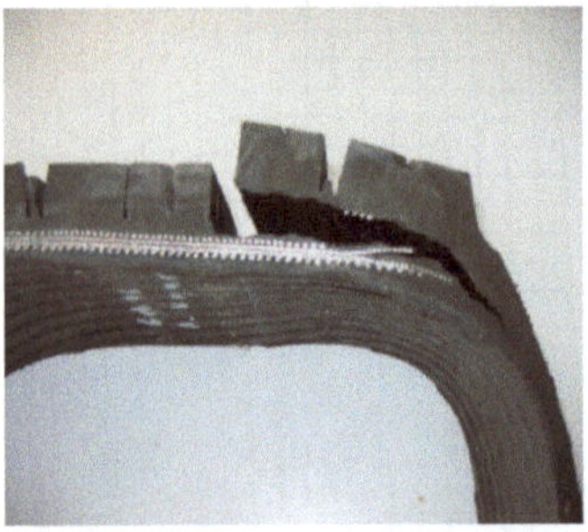

b) 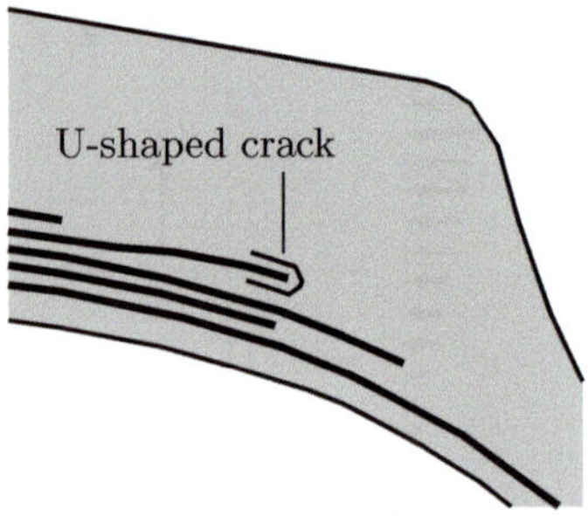

Fig. 33 Part of a tire's cross-section: a) characteristic failure mode by delamination; b) cross-section of tire model with a U-shaped crack at belt edges

crack at the belt edge as depicted in Fig. 33 b), the energy release rate as a function of the tire's circumference is given in Fig. 34 for two different tire designs A and B. The peak in the energy release rate corresponds to the footprint area around $\varphi = 0°$.

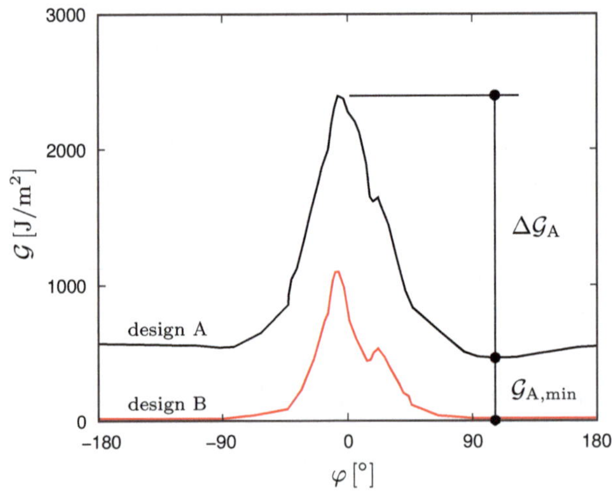

Fig. 34 Energy release rate $\mathcal{G}$ as function of the tire's circumference φ for two different tire designs A and B

For a rolling tire, the relation of the energy release rate given in Fig. 34 can be interpreted as the evolution of the energy release rate for a point at the crack tip during one cycle. $\Delta\mathcal{G} = \mathcal{G}_{max} - \mathcal{G}_{min}$ is the energy release rate amplitude which is one of the characteristic values for crack growth beside the important limits $\mathcal{G}_{max}$ and $\mathcal{G}_{min}$ which strongly govern the type of crack growth and related effects like SIC as discussed later. If the energy release rate amplitude is plotted together with the crack growth rate $\Delta a/\Delta N$, where Δa is the crack length increment observed during ΔN load cycles, in a Paris-plot with double logarithmic scale, a linear relation between $\Delta\mathcal{G}$ and $\Delta a/\Delta N$ can be recognized. In total, three domains can be observed in a schematic Paris-plot as depicted in Fig. 35 a), where domain I corresponds to the initiation of crack growth, domain II is related to stable crack propagation and domain III is linked with a fast crack growth leading to failure.

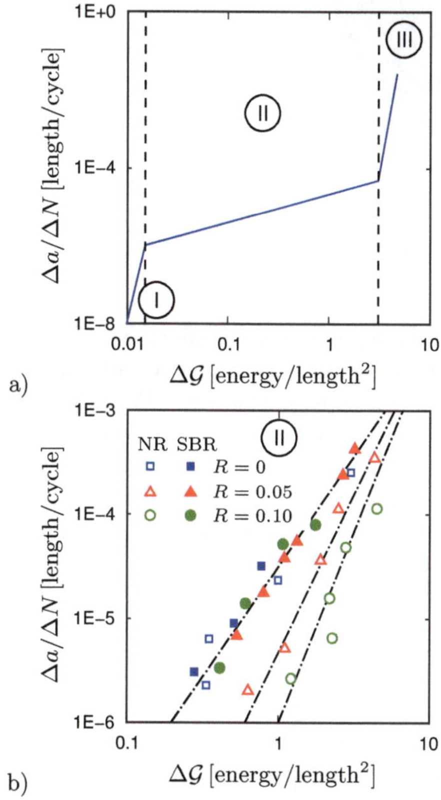

Fig. 35 Schematic representation of crack growth rate as a function of the energy release rate: a) qualitative illustration of different domains I to III of crack growth; b) qualitative dependence of crack growth rate on loading ratio R for filled NR and filled SBR

As already mentioned before, the limit values $\mathcal{G}_{\text{max}}$, $\mathcal{G}_{\text{min}}$ and their ratio $R = \mathcal{G}_{\text{min}}/\mathcal{G}_{\text{max}}$ have to be considered as well as the amplitude $\Delta\mathcal{G}$ in-between for durability analysis. As an example, values of $R > 0$ can lead to SIC in rubber which increases the crack growth resistance at the crack tip. In contrast to filled NR, filled SBR does not show pronounced SIC effects as observed from a comparison of the corresponding Paris-plot given in Fig. 35 b).

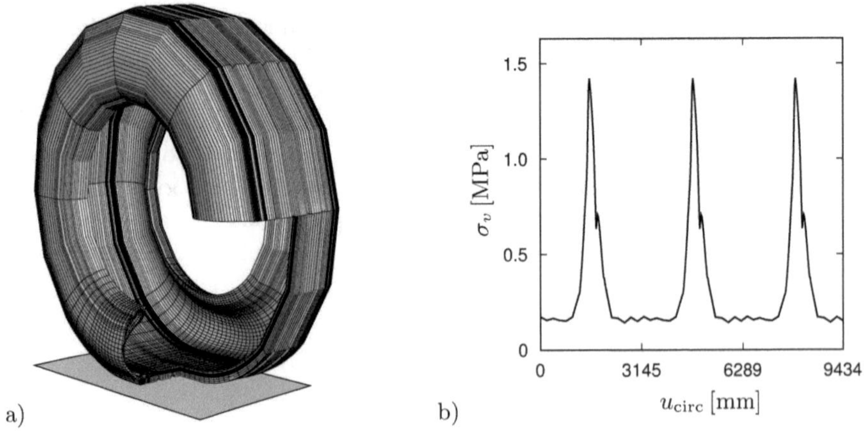

Fig. 36 Dwell-effect for a tire: a) finite element discretized tire structure of size 385/65 R22.5; b) von Mises stress σ_v of a material point in the tire as a function of the circumferential length u_{circ}

Another influence factor is the dwell-effect (see Näser et al. [71]) which is the relation between the time period of two subsequent load peaks, called dwell-time, and the crack propagation rate. With increasing dwell-time, the rate of crack growth per load application also increases. This phenomenon has been observed in experiment and simulation as briefly outlined in the following. The finite element model of a truck tire depicted in Fig. 36 a) has been used to compute the stress state of a material point near the belt edge during the revolution of the tire.

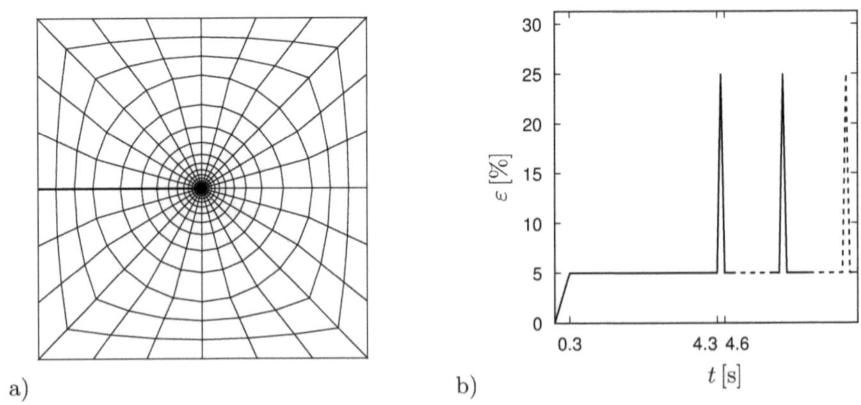

Fig. 37 Model problem consisting of an initially notched rubber plate: a) undeformed geometry with finite element mesh; b) time-dependent strain controlled loading ε of the rubber plate as function of time t

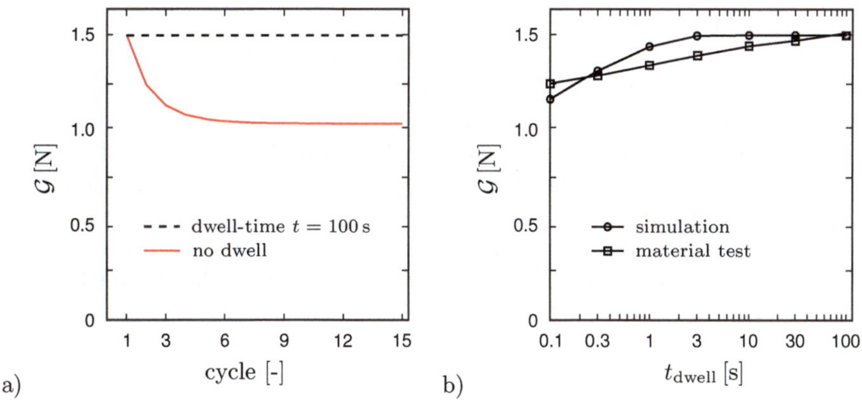

Fig. 38 Energy release rate with and without dwell-effect: a) evolution of energy release rate as function of cycles with dwell-time $t = 0$ s and $t = 100$ s; b) comparison of simulated energy release rate including dwell-effect with experimental data from material test

The stress-circumference relation depicted in Fig. 36 b) is due to the cyclic loading which shows dwell-times between the peaks. With increasing dwell-time, the energy release rate increases which has been observed during experiments by Harbour et al. [72] and has been numerically analyzed for a model problem shown in Fig. 37 a). The notched square plate of dimensions 150 mm × 150 mm × 1 mm is composed of 3D-mixed-continuum elements and subjected to tension loading. The numerical results of the model problem have been validated by an experiment on a pure-shear specimen. For the numerical simulation, the visco-elastic material of the specimen (styrene-butadiene rubber) is modeled using the developed model presented in Sect. 2.4 with the parameters $\mu = 0.4$ MPa, $N = 4.0$, $\mu_v = 0.8$ MPa, $N_v = 2.9$, $\dot{\gamma}_0/\hat{\tau}^m = 1.6\,\text{s}^{-1}\text{MPa}^{-m}$, $c = 0$ and $m = 1$ for the considered strain rate. The mechanical loading of the specimen given in Fig. 37 b) has been derived from the load history of a tire's material point in the tire simulation. Logarithmic time increments in the interval from 0 s to 100 s have been used as dwell-times between the load peaks of 20 % strain amplitude. The energy release at the crack tip of the rubber plate has been computed with the help of the material force method as outlined in Sect. 3.3 by considering a radius of 50 mm around the crack tip. The limit cases of 0 s and 100 s dwell-time are depicted in Fig. 38 as well as the comparison with the experimentally obtained energy release rates for the different dwell-times in the interval.

By using the material force method, the energy release rate along the circumference of a tire with initially U-shaped cracks at the belt edge is numerically computed. If the tire is in a steady state of rolling, centrifugal forces due to the rotation induced inertia effects occur and can be taken as physical body forces into account. In the present case of a truck tire, the influence of the centrifugal forces is negligible due to a low rolling velocity, high inner pressure and axle loads. Therefore, a static analysis has been carried out to compute the nodal material forces. In Fig. 39 a), the discrete nodal material forces are depicted for the whole finite element discretized tire structure. To obtain the continuous function of the energy release rate depicted in Fig. 39 b), the nodal material forces have been converted to line forces along the circumference. Again, the peak in the energy release rate corresponds to the footprint area around $\varphi = 0°$. Neglecting inertia and time-dependent effects, the change as well as the limits of the obtained energy release rate can be used for a durability investigation of a given tire design. Further details with respect to experimental investigations are discussed in Näser et al. [73].

a)

b)

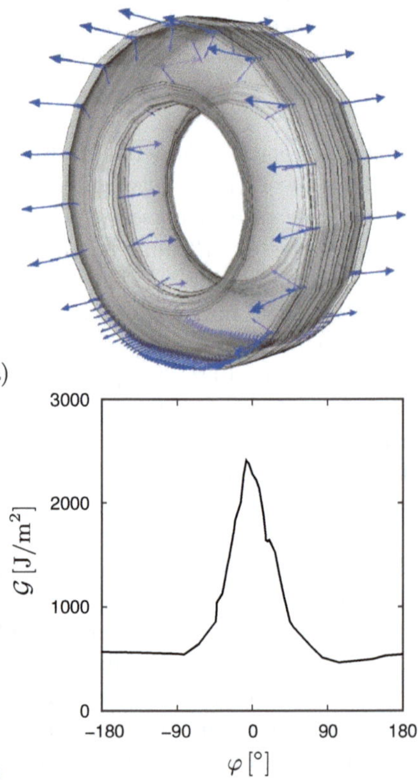

Fig. 39 Computation of energy release rate by material forces: a) discretized tire structure of size 385/65 R22.5 with discrete nodal material forces due to U-shaped crack at belt edge; b) energy release rate as a function of the tire's circumference

6 Conclusion and Outlook

In the present work, the numerical simulation of elastomers including durability analysis and fracture mechanical investigations has been built up by focusing on the micro-, meso- and macroscale to derive numerical models formulated on the macroscale. Theoretical considerations with respect to modeling of significant phenomena and their numerically efficient and consistent implementation in the framework of the FEM have been carried out. In the model finding phase and the validation phase, experimental input data and comparison with experimental observations have been indispensable. In a first step, theoretical considerations with respect to the microstructure of filled elastomers have enabled the development of material models formulated on the macroscale. In a second step, intensive research has been carried out in order to propose numerically efficient simulation strategies within the FEM to realistically model the material and fracture mechanical behavior of filled elastomers. The latter required substantial modifications of the global finite element program structure compared to today's standard finite element techniques. In a third step, numerical simulations of elastomeric structures have been carried out to predict arbitrary crack propagation and to identify their durability characteristics with respect to fracture.

Material Modeling. Finite visco-elastic and finite endochronic elasto-plastic material models for filled elastomers based on an elastic volumetric and elastic as well as inelastic isochoric material response have been derived and implemented by using an elastic predictor and inelastic corrector algorithm together with consistently derived expressions for stresses and material tangents. A parameter study with respect to the model parameters revealed their influence on the simulated material response. The implemented models have been verified with respect to convergence and stability aspects of the proposed algorithm. The parameter identification enabled finally the application of the models to understand and predict the material behavior observed during experiment in the validation phase.

Crack and Durability Properties. Failure criteria for crack propagation on the macroscale have been derived by first considering the material alteration on the microscale using a continuum failure model. The energy-entropic model with a coupled representation of the volumetric and isochoric material response is able to represent the rubberlike elasticity as well as rubber failure caused by bond rupture derived from micromolecular considerations. The failure envelops obtained from material tests have been well captured by the model. Applied to ageing-induced surface cracking, the model is able to capture crack initiation. Fracture sensitivity is assessed by the further derived material force method to take also the elastomer specific characteristics during crack propagation into account. As a result, the numerically interesting purely crack driving force could be separated from the global energy release rate for dissipative materials.

Arbitrary Crack Propagation. The formulation of rate-dependent cohesive elements within the standard FEM enabled the representation of finite cracks. By the example of a peel test, the rate-dependent traction-separation characteristic of the cohesive zone has been demonstrated and compared to the experimentally observed one. Together with averaged stress and energy based crack propagation criteria for crack growth and crack direction, arbitrary crack patterns in elastomeric structures could be tracked by using an adaptive meshing strategy and the adaptive implementation of cohesive elements.

Structural Investigations. For different characteristic tensile and pure shear rubber specimens, the averaged stress and energy based criteria have been studied. It can be concluded that the combination of averaged stress and energy based criteria will increase the robustness with respect to distorted mesh configurations and, as a consequence, will improve the prediction of crack paths. The durability analysis of a tire design by means of the material force method has shown the various fields of application for the developed simulation approaches.

Outlook. However, future research work will still be required to transfer the gained fundamental research results to industry for the application in the design process of elastomeric components. A direct cooperation with industry will lead to user-specific solutions for the improvement of quality, the increase of productivity as well as the product's sustainability. Therefore, experimental and numerical investigation methods accounting for all significant phenomena are required. The aim is to intensify the cooperation with industry partners to transfer the gained knowledge, to extent the computer code to practical large scale applications and to already evaluate the product's safety and risk of failure during the design process. The afore-described steps of the numerical reproduction of the material and fracture behavior of elastomers are recently carried out in the ongoing work of the DFG transfer project "Thermo-mechanical durability analysis for the design optimization of elastomeric components in industrial research and development". In addition, the extension to other interaction phenomena (e.g. tire-pavement interaction, vehicle-tire-pavement interaction etc.) are currently subjects of research and will considerably enlarge the numerical model approaches in the future.

Acknowledgments. The authors would like to gratefully acknowledge the financial support by the Deutsche Forschungsgemeinschaft (DFG) under contract KA1163/3-1 and KA1163/3-2. The current financial support by the DFG for project KA1163/16-1 to bring the gained knowledge to industrial application is highly appreciated.

References

1. Grambow, A.: Determination of Material Parameters for Filled Rubber Depending on Time, Temperature and Loading Condition. PhD thesis, Fakultät für Maschinenwesen, Rheinisch-Westfälische Technische Hochschule Aachen (2002)
2. Mooney, M.: A theory of large elastic deformation. Journal of Applied Physics 11, 582–592 (1940)
3. Yeoh, O.: Characterization of elastic properties of carbon-black-filled rubber vulcanizates. Rubber Chemistry and Technology 63, 792–805 (1990)
4. Arruda, E., Boyce, M.: A three-dimensional constitutive model for the large stretch behavior of rubber elastic materials. Journal of the Mechanics and Physics of Solids 41, 389–412 (1993)
5. Miehe, C., Göktepe, S., Lulei, F.: A micro-macro approach to rubber-like materials. Part I: The non-affine micro-sphere model of rubber elasticity. Journal of the Mechanics and Physics of Solids 52, 2617–2660 (2004)
6. Marckmann, G., Verron, E.: Comparison of hyperelastic models for rubberlike materials. Rubber Chemistry and Technology 79, 835–858 (2006)
7. Kaliske, M., Heinrich, G.: An extended tube-model for rubber elasticity: Statistical-mechanical theory and finite element implementation. Rubber Chemistry and Technology 72, 602–632 (1999)
8. Dal, H., Kaliske, M.: Bergström-Boyce model for nonlinear finite rubber viscoelasticity: Theoretical aspects and algorithmic treatment for the FE method. Computational Mechanics 44, 809–823 (2009)
9. Behnke, R., Dal, H., Kaliske, M.: An extended tube model for thermoviscoelasticity of rubberlike materials: Parameter identification and examples. Proceedings in Applied Mathematics and Mechanics 11, 353–354 (2011)
10. Behnke, R., Dal, H., Kaliske, M.: An extended tube model for thermoviscoelasticity of rubberlike materials: Theory and numerical implementation. In: Jerrams, S., Murphy, N. (eds.) Constitutive Models for Rubber VII, pp. 87–92. CRC Press, Taylor & Francis Group (2011)
11. Kaliske, M., Zopf, C., Brüggemann, C.: Experimental characterization and constitutive modeling of the mechanical properties of uncured rubber. Rubber Chemistry and Technology 83, 1–15 (2010)
12. Dal, H., Kaliske, M., Zopf, C.: Theoretical and numerical modelling of unvulcanized rubber. In: Jerrams, S., Murphy, N. (eds.) Constitutive Models for Rubber VII, pp. 99–106. CRC Press, Taylor & Francis Group (2011)
13. Green, M., Tobolsky, A.: A new approach to the theory of relaxing polymeric media. The Journal of Chemical Physics 14, 80–92 (1946)
14. Doi, M., Edwards, S.F.: The Theory of Polymer Dynamics. Clarendon Press, Oxford (1986)
15. Bergström, J., Boyce, M.: Constitutive modeling of the large strain time-dependent behavior of elastomers. Journal of the Mechanics and Physics of Solids 46, 931–954 (1998)
16. Reese, S., Govindjee, S.: A theory of finite viscoelasticity and numerical aspects. International Journal of Solids and Structures 35, 3455–3482 (1998)
17. Miehe, C., Keck, J.: Superimposed finite elastic-viscoelastic-plastoelastic stress response with damage in filled rubbery polymers. Experiments, modelling and algorithmic implementation. Journal of the Mechanics and Physics of Solids 48, 323–365 (2000)

18. Miehe, C., Göktepe, S.: A micro-macro approach to rubber-like materials. Part II: The micro-sphere model of finite rubber viscoelasticity. Journal of the Mechanics and Physics of Solids 53, 2231–2258 (2005)
19. Cohen, A.: A Padé approximant to the inverse Langevin function. Rheologica Acta 30, 270–273 (1991)
20. Amin, A., Lion, A., Sekita, S., Okui, Y.: Nonlinear dependence of viscosity in modeling the rate-dependent response of natural and high damping rubbers in compression and shear: Experimental identification and numerical verification. International Journal of Plasticity 22, 1610–1657 (2006)
21. Dal, H.: Approaches to the Modeling of Inelasticity and Failure of Rubberlike Materials. Theory and Numerics. PhD thesis, Institut für Statik und Dynamik der Tragwerke, Technische Universität Dresden (2012)
22. Kaliske, M., Rothert, H.: Constitutive approach to rate-independent properties of filled elastomers. International Journal of Solids and Structures 35, 2057–2071 (1998)
23. Valanis, K.: A theory of viscoplasticity without a yield surface. Archives of Mechanics 23, 517–533 (1971)
24. Kilian, H., Strauss, M., Hamm, W.: Universal properties in fillerloaded rubbers. Rubber Chemistry and Technology 67, 1–16 (1994)
25. Netzker, C., Dal, H., Kaliske, M.: An endochronic plasticity formulation for filled rubber. International Journal of Solids and Structures 47, 2371–2379 (2010)
26. Besdo, D., Ihlemann, J.: A phenomenological constitutive model for rubberlike materials and its numerical applications. International Journal of Plasticity 19, 1019–1036 (2003)
27. Maugin, G.: Material Inhomogeneities in Elasticity. Chapman Hall, London (1993)
28. Gurtin, M.: Configurational Forces as Basic Concept of Continuum Physics. Springer, New York (2000)
29. Kienzler, R., Herrmann, G.: Mechanics in Material Space – with Application to Defect and Fracture Mechanics. Springer, Berlin (2000)
30. Morse, P.: Diatomic molecules according to the wave mechanics. II. Vibrational levels. Physical Review 34, 57–64 (1929)
31. Dal, H., Kaliske, M.: A micro-continuum-mechanical model for failure of rubber-like materials: Application to ageing-induced fracturing. Journal of the Mechanics and Physics of Solids 57, 1340–1356 (2009)
32. Treloar, L.: Stress-strain data for vulcanised rubber under various types of deformation. Transactions of the Faraday Society 40, 59–70 (1944)
33. Hamdi, A., Naït Abdelaziz, M., Aït Hocine, N., Heuillet, P., Benseddiq, N.: Fracture criterion of rubber-like materials under plane stress conditions. Polymer Testing 25, 994–1005 (2006)
34. Griffith, A.: The phenomena of rupture and flow in solids. Philosophical Transactions of the Royal Society of London. Series A 221, 163–198 (1920)
35. Rice, J.: A path independent integral and the approximate analysis of strain concentration by notches and cracks. Journal of Applied Mechanics 35, 379–386 (1968)
36. Eshelby, J.: The elastic energy-momentum tensor. Journal of Elasticity 5, 321–335 (1975)

37. Netzker, C., Horst, T., Reincke, K., Behnke, R., Kaliske, M., Heinrich, G., Grellmann, W.: Analysis of stable crack propagation in filled rubber based on a global energy balance. International Journal of Fracture (2013), doi:10.1007/s10704-013-9816-5
38. Näser, B., Kaliske, M., Müller, R.: Material forces for inelastic models at large strains: Application to fracture mechanics. Computational Mechanics 40, 1005–1013 (2007)
39. Näser, B.: Zur numerischen Bruchmechanik dissipativer Materialien. PhD thesis, Institut für Statik und Dynamik der Tragwerke, Technische Universität Dresden (2009)
40. Behnke, R., Kaliske, M.: Computation of material forces for the thermo-mechanical response of dynamically loaded elastomers. Proceedings in Applied Mathematics and Mechanics 12, 299–300 (2012)
41. Mueller, R., Kolling, S., Gross, D.: On configurational forces in the context of the finite element method. International Journal for Numerical Methods in Engineering 53, 1557–1574 (2002)
42. Steinmann, P., Ackermann, D., Barth, F.: Application of material forces to hyperelastostatic fracture mechanics. II. Computational setting. International Journal of Solids and Structures 38, 5509–5526 (2001)
43. Liebe, T., Denzer, R., Steinmann, P.: Application of the material force method to isotropic continuum damage. Computational Mechanics 30, 171–184 (2003)
44. Menzel, A., Denzer, R., Steinmann, P.: On the comparison of two approaches to compute material forces for inelastic materials. Application to single-slip crystal-plasticity. Computer Methods in Applied Mechanics and Engineering 193, 5411–5428 (2004)
45. Zimmermann, D.: Material Forces in Finite Inelasticity and Structural Dynamics: Topology Optimization, Mesh Refinement and Fracture. PhD thesis, Institut für Mechanik (Bauwesen), Lehrstuhl I, Universität Stuttgart (2008)
46. Kaliske, M., Dal, H., Fleischhauer, R., Jenkel, C., Netzker, C.: Characterization of fracture processes by continuum and discrete modelling. Computational Mechanics 50, 303–320 (2012)
47. Hiermaier, S.: Structures under Crash and Impact. Springer, Berlin (2008)
48. Belytschko, T., Black, T.: Elastic crack growth in finite elements with minimal remeshing. International Journal for Numerical Methods in Engineering 45, 601–620 (1999)
49. Neumann, J.: Anwendung von adaptiven Finite Element Algorithmen auf Probleme der Strukturdynamik. PhD thesis, Fakultät für Bauingenieur-, Geo- und Umweltwissenschaften, Institut für Mechanik, Universität Karlsruhe (2004)
50. Song, J.H., Wang, H., Belytschko, T.: A comparative study on finite element methods for dynamic fracture. Computational Mechanics 42, 239–250 (2008)
51. Dugdale, D.: Yielding of steel sheets containing slits. Journal of the Mechanics and Physics of Solids 8, 100–104 (1960)
52. Barenblatt, G.I.: The mathematical theory of equilibrium cracks in brittle fracture. Advances in Applied Mechanics 7, 55–129 (1962)
53. Needleman, A.: A continuum model for void nucleation by inclusion debonding. Journal of Applied Mechanics 54, 525–531 (1987)
54. Geißler, G.: The Cohesive Crack Tip Model within the Finite Element Method. Implementations, Enhancements, Applications. PhD thesis, Institut für Statik und Dynamik der Tragwerke, Technische Universität Dresden (2009)

55. Geißler, G., Kaliske, M.: Time-dependent cohesive zone modelling for discrete fracture simulation. Engineering Fracture Mechanics 77, 153–169 (2010)
56. Kaliske, M., Rothert, H.: Formulation and implementation of three-dimensional viscoelasticity at small and finite strains. Computational Mechanics 19, 228–239 (1997)
57. Geißler, G., Kaliske, M., Nase, M., Grellmann, W.: Peel process simulation of sealed polymeric film: Computational modelling of experimental results. Engineering Computations 24, 586–607 (2007)
58. Nase, M., Langer, B., Baumann, H., Grellmann, W., Geißler, G., Kaliske, M.: Evaluation and simulation of the peel behavior of polyethylene/polybutene-1 peel systems. Journal of Applied Polymer Science 111, 363–370 (2009)
59. Hillerborg, A., Modéer, M., Petersson, P.E.: Analysis of crack formation and crack growth in concrete by means of fracture mechanics and finite elements. Cement and Concrete Research 6, 773–781 (1976)
60. Camacho, G., Ortiz, M.: Computational modelling of impact damage in brittle materials. International Journal of Solids and Structures 33, 2899–2938 (1996)
61. Pandolfi, A., Ortiz, M.: Solid modeling aspects of three-dimensional fragmentation. Engineering with Computers 14, 287–308 (1998)
62. Pandolfi, A., Ortiz, M.: An efficient adaptive procedure for three-dimensional fragmentation simulations. Engineering with Computers 18, 148–159 (2002)
63. Papoulia, K., Sam, C.H., Vavasis, S.: Time continuity in cohesive finite element modeling. International Journal for Numerical Methods in Engineering 58, 679–701 (2003)
64. Wells, G., Sluys, L.: A new method for modelling cohesive cracks using finite elements. International Journal for Numerical Methods in Engineering 50, 2667–2682 (2001)
65. Pidaparti, R., Yang, T., Soedel, W.: Plane stress finite element prediction of mixed-mode rubber fracture and experimental verification. International Journal of Fracture 45, 221–241 (1990)
66. Gürses, E.: Aspects of Energy Minimization in Solid Mechanics: Evolution of Inelastic Microstructures and Crack Propagation. PhD thesis, Institut für Mechanik (Bauwesen), Lehrstuhl I, Universität Stuttgart (2007)
67. Kaliske, M., Özenç, K., Dal, H.: Aspects of crack propagation in small and finite strain continua. In: Jerrams, S., Murphy, N. (eds.) Constitutive Models for Rubber VII, pp. 137–142. CRC Press, Taylor & Francis Group (2011)
68. Kaliske, M., Näser, B., Meiners, C.: Inelastic fracture mechanics for tire durability simulations. Tire Science and Technology 35, 239–250 (2007)
69. Nasdala, L., Wei, Y., Rothert, H., Kaliske, M.: Lifetime prediction of tires with regard to oxidative aging. Tire Science and Technology 36, 63–79 (2008)
70. Previati, G., Kaliske, M.: Crack propagation in pneumatic tires: Continuum mechanics and fracture mechanics approaches. International Journal of Fatigue 37, 69–78 (2012)
71. Näser, B., Kaliske, M., Dal, H., Netzker, C.: Fracture mechanical behaviour of visco-elastic materials: Application to the so-called dwell-effect. Zeitschrift für Angewandte Mathematik und Mechanik 89, 677–686 (2009)
72. Harbour, R., Fatemi, A., Mars, W.: The effect of a dwell period on fatigue crack growth rates in filled SBR and NR. Rubber Chemistry and Technology 80, 838–853 (2007)
73. Näser, B., Kaliske, M., Mars, W.: Fatigue investigation of elastomeric structures. Tire Science and Technology 38, 194–212 (2010)

Technical Material Diagnostics – Fracture Mechanics of Filled Elastomer Blends

Wolfgang Grellmann[1] and Katrin Reincke[2]

[1] Martin Luther University of Halle-Wittenberg,
Centre of Engineering Sciences, D-06099 Halle
wolfgang.grellmann@iw.uni-halle.de
[2] Formerly Martin Luther University of Halle-Wittenberg,
Polymer Service GmbH Merseburg, Geusaer Str., Geb. 131, D-06217 Merseburg
katrin.reincke@psm.uni-halle.de

Abstract. This publication contains results of different mechanical and fracture mechanics investigations with elastomeric materials. The aims of the work were the characterization of the influence of the structure on the mechanical and fracture mechanical properties as a basis for the establishment of practically usable structure–property relationships as well as the further clarification of the crack processes themselves. The experimental investigations included fracture tests under quasi-static and under impact-like loading conditions. For the experiments, different elastomeric materials were selected: filler-reinforced styrene–butadiene rubber, ethylene–propylene–diene rubber and natural rubber model vulcanizates as well as related blends. Beside various mechanical tests such as tensile test, dynamic mechanical analysis, hardness testing and tear tests, fracture mechanics investigations under quasi-static and impact-like loading conditions were performed. Here, especially the impact tests with variation of the test temperature are new in this field. The quasi-static tests included single- and multi-specimen tests with the aim of recording crack resistance curves. From these results, the materials' resistance against stable crack initiation and propagation was determined. Additionally, comprehensive microfractographic investigations were performed with the aim of a reliable and reproducible quantitative description of the fracture surfaces as the basis for an adequate characterization of the fracture process zone for example.

1 Introduction

The application of elastomer products ranges among others from tyres and conveyor belts over seals to springs and dampers. The materials can be exposed e.g. to mechanical loading, UV radiation, liquids, ozone etc. or a combination of these factors what requires a defined characteristic and property level of the material and can additionally result in a more or less strong change of material

W. Grellmann et al. (Eds.): *Fracture Mechanics & Statistical Mech.*, LNACM 70, pp. 227–268.
DOI: 10.1007/978-3-642-37910-9_6

parameters. Therefore, only with knowledge of the complex interrelationship between the structure of the material and its resulting physical properties, one is able to develop materials which meet the technical requirements of the aspired application. Beside the economic insert of raw materials and additives, simultaneously a high mechanical property level, a long life time, an economic usage property for example in the case of tyres and last but not least a low price are necessary. Regarding the life time of the components, the fracture properties are of outstanding importance and there are still a lot of open questions in this field. Mechanisms of crack initiation and propagation in an elastomeric material are subject of scientific interest, because it is partly still unknown how these processes start and especially proceed.

By applying fracture mechanical methods, one can obtain deepened knowledge on materials' deformation and fracture behaviour, because fracture mechanical methods have been proven as a useful tool for establishment of structure–property correlations for polymeric materials because of their sensitivity to structural changes [1, 2]. The starting point of the work was the thought on the wear phenomenon 'fatigue wear' mainly appearing in the case of passenger car tires and leading to abrasion losses within the tyre tread. Fatigue wear is caused by initiation and propagation of cracks, from which the application of fracture mechanics concepts for assessment of the materials was derived with the aim of increasing the lifetime. Hence, fracture mechanics testing of elastomers has a very high economic significance and one of the main aims of this publication is to show possibilities of experimental methods of polymer diagnostics and fracture mechanics for quantitative characterization of crack initiation and propagation properties of elastomeric materials.

Generally, by the application of fracture mechanics it is assumed that materials fail as a result of initiation and propagation of cracks [3]. Conditions of it are investigated with the result of quantitative connections between external loading of a notched specimen and size and kind of the resulting crack. For the purpose of derivation of failure criteria, various fracture mechanics concepts have been developed. Linear-elastic fracture mechanics (LEFM) can be applied for the characterization of materials having a linear-elastic deformation behaviour. An extension of the LEFM represents the linear-elastic fracture mechanics with small-scale yielding that considers small plastic deformation of the material in front of a crack tip.

When applying the LEFM for the assessment of a material, it is possible to use the stress-intensity factor K or the energy release rate G for a description of the stress field in front of a crack tip. The main fracture criterion of the LEFM is a critical stress-intensity factor called fracture toughness $K_{\mathrm{I,II,III}}$ (indices denoting the crack opening mode) basing on the description of the stress state near the crack tip (see Eq. 1). When the fracture toughness is exceeded, an inactive crack begins to grow in an unstable manner releasing stored elastic energy. The crack velocity is nearly sound velocity and stopping the crack growth is not possible.

$$\sigma_{ij} = \frac{K}{(2\pi r)^{1/2}} g_{ij}(\Theta) \tag{1}$$

with σ_{ij} – normal and shear stresses; r; Θ – polar coordinates with the crack tip as the origin and g_{ij} – dimensional function

The experimentally determined value of the fracture toughness is a geometry-independent material parameter provided that a plane-strain stress state is ensured. This means, specimens with suitable dimensions must be used.

For the determination of a second fracture mechanics parameter of the LEFM – the energy release rate $G_{I,II,III}$ – it is assumed that for the propagation of a crack (this means the extension of the crack by a certain increment) certain energy is necessary, which comes either from the applied load or from the release of deformation energy in a deformed body. The energy release rate is defined as

$$G = -\frac{1}{B}\frac{\partial U}{\partial a} \tag{2}$$

with B – thickness of the specimen; U – elastically stored or potential energy and a – crack length

Concepts of elastic–plastic fracture mechanics (EPFM) have reached greatest importance within fracture mechanics of polymers. Here, the crack-tip-opening displacement (CTOD) and the J-integral concept are mostly used for evaluation of crack toughness behavior of various polymeric materials, although these materials are more or less viscoelastic and therefore, load–deformation curves are non-linear. However, during related experimental investigations with monotonic increase of the loading, the viscoelasticity stays mostly disregarded.

The J-integral as fracture criterion bases on an energetic interpretation of the fracture process: A path-independent line integral encloses the plastically deformed area and is used to describe the stress–strain field around the crack tip. Generally, J describes this part of energy flowing under deformation through the closed area around the crack tip, which is not stored but dissipated by crack extension.

Considering a material showing stable crack propagation, the crack resistance curve (R-curve) concept of the EPFM can be used for a quantitative description of the stable fracture process including the crack-tip blunting, the crack initiation and the crack propagation. Fig. 1 shows a schematic representation of a crack resistance curve (R-curve) with the loading parameter J and the damage parameter Δa being the size of the stable crack growth. In the beginning of the stable crack process, the blunting of the crack tip with increasing loading takes place. This process is characterized by a linear connection between J and Δa namely the blunting line. At a critical point, the crack is initiated, this means, the crack growth begins. For some thermoplastic polymers, a stretch zone could be proven. In this case, the stretch zone width (SZW) can be determined as a critical point where the crack begins to expand and so, a physical crack initiation value J_{iphys} can be determined. For such materials without forming a stretch zone, the ESIS protocol [4], allows for the determination of the technical crack initiation value $J_{0.2}$ at a stable crack-growth size of 0.2 mm as an alternative crack initiation value. After the crack initiation, the crack resistance curve is non-linear and flattens often with increasing crack growth.

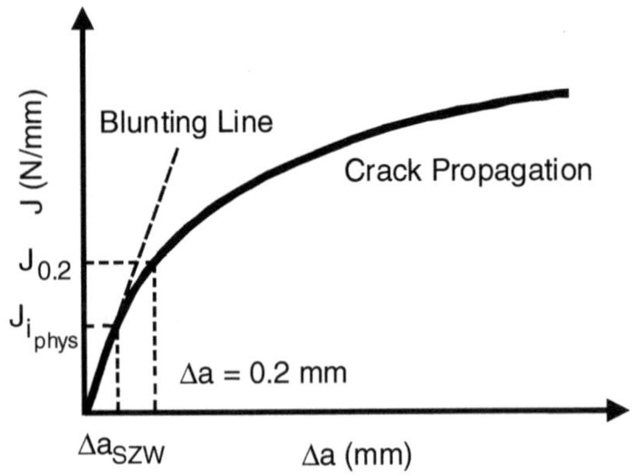

Fig. 1 Schematic representation of a crack resistance curve (R-curve)

For obtaining R-curve data, various experimental methods can be applied including single-specimen or the multi-specimen methods under quasi-static or impact-like loading conditions. For thermoplastic polymers, there are several well-known standards for R-curve determination. One example is the ESIS TC 4 standard draft "A Testing Protocol for Conducting *J*-Crack Growth Resistance Curve Tests on Plastics" [1–5].

In the field of fracture mechanics for elastomers, the fundamental work of Rivlin and Thomas from 1953 [6] is the basis of most works published. Here, also starting from the energy balance, the tearing energy *T* was deduced as a fracture mechanics parameter for elastomeric materials. According to Eq. 3, *T* is the change of the deformation energy due to the energy necessary for the creation of a new crack surface and therefore, it is formally identical to the definition of the *J* integral and the energy release rate *G:*

$$G \equiv J \equiv T = -\frac{1}{B}\frac{\partial U}{\partial a} \tag{3}$$

Rivlin and Thomas [6] used three types of specimens: the single-edge-notched tension (SENT), the trouser specimen and the pure shear specimen (PS) as it is shown in Fig. 2 and derived Eqs. (4) to (6) for the determination of the tearing energy *T*.

For SENT specimens:

$$T = 2\frac{\pi}{\sqrt{\lambda}}W_0 a \tag{4}$$

with λ – stretch ratio; W_0 – elastic energy density

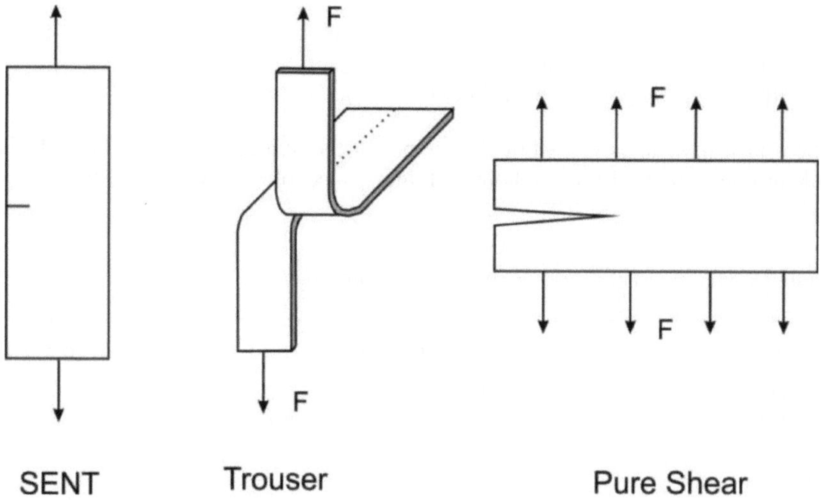

Fig. 2 Schematic representation of typical specimens for the experimental determination of fracture mechanical parameters for elastomers

For trouser specimens:

$$T = \frac{W_0 A_0 - 2\lambda F}{B} \tag{5}$$

with A_0 – cross-section area of the undeformed specimen; F – load

For PS specimens:

$$T = W_0 l_0 \tag{6}$$

with l_0 – initial height of the specimen (corresponds to the initial distance of the clamps)

Within the scientific literature, the use of the SENT and the PS specimens has been mostly described. The advantage of the PS specimen can be seen from Eq. 4 and lies in the independence of the current crack length. While for the SENT specimens, the current crack length a must be known, this is not the case for PS specimens. This is of special interest for fracture mechanics tests under cyclic loading (fatigue tests).

2 Materials

2.1 Composition

During the project work, three series of materials were investigated. As matrices different polymers were selected and model vulcanizates were produced. The elastomeric materials which were examined were on the basis of

- solution styrene–butadiene rubber (S-SBR),
- natural rubber (NR) and
- ethylene–propylene–diene rubber (EPDM).

The following Tables 1–4 show the recipes of the related rubber mixtures. Beside vulcanizates on the basis of homopolymers, also blends were investigated (see Table 5).

Table 1 Composition of the materials of series 1; all data in phr*

S-SBR	100	100	100	100	100	100	100	100	100
CB N339	–	20	40	60	–	–	–	–	–
CB N115	–	–	–	–	60	–	–	–	–
CB N550	–	–	–	–	–	60	–	–	–
Silica Ultrasil GR7000®	–	–	–	–	–	–	20	40	60
Silane	–	–	–	–	–	–	1.7	3.3	5
Ageing Protection IPPD**	1.5	1.5	1.5	1.5	1.5	1.5	1.5	1.5	1.5
Stearic acid	1.0	1.0	1.0	1.0	10.	1.0	1.0	1.0	1.0
Zinc oxide	3.0	3.0	3.0	3.0	3.0	3.0	3.0	3.0	3.0
CBS***	2.5	2.5	2.5	2.5	2.5	2.5	2.5	2.5	2.5
Sulphur	1.7	1.7	1.7	1.7	1.7	1.7	1.7	1.7	1.7

* phr – parts per hundred parts of rubber
** IPPD – N-isopropyl-N-phenyl-p-phenylendiamine
*** CBS – N-cyclohexyl-2-benzothiazolesulfenamide

Table 2 Composition of the SBR materials of series 2

S-SBR	100	100	100	100
CB N234	–	20	40	60
Ageing Protection IPPD	1.5	1.5	1.5	1.5
Stearic acid	1.0	1.0	1.0	1.0
Zinc oxide	3.0	3.0	3.0	3.0
DPG*	–	–	–	–
CBS	2.5	2.5	2.5	2.5
Sulphur	1.7	1.7	1.7	1.7

* DPG – diphenyl guanidine

Table 3 Composition of the NR materials of series 2

NR	100	100	100	100
CB N234	–	20	40	60
Ageing Protection IPPD	1.5	1.5	1.5	1.5
Stearic acid	1.0	1.0	1.0	1.0
Zinc oxide	3.0	3.0	3.0	3.0
DPG	–	–	–	–
CBS	2.5	2.5	2.5	2.5
Sulphur	1.7	1.7	1.7	1.7

Table 4 Composition of the EPDM materials of series 2

EPDM	100	100	100	100
CB N234	–	20	40	60
Ageing Protection IPPD	1.5	1.5	1.5	1.5
Stearic acid	1.0	1.0	1.0	1.0
Zinc oxide	3.0	3.0	3.0	3.0
DPG	1.5	1.5	1.5	1.5
CBS	2.5	2.5	2.5	2.5
Sulphur	1.7	1.7	1.7	1.7

Table 5 Composition of the materials of series 3

S-SBR	50	50	50	50
NR	50	50	–	–
EPDM	–	–	50	50
CB N234	–	60	–	60
Ageing Protection IPPD**	1.5	1.5	1.5	1.5
Stearic acid	1.0	1.0	1.0	1.0
Zinc oxide	3.0	3.0	3.0	3.0
DPG	–	–	1.5	1.5
CBS	2.5	2.5	2.5	2.5
Sulphur	1.7	1.7	1.7	1.7

The solution-styrene–butadiene polymer was S-SBR type Buna VSL 5025-0 with a vinyl content of 50 w.-% and a styrene content of 25 w.-%, EPDM was Keltan 512 and NR CV 50. As can be seen, the content of the fillers carbon black (CB) and silica was varied between 20 and 60 phr. Blends contained 60 phr CB.

The materials were compounded in an intermeshing mixer of industrial type (Werner & Pfleiderer GK 1,5 E). The vulcanization system was semi efficiently with sulphur and the accelerator CBS and the vulcanization took place in a heat press at up to 90 % of the torque maximum (t_{90}) of the vulcanization curve measured during vulcameter tests. All the mixtures contained the processing additives stearic acid and zinc oxide ZnO and the ageing protection substance IPPD. The silica mixtures (Table 1) contained additionally silane as a compatibilisator.

Table 6 Details on carbon black [7, 8]

Carbon Black Type	Specific Surface N_2SA (m^2/g)	Particle Diameter (nm)
N115	145	18
N234	126	20
N339	91	30
N550	42	56

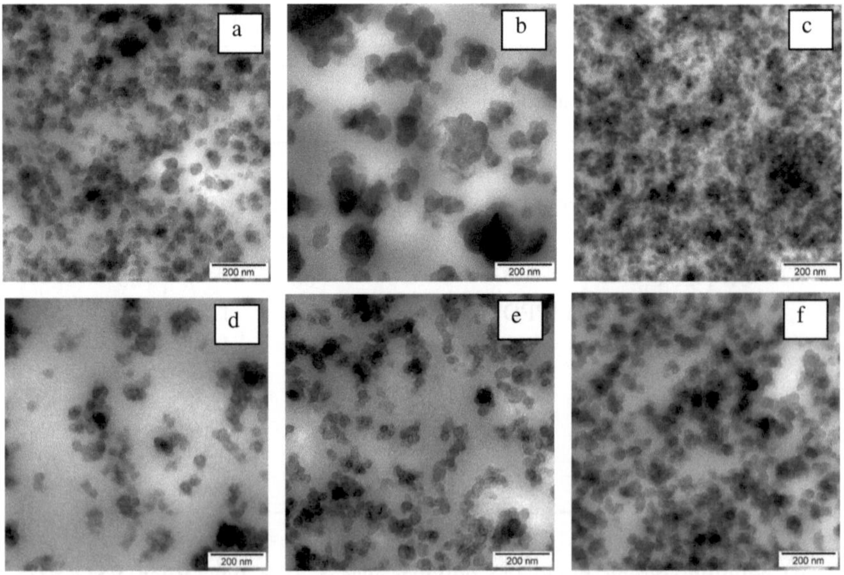

Fig. 3 TEM images of S-SBR vulcanizates with 60 phr N115 (a), 60 phr N550 (b), 60 phr Ultrasil GR7000® (c), 20 phr N 339 (d), 40 phr N339 (e) and 60 phr N339 (f) [9]

The fillers used were different types of CB: N115, N234, N339 and N550. The lower the number within the specification of the carbon black type the higher is the activity. This means, the primary particles as well as the aggregates are smaller, the specific surface characterized by nitrogen absorption (N_2SA) is larger and therefore, more active places on the surface of the filler particles are available for the contact with the polymer chains. It is assumed that with increasing activity the polymer–filler interaction increases too and thus, the macroscopic properties depend on the filler type. However, decreasing filler particle size and specific surface area can be also combined with increasing agglomeration of the filler aggregates also influencing the resulting properties. The following TEM images of some materials from series 1 (Fig. 3) and of NR vulcanizates from series 2 (Fig. 4) illustrate the differences of the various filler types, the increasing concentration and dispersion of the filler particles and aggregates of CB N339 and CB N234.

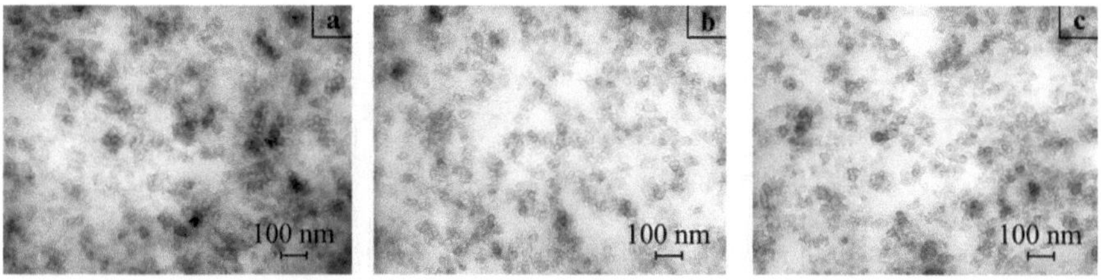

Fig. 4 TEM images of NR vulcanizates with 20 phr (a), 40 phr (b) and 60 phr (c) of CB N234 [10]

2.2 Mechanical Properties

The conventional tensile test is one of the most important methods of polymer testing and also for elastomeric materials it is one of the basic tests for the quantitative characterization of *strength* and *deformability*. The tests were performed according to ISO 37 [11] by using the universal testing machine ZWICK Z020 by the company Zwick GmbH, Ulm (Germany). The test speed was 100 mm/min and S2 specimens were punched from plates with a metal cutter. Per material/series 5 specimens were tested. From the recorded stress–strain diagrams (σ–ε diagram) various material parameters were determined: stress values at defined strain values (σ_{100}, σ_{200}, σ_{300}), the tensile stress at maximum (tensile strength) σ_{max} and the related strain value ε_{max}. If the maximum of the stress is also the point of fracture, σ_{max} is equal to the fracture stress σ_R and ε_{max} is equal to ε_R.

In Figs. 5–7, results of the tensile tests are shown. The tensile strength being equal to the tensile stress at fracture σ_R is increased with increasing filler content for all polymers and filler types investigated. However, the type of the polymer and also the type of the filler can influence the reinforcement level. For example, more active carbon black N234 results in higher σ_R values at comparable filler loadings (see Fig. 5 a and b). On the other hand, the materials with CB N339 have nearly the same strength level as the vulcanizates with precipitated silica. Furthermore, by using natural rubber, the strength of the unfilled vulcanizate is much higher than that of the SBR due to the strain crystallization of the NR leading to a self-reinforcement, especially at high strain. For all polymer–filler combinations, the tensile strength remains at nearly the same level when the filler content is increased from 40 to 50 phr. Using a blend instead of a homopolymer as the basis of the compound, the strength is reduced (SBR/EPDM with 60 phr CB) or slightly increased (unfilled SBR/NR and SBR/EPDM as well as SBR/NR with 60 phr CB).

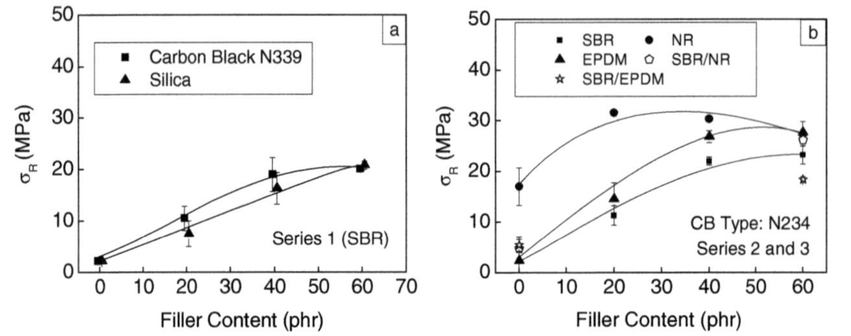

Fig. 5 Tensile (fracture) strength σ_R of the materials of series 1 basing on SBR (a) and of series 2 and 3 basing on different polymers and reinforced with carbon black N234 (b) as a function of the filler content

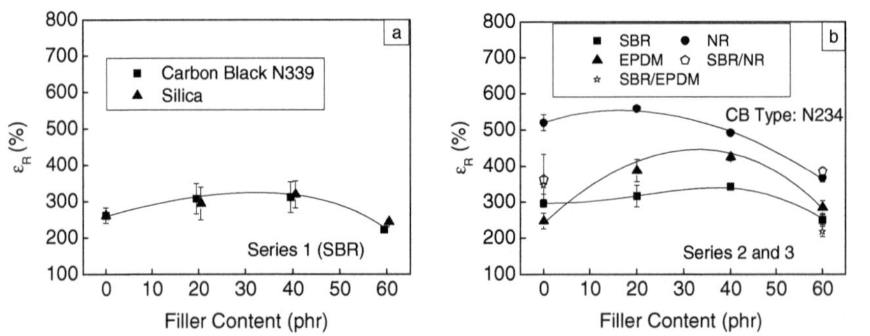

Fig. 6 Tensile (fracture) strain ε_R of the materials of series 1 basing on SBR (a) and of series 2 and 3 basing on different polymers and reinforced with carbon black N234 (b) as a function of the filler content

Also the deformability, expressed by the tensile strain at maximum stress/fracture ε_R, depends on the composition of the elastomers. Figs. 6 a and b show that except for EPDM vulcanizates the tensile strain at fracture remains relatively constant up to a filler content of 40 phr. With 60 phr of CB and silica, the filler concentration is so high leading e.g. to reduced flexibility of the polymer chains in between and therefore to a reduced macroscopic deformability. Locally, the tensile stress of the chains is reached at lower macroscopic deformation because of the small distances between the filler aggregates. Fig. 3 illustrates the density of the particles/agglomerates within the polymer matrix with 60 phr of the filler.

In Figs. 7 a and b, the results of the tensile test show the influence of the type of carbon black on the strength and deformability. The strength is increased with increasing specific surface and decreasing primary particle size of the carbon black (see Table 6), as expected. However, N115 leads to a slightly reduced strength what can be traced back e.g. to filler agglomerations which may act as defects. The tensile strain has a minimum value for the vulcanizate with CB N339.

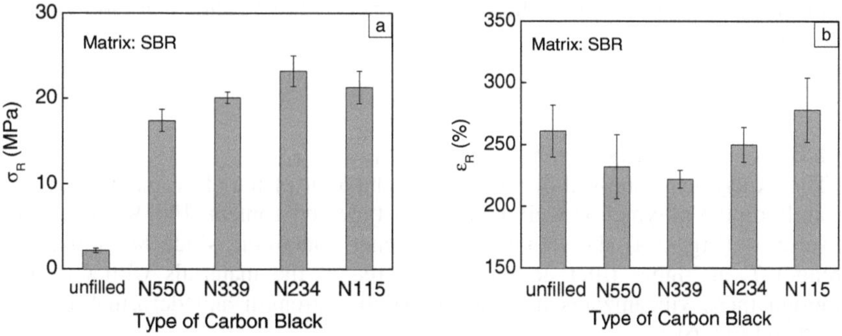

Fig. 7 Tensile (fracture) stress σ_R (a) and tensile strain ε_R (b) of the SBR vulcanizates reinforced with 60 phr of different types of carbon black

For all materials investigated, Shore A hardness experiments were performed according to [12] by using 6 mm thick specimens and the durometer Zwick 3150 by the company Zwick GmbH, Ulm (Germany). For each material/series, 10 single measurements were done. Results are shown in Fig. 8. As it is known, the hardness depends linearly on the filler content for all polymer–filler combinations. Similar to the strength, also the hardness is slightly changed for the polymer blends.

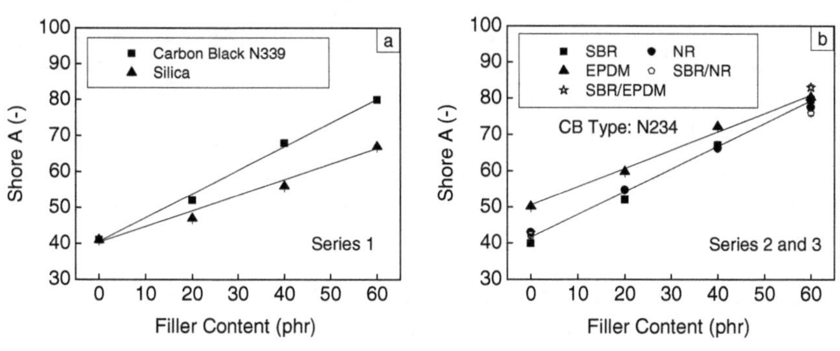

Fig. 8 Shore A hardness values of series 1 basing on SBR (a) and of series 2 and 3 basing on different polymers and reinforced with carbon black N234 (b) as a function of the filler content

However, while the strength was reduced for the filled SBR/EPDM blend, the hardness is a little bit higher compared to that of the homopolymer-based compounds. The Shore A value of the filled SBR/NR blend is decreased in comparison with the homopolymer.

By means of tear tests with trouser specimens (see Fig. 1), the behaviour of the materials in the presence of a sharp cut was characterized under quasi-static loading. The applied standard was ISO 34-1 [13]. The universal testing machine Zwicki by Zwick GmbH, Ulm (Germany) was used and a testing speed of 100 mm/min was realized. During the tests, the load–displacement diagrams were recorded, from which maximum load F_{max} or the median of load F_{med} was obtained. According to Eq. 7, the tear resistance T_s (also called tear strength) was calculated.

$$T_s = \frac{F}{B} \tag{7}$$

with B – specimen thickness; F – F_{max} or F_{med}.

The experiments showed different tear behaviour of the elastomers, depending strongly on the polymer as well as the filler type and content (Fig. 9). The results of series 1 (Fig. 9 a) show that the elastomer containing silica has higher tear strength T_s at higher filler contents compared to the materials with CB N339 although these vulcanizates have a comparable strength and deformability (see Fig. 5 and Fig. 6).

The NR compounds have the highest tear strength increasing with the filler content as it was expected, and the EPDM vulcanizates the lowest one. EPDM is known for a high notch sensitivity and this fact is reflected by the tear strength values, also of the blends containing EPDM as one matrix component.

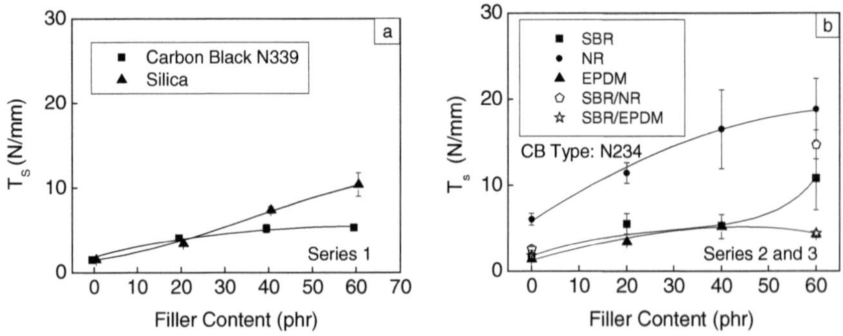

Fig. 9 Tear strength T_s of series 1 basing on SBR (a) and of series 2 and 3 basing on different polymers and reinforced with carbon black N234 (b) as a function of the filler content

The dynamic–mechanical (thermal) analysis (DM(T)A) can be used to quantify the *viscoelastic behaviour* of polymers. Here, the cyclic, sinusoidal loading of the specimens was a tension loading with a frequency of 1 Hz. The tests were performed as a temperature sweep with a temperature range between –100 °C and 100 °C. As a result, the storage modulus *E*', the loss modulus *E*'' and the mechanical loss factor tan δ were determined. Fig. 10 shows the dependence of the storage modulus of the materials of series 2 as an example, where the typical behaviour of an elastomeric material can be seen. Up to the glass transition area, the storage modulus has high values because the materials are in the energy elastic range. Here, the polymer chains have only a low mobility and the filler content or the polymer type have only a small influence on the height of E'.

The temperature of the glass transition area is defined by the polymer type and is nearly independent of the filler content (see Table 7). Remarkable differences of E' are visible as expected in the entropy elastic area above the glass transition. With increasing filler content, the storage modulus has a higher level and also a different (negative) slope when the temperature increases.

Table 7 Glass transition temperatures T_g of the materials of series 2 (type of CB: N234)

	T_g (°C)		T_g (°C)		T_g (°C)
NR	–51	SBR	–3.5	EPDM	–42
NR + 20 phr CB	–51	SBR + 20 phr CB	–3.8	EPDM + 20 phr CB	–42
NR + 40 phr CB	–51	SBR + 40 phr CB	–4.8	EPDM + 40 phr CB	–43
NR + 60 phr CB	–53	SBR + 60 phr CB	–6.2	EPDM + 60 phr CB	–43

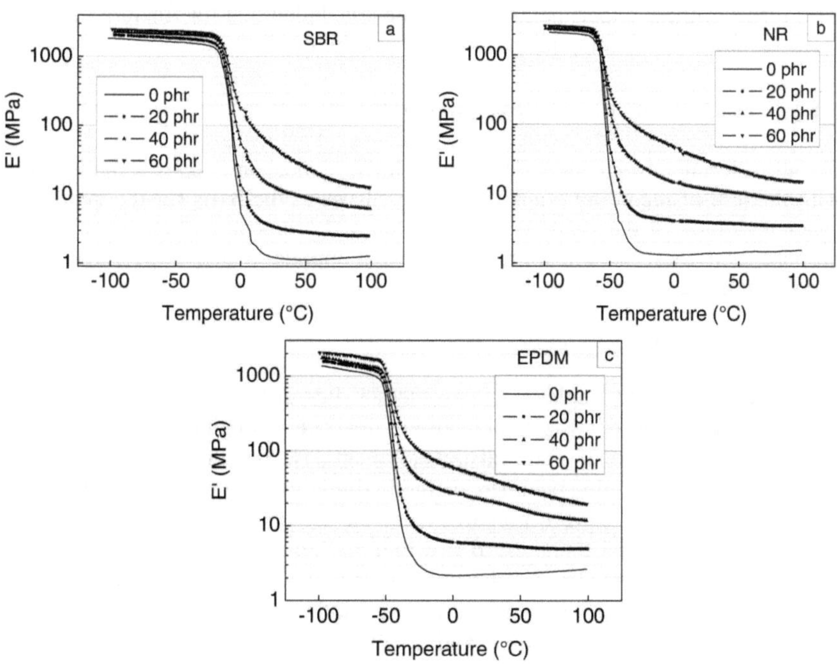

Fig. 10 Storage modulus E' of the materials of series 2 (S-SBR, NR and EPDM vulcanizates with different amounts of carbon black N234)

3 Fracture Mechanics Characterization

3.1 *Experimental*

3.1.1 Specimen Preparation

The specimens for the fracture mechanics experiments were prepared by cutting from vulcanized plates with various thicknesses. For this purpose, cutters with the related geometry were used. Details on the specimen geometry can be found within the following sections.

Generally, the notching of specimens for fracture mechanics investigations requires particular attention. At first, the notch radius should be as small as possible. This means, a sharp, thin metal blade must be used. Furthermore, the process of notching should be realized in this way that the notch is perpendicular to the specimen edge without any distortion. Our investigations regarding the influence of the notch insertion showed the influence on the stress field in front of the notch and therefore on the stress-intensity factor when the notch is distorted. When the crack is than initiated, the crack propagation mostly takes place perpendicularly to the direction of the external loading, but depending on the notch distortion, different crack profiles are obtained. Further details can be found in [14, 15] and in the chapter „Analysis of Dynamic Crack Propagation in Elastomers by Simultaneous Tensile- and Pure-Shear-Mode Testing“ of this book. As a result of these experiments, it could be concluded that for fracture mechanics investigations the notching procedure is of special importance.

3.1.2 Quasi-static Fracture Mechanics Tests

For all materials investigated, quasi-static fracture mechanics tests were performed with the aim of recording crack resistance curves as the basis for the quantitative characterization of the stable crack initiation and propagation behaviour. For this purpose, two quasi-static test procedures were applied: a single-specimen method and the multiple-specimen technique [9–20].

Single-Specimen Method (SSM)

The advantage of the single-specimen method lies in the reduced material insert and therefore reduced costs which allows for crack toughness characterization during an early stage of material development [1]. For the R-curve tests with application of the single-specimen method, single edge notched tension (SENT) specimens (see Fig. 1) were prepared with the dimensions length L = 100 mm, width W = 25 mm and thickness B = 6 mm and an a/W ratio of at least 0.2 was realized. Earlier results [16, 21] showed that specimens with a minimum thickness of 6 mm should be used for such tests to exclude or at least to reduce a plane-stress state within the specimen and to obtain geometry-independent fracture mechanics parameters. The notches were produced by using a thin metal blade.

For the tests, the universal testing machine Zwick Z020 by the company Zwick GmbH, Ulm (Germany) was used. The test speed was 10 mm/min and the temperature was 23 °C. For each material, three specimens were tested, whereupon only one specimen was regarded for recording a crack resistance curve. The initial gauge length l_0 was 40 mm. The specimens were loaded with a monotonically increasing load in tension. This means, the crack opening mode I was realized.

Before starting the test, the notch tip was prepared with TiO_2 powder to enable the examination of the crack initiation. During the test, the notch tip/opened crack was observed and pictures of it were taken at different stages of the deformation and crack process. After the crack initiation, the distance between the edges of the notch, becomes larger with the deformation of the specimen (see Fig. 11). According to [22], this distance was defined as crack opening l_R. Investigations indicated that a linear connection between the real crack size Δa and l_R exists (see Fig. 12), so that the latter can be used as x value of the R-curve.

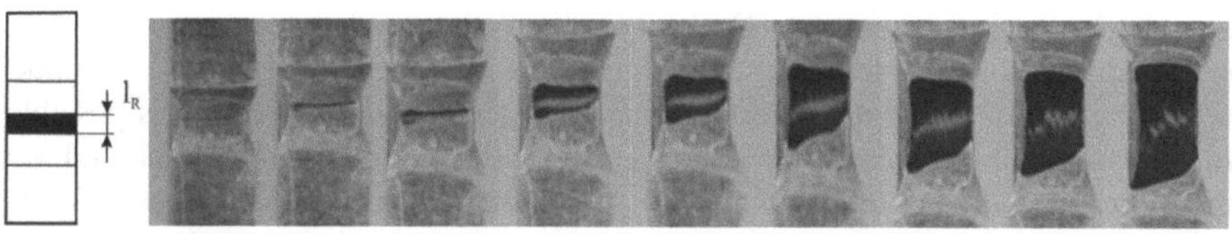

Fig. 11 Increasingly loaded specimen during a fracture mechanics test and definition of the damage parameter crack opening l_R [9]

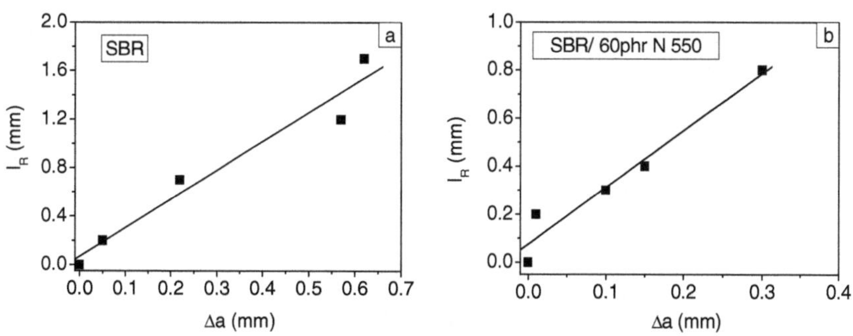

Fig. 12 Interrelationship of crack opening l_R from quasi-static single-specimen test and crack growth Δa from quasi-static multiple-specimen test [23]

The related y value of the R-curve J was obtained on the basis of the recorded load–displacement (F–l) diagram. The deformation energy U was determined by integrating the F–l diagram up to certain deformation values. The intervals must correspond to the times of taking the pictures of the crack opening. J values were calculated according to:

$$J = \frac{\eta U}{B(W-a)} \tag{8}$$

with η – geometry function

The J–l_R data pairs were plotted as a crack resistance (R-) curve and two fracture parameters were obtained as a result of the experiment. The first one was the physical crack initiation value J_i, which was determined through observation of the notch tip and therefore, a subjective influence cannot be excluded. The second was a crack propagation value T_J^* coming from the slope of the R-curve at the point of $l_R = 0.5$ mm, this means in the beginning of the crack propagation process.

Multiple-Specimen Method (MSM)

By applying the multiple-specimen method, a certain number of equal SENT specimens with the same dimensions as described for the single-specimen method were loaded in tension one time up to different load levels. In this way, different amounts of stable crack propagation resulted and the specimens were not loaded up to the total fracture. The experimental conditions were comparable to that of the single-specimen method. During loading, the load–extension diagram is recorded including loading and unloading cycle (Fig. 13) so that beside J values according to Eq. 8 also the determination of J_{diss} according to Eq. 9 was possible. In this way, the large amount of the elastic deformation stays disregarded and therefore, it was assumed that J_{diss} is a more precise parameter for the quantification of the crack resistance.

$$J_{diss} = \frac{\eta A_{diss}}{B(W-a)} \tag{9}$$

with A_{diss} – energy corresponding to the area between the loading and the unloading cycle

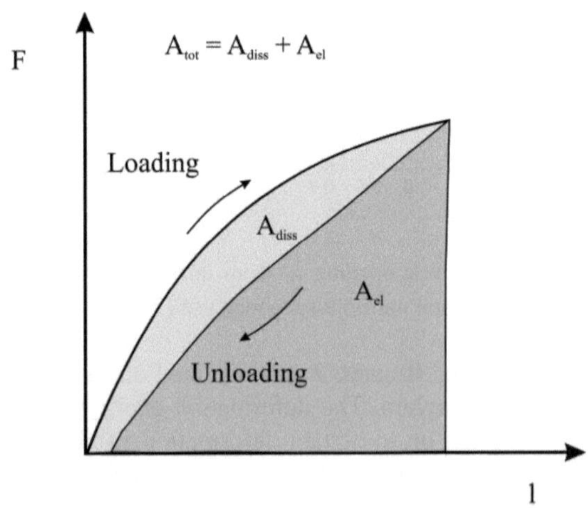

Fig. 13 Schematic load (F)–extension (l) diagram from a quasi-static fracture mechanics test (multiple-specimen method); A_{tot} – total deformation energy corresponds to the complete area under the loading curve, A_{el} – elastic deformation energy

After the test, the specimens were cut and the sizes of the stable crack growth Δa were determined by using a light microscope. So, pairs of J and Δa were available for the construction of a crack resistance curve. The analysis of these curves was similar to that of the single-specimen method: T_J^* values as a quantitative measure for the stable crack propagation resistance were determined at a crack growth $\Delta a = 0.1$ mm.

3.1.3 Instrumented Notched Tensile-Impact Test

Starting from the valid standard for conventional tensile-impact tests DIN EN ISO 8256 „Plastics – Determination of Tensile-impact Strength" [24], some instructions for performance of instrumented tensile-impact tests have been adopted. Finally, an own procedure for performance and evaluation of instrumented notched tensile-impact tests was developed [25]. According to this procedure, all materials were examined by using double edge notched specimens (DENT specimens) as shown in Fig. 14. Following dimensions of the DENT specimens were realized: length $L = 64$ mm, width $W = 10$ mm and initial crack length a should be 2 mm. The thickness of the specimens follows from the thickness of the plates, from which the specimens were punched. Because the notch radius must be very small, the notches were made by a thin metal blade.

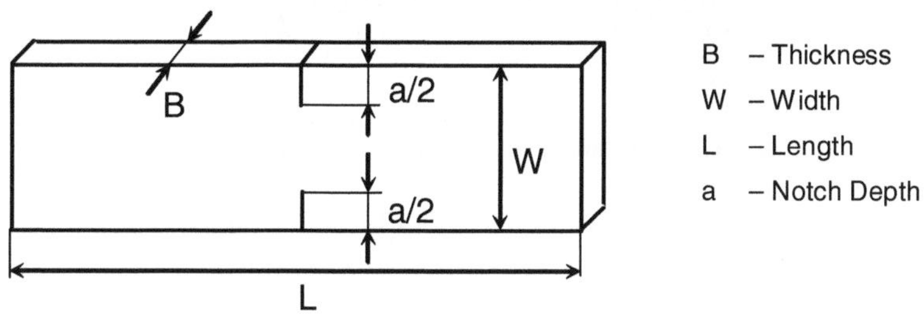

Fig. 14 DENT specimen for instrumented tensile-impact tests

The tests were performed with the instrumented pendulum device Resil Impactor Junior (Ceast, Italy) and pendulum hammer energy of 7.5 J or 15 J, depending on the energy uptake of the material. The test speed was 3.7 m/s resulting from the maximum falling angle of 160°. Additionally, for selected materials, the test speed was varied in the range between 1.3 m/s and 3.7 m/s.

Fig. 15 shows a schematic representation of the test set-up and a load–extension diagram of an elastomeric material. Additionally, in Fig. 16 one can see how a specimen is fixed between the crosshead and the fixed clamp containing the piezo load cell with a load working range up to 2.2 kN. For starting the test, the pendulum hammer was triggered from its 160° starting position, then described a downwards circle motion, takes with the unsecured crosshead together with the specimen from the lowest position (0° stage) and follows its circle motion upwards. In this way, the specimen is strained in its longitudinal direction. At the

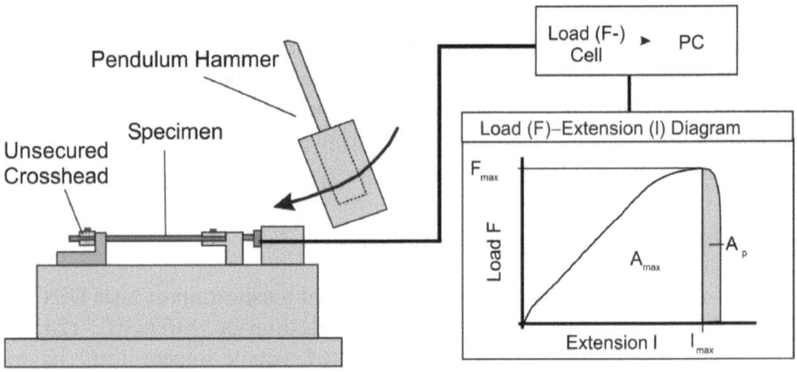

Fig. 15 Test set-up of the instrumented tensile-impact test and analysis of a load–extension diagram of an elastomeric material; F_{max} is the maximum load and l_{max} the related extension, A_{max} is the energy up to F_{max} and A_p is the crack propagation energy

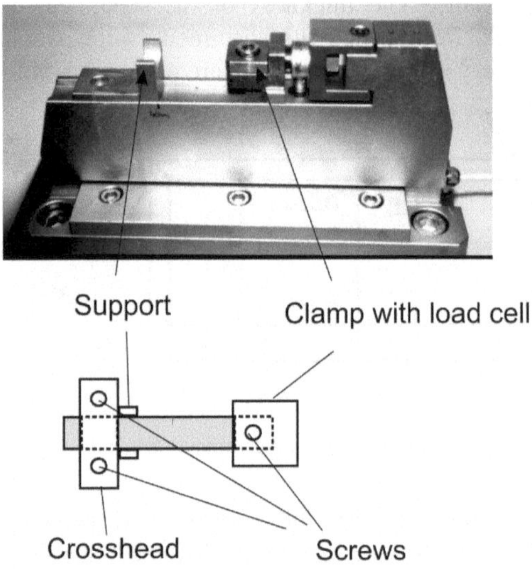

Fig. 16 Instrumented tensile-impact unit of the pendulum device and schematic representation of the specimen clamping

same time, a load–time diagram is recorded, which is transferred into a load–extension (F–l) diagram. For the examinations in dependence on the temperature, special specimen holders were used, which required bulges at the ends of the specimens.

From the analysis of the diagrams, various energy values can be obtained (shown schematically in Fig. 15) from which J_d values as a measure of the crack toughness related to the resistance against initiation and propagation of an unstable crack are calculated according to Eq. 10. These J_d values are related to the resistance of the materials against unstable crack propagation. The index "d" means "dynamic" and was introduced to make possible a clear distinction between J values from impact-like ("dynamic") and quasi-static tests.

$$J_d = \frac{\eta A_{max}}{B(W-a)} \tag{10}$$

with the geometrical function η [3]:

$$\eta = -0.06 + 5.99\left(\frac{a}{W}\right) - 7.42\left(\frac{a}{W}\right)^2 + 3.29\left(\frac{a}{W}\right)^3$$

with A_{max} – energy up to F_{max}

Because the temperature-dependent fracture behaviour is of large practical interest, instrumented notched tensile-impact tests were performed at different temperatures. However, an internal tempering is not possible for such tests. For this reason, an external tempering was realized with special mountings and specimens with bulges at both ends (see Fig. 17). So, the test temperature could be varied between –70 and 80 °C. Before the test, the specimens were conditioned for at least 30 minutes in a temperature chamber belonging to the falling dart tester Fractovis by the company Ceast (Italy), located near the pendulum device. The time spent for positioning the specimen in the tensile-impact unit of the pendulum device is short enough to prevent remarkable changes in the temperature of the specimen. Moreover, the time for positioning the specimen in the impact machine is reduced by the use of the specimen mounting shown in Fig. 17.

Fig. 17 Mountings (with specimen) for instrumented tensile-impact tests in dependence on temperature as well as a specimen with bulges

3.2 *Results and Discussion*

3.2.1 Structural Influence on the Fracture Behaviour

The fracture behaviour of the materials described in section 2.1 was characterized by using quasi-static and impact-like tests. In the following graphs, selected results of quasi-static tests are shown. At first, Fig. 18 contains crack resistance curves of the materials of series 1 with different contents of carbon black type N339 as well as with different carbon black types. The addition of more filler leads to an increasing level of J values (Fig. 18 a). This was expected because of the reinforcement effect, which can be seen also from the mechanical properties discussed in section 2.2.

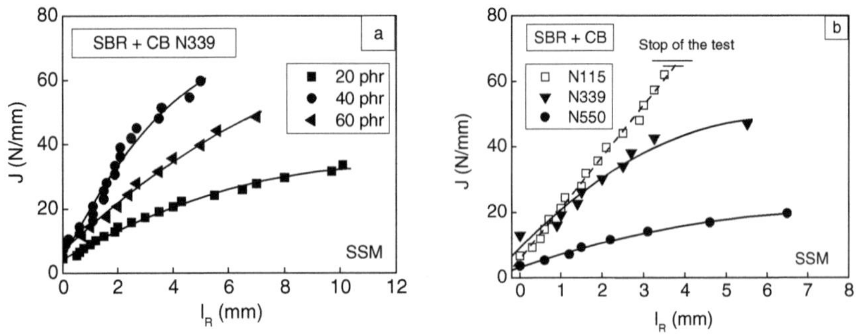

Fig. 18 Crack resistance curves of SBR vulcanizates with different amounts of carbon black N339 (a) and with 60 phr of different carbon black types (b) determined by using fracture mechanics tests in single-specimen method [9, 18, 26]

The qualitative comparison of the stable crack growth behaviour of SBR vulcanizates with different carbon blacks (Fig. 18 b) shows a strong influence of carbon-black activity, particle size and structure. The lower the activity, the lower is the level of the J values as well as the slope of the crack resistance curves. The elastomer with carbon black N115 has a very high resistance against stable crack propagation. This led to the deviation of the crack in the direction of the load requiring a stop of the test.

In Fig. 19, the fracture mechanics parameters J_i and T_J^* as quantitative measures for the crack initiation and crack propagation resistance are shown for the materials of series 1.

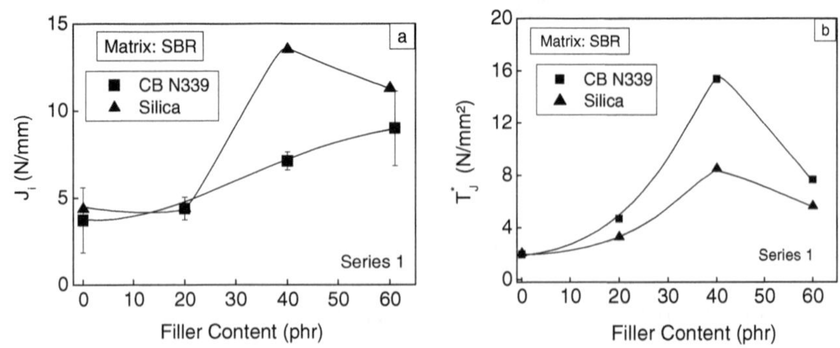

Fig. 19 Crack initiation and propagation values J_i and T_J^* of S-SBR vulcanizates with different contents of carbon black (a) and crack propagation values T_J^* of S-SBR vulcanizates with different contents of silica determined at different sizes of crack opening (b); determined in the quasi-static fracture mechanics test [18]

Generally, the crack initiation value J_i (Fig. 19 a) increases with the carbon-black as well as with the silica content. Similar to the crack propagation value T_J^* (Fig. 19 b), J_i of the silica-filled materials has a maximum at a filler content of 40 phr. This maximum of the fracture resistance was found and discussed in an earlier study for SBR/N330 compounds [21] and was now confirmed for other polymer/filler combination. One important reason for the reduction of the fracture resistance above a filler content of 40 phr in such non-plasticized compounds is the lowering of the deformability and flexibility of the polymer chains due to the large amount of filler. Fig. 20 a shows the network density determined from swelling experiments. It can be seen a strong increase in network density connected with smaller molecular weight between two net nodes. Therefore, mean macroscopic deformation can result in large strains of the polymer chains on the molecular length scale, leading to earlier fracture compared to materials with higher chain flexibility. A filler agglomeration at higher filler contents acting as crack starting sites can be excluded as a reason for the reduced fracture resistance with 60 phr filler. This is illustrated by Fig. 20 b, where the macro dispersion index is shown. The higher this value, the better is the macro filler dispersion.

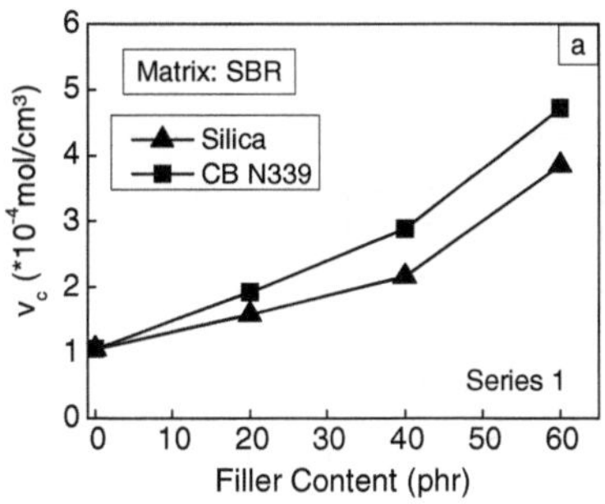

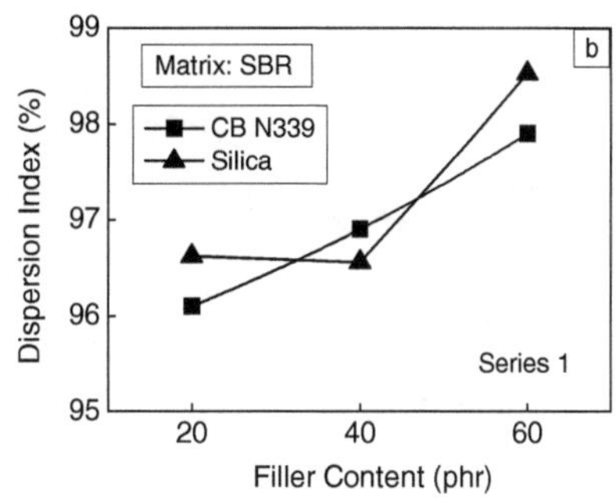

Fig. 20 Network density (a) and dispersion index (b) [26]

The crack initiation and propagation values in Table 8 illustrate the influence of the carbon-black type. The increased specific surface area and the smaller primary particles are connected with higher crack resistance. Especially with N115, the most active carbon black, the parameters reflect a high fracture resistance owing to a strong polymer–filler interaction.

Table 8 Crack initiation and propagation values J_i and T_J^* of S-SBR vulcanizates with 60 phr of different carbon blacks [27]

	J_i (N/mm)	T_J^* (N/mm^2)
SBR + CB N115	16.9	16.0
SBR + CB N339	9.0	11.6
SBR + CB N550	5.7	4.0

Further crack resistance curves from single-specimen fracture mechanics tests are shown in Fig. 21 for SBR and NR materials of series 2. Firstly, again a dependence on the filler content, in this case CB N234, is visible. Without filler, a very flat R-curve resulted, indicating a low crack propagation resistance. Comparing the crack resistance curves of SBR and NR vulcanizates, one can see a very large difference. The J level for NR with filler is much higher. This is attributed to the strain-induced crystallization (SIC) increasingly appearing at higher strains. According to [28], with increasing strain, the number of short, stretched polymer chains acting as nucleation sites becomes larger. The distance of the chains decreases and at a certain point, the aligned chains form crystallites. These crystallites contribute to the mechanical response like filler particles or network nodes and a self-reinforcement is obtained which is connected with an increase in strength, wear or tear strength. As it can be seen from Fig. 21 b, also the crack propagation resistance is enhanced. Further experimental results on the SIC effect can be found in chapter "Online characterization of rubber composites during deformation and fracture".

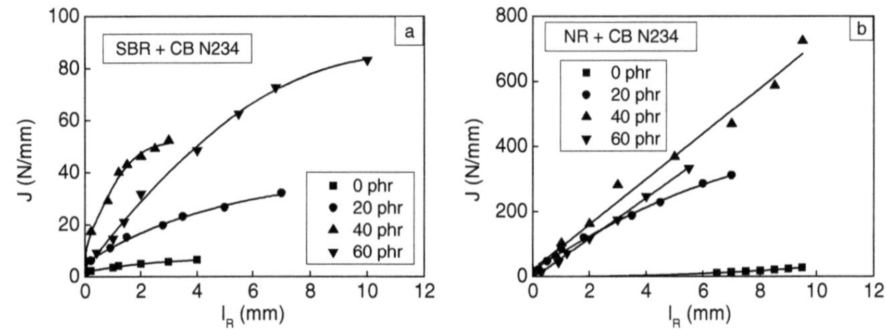

Fig. 21 Crack resistance curves from the fracture mechanics test with single-specimen method for SBR (a) and NR elastomers (b) with increasing filler content [29]

One aim of the project work was to proof the applicability of the single-specimen method for a reliable and reproducible, quantitative description of the fracture behaviour also for other elastomers. For SBR, this was done in the past and was published in [30]. In Fig. 22, crack resistance curves from the single-specimen and from the multiple-specimen method for EPDM with different amounts of CB N234 are shown.

The direct comparison shows the influence of the filler addition again and similar crack propagation behaviour by using the different methods. The same qualitative comparability was found for NR and SBR materials with CB N234. Therefore, it can be stated that both tests can be used for material ranking regarding stable crack resistance. For this purpose, the single-specimen test is to prefer because it needs only one specimen and therefore, the expense in time and

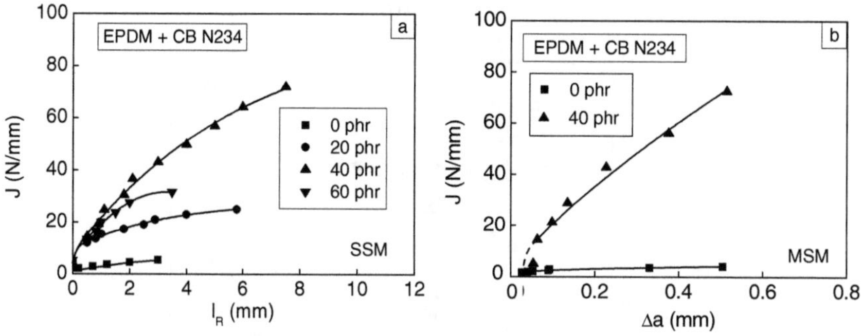

Fig. 22 Crack resistance curves from single (a) and multiple-specimen tests (b) for EPDM with different contents of carbon black N234

material is much lower. However, under some points of view, the MSM has to be preferred.

The damage parameter stable crack growth size Δa has the advantage of being uninfluenced by large elastic deformation as it is the case for the crack opening l_R. Furthermore, the application of the multiple-specimen method allows for the splitting of the deformation energy by registration the complete loading and unloading cycle (see Fig. 13). The deformation during such a test can be connected with extensive elastic deformation within each unit of volume, also far away from the crack tip, especially at higher strain ratios. This means, total deformation energy A_{tot}, determined from the area under the load–extension diagram, contains a large part A_{el} because of elastic deformation (Fig. 13). Furthermore, in contrast to other materials, especially in reinforced elastomers, energy dissipation takes place to a higher extent due e.g. to friction and slide processes of polymer chains at the strongly structured filler surface or breakup of filler–filler bonds. It is assumed that most of the dissipation processes concentrate in a limited zone in front of the crack tip where large stresses and strains appear. Moreover, when crack propagation occurs, the total deformation energy from a load–extension diagram includes additionally a certain amount A_{crack}, which was necessary to produce a new surface. The energy denoted as A_{diss} in Fig. 13 contains both: A_{crack} as well as dissipated energy. However, this splitting of the A_{el} and A_{diss} is only possible when loading and unloading cycle are measured during a multiple-specimen test, which should be regarded as a benefit of this methodology. Fig. 23 shows crack resistance curves for the SBR material with 60 phr N550 (series 1) recorded by using the SSM and the MSM. It can be seen that there are no principal differences between the crack resistance curves from the multiple-specimen test by using A_{tot} for calculation of J and that of the single-specimen method. However, the R-curves J–Δa and J_{diss}–Δa from the multiple-specimen method differ owing to the fact that the elastic deformation energy is disregarded for J_{diss}.

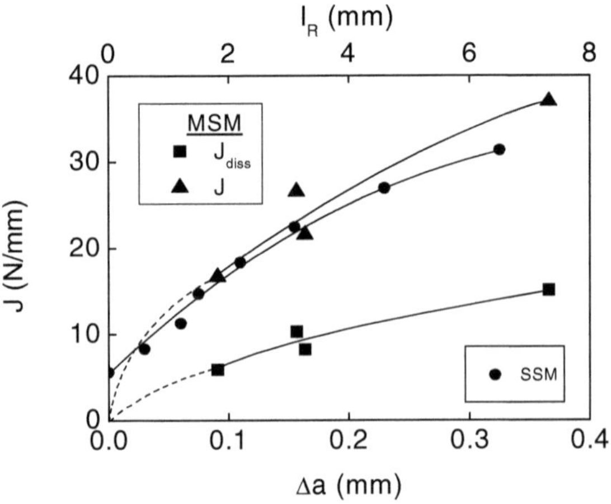

Fig. 23 Crack resistance curves from single- and multiple-specimen tests for S-SBR with 50 phr N550

Fig. 24 shows J_{diss}–Δa curves from fracture mechanics tests with MSM of all materials of series 2. Again, the influence of the polymer type and the filler content can be seen. Without filler or with low filler content of 20 phr, the materials have a relatively low resistance against stable crack propagation, and for

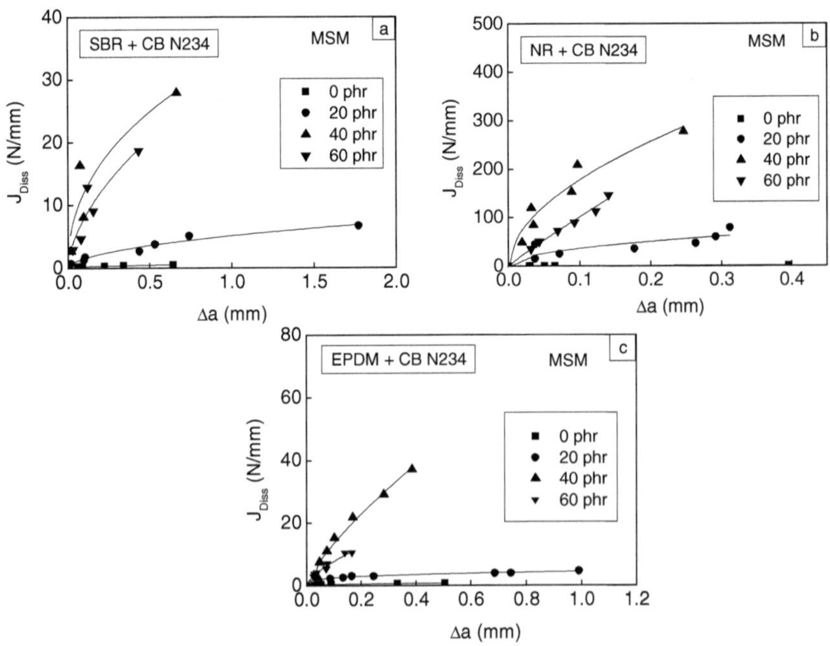

Fig. 24 Crack resistance curves from multiple-specimen tests for S-SBR (a), NR (b) and EPDM (c) with different amounts of carbon black N234

every polymer type, the compound with 40 phr carbon black has the highest J level. The belonging crack propagation values determined from the R-curves resulting from the tests with SSM and MSM are represented in Fig. 25 and Fig. 26, respectively.

Independent of the method, the NR vulcanizates have the largest resistance against crack propagation which is mainly attributed to the strain-induced crystallization as discussed above. Furthermore, the crack resistance maximum at 40 phr was found for all polymers. Reasons for this behaviour were discussed in [21] and above. Under the practical point of view such elastomers should be designed in this way that a certain deformability/flexibility is given. So, also on the molecular length scale, the chains are able to compensate a higher external loading without a fracture by elastic deformation. Thus, because the fracture of a polymer begins at the molecular level, also the macroscopic fracture appears at higher external loadings.

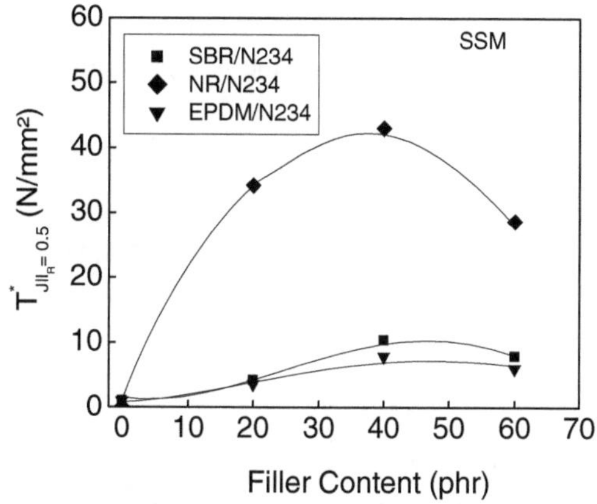

Fig. 25 Crack propagation values T_J^* determined from crack resistance curves (SSM) at the point $l_R = 0.5$ mm for the materials of series 2

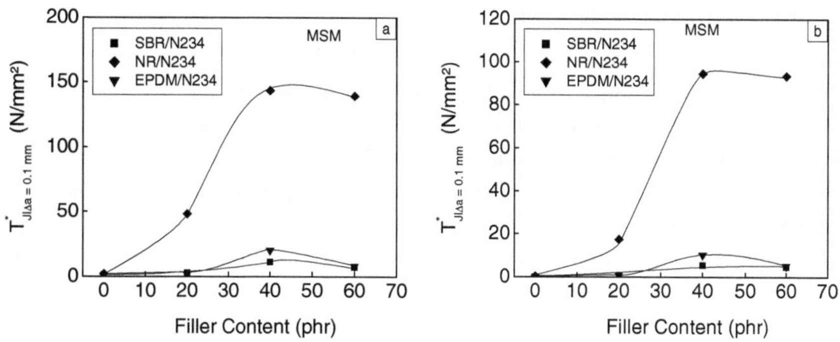

Fig. 26 Crack propagation values T_J^* determined from crack resistance curves (MSM) J–Δa (a) and J_{diss}–Δa (b) at the point $\Delta a = 0.1$ mm for the materials of series 2

Further subject of our investigations were some elastomers basing on polymer blends. Beside the mechanical basic characterization as it was shown in Section 2.2, also fracture mechanics tests were performed. In Fig. 27, some examples of crack resistance curves of these materials are shown. The tests were performed with application of the SSM. However, to get information on the reproducibility of such tests, for each material 2 or 3 specimens were used. It can be seen that a relatively good agreement of the single tests was obtained. From the experience and from these results, it can be stated that this experimental method is a valuable possibility to get fast and reliable data for the characterization of the stable fracture behaviour.

For the SBR/NR blend (Fig. 27 b), a typical stick-slip fracture behaviour was observed characterized by jumps in the R-curve. At these points, the crack opening l_R is suddenly enlarged while nearly no energy is consumed by the specimen. Such behaviour was interpreted as a typical sign of SIC and in the case of NR it is connected with a high crack resistance. Surprisingly, it was only observed for the blend, but not for the homopolymer-based NR vulcanizates (Fig. 21).

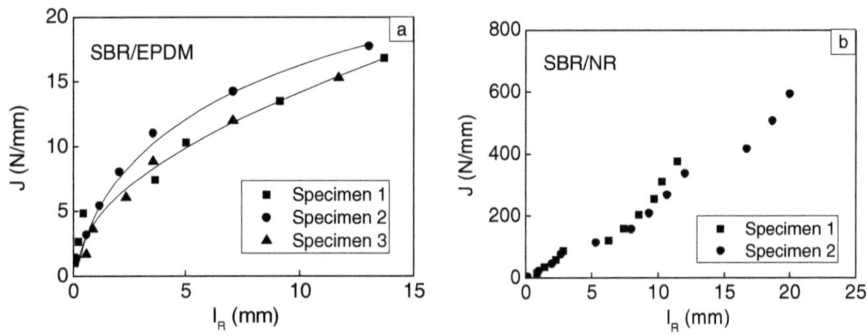

Fig. 27 Crack resistance curves from the fracture mechanics test with SSM for two elastomer blends with 60 phr N234

The crack initiation and crack propagation values J_i and T_J^* of the blend materials are shown in Table 9. Similar to the vulcanizates basing on single polymers, the addition of 60 phr of the active carbon black N234 leads to an increase in the crack resistance. However, the comparison of the initiation and propagation values with these of the comparable homo-polymer elastomers reveals mostly a decrease. It can be assumed a non-optimized blend composition and/or a disadvantageous filler distribution as reasons.

Table 9 Crack initiation and propagation values J_i and T_J^* of the investigated vulcanizates basing on polymer blends

Elastomers	J_i (N/mm)	T_J^* (N/mm²)
SBR/EPDM	1.3	0.4
SBR/EPDM + 60 phr	2.9	2.9
SBR/NR	1.6	0.8
SBR/NR + 60 phr	5.0	28.0

The following graphs contain results of the instrumented notched tensile-impact test illustrating the structural influence, in this case again the filler content and the filler and polymer type. It can be seen from Fig. 28 a, the crack toughness values J_d of the SBR vulcanizates of series 1 increase with increasing content of CB N339 and silica, this means the materials resistance against unstable crack propagation becomes larger with the filler addition. However, the course of the tear strength from quasi-static tear test – in practice often a measure for the cut growth resistance – shows a remarkable difference for the two fillers. Adding larger contents of silica leads to much higher tear strength than carbon black (see Fig. 9).

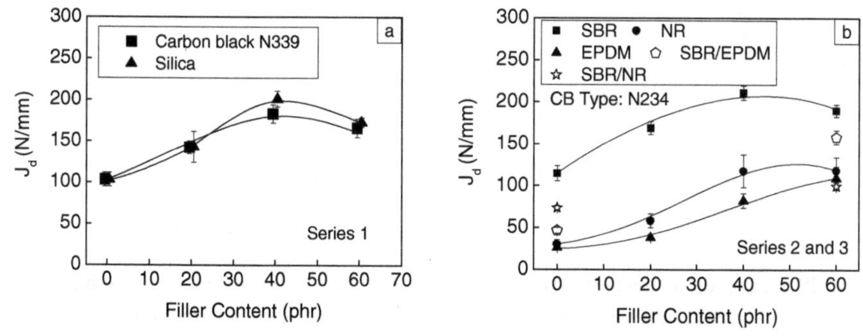

Fig. 28 Crack toughness values J_d of series 1 basing on SBR (a) [18] and of series 2 and 3 basing on different polymers and reinforced with carbon black N234 (b) as a function of the filler content

Thus, an important conclusion of this result for industrial practice is that tear strength cannot replace the application of test methods for the characterization of crack toughness, because T_s is not a toughness parameter.

For the materials of series 2 and 3 (Fig. 28 b), also an increase in crack toughness with the filler content was found. Similar to the results of the quasi-static fracture mechanics test, also a maximum of J_d was observed at 40 phr filler content, apart from the EPDM elastomer. From the measuring values of the carbon-black filled samples F_{max} and l_{max} it could be derived that there is a change in the fracture behaviour [18]. While up to 40 phr, the fracture is load-controlled, at the highest filler content of 60 phr, the deformation ability is more important. Latter decreases linearly with increasing filler content as can be seen from course of l_{max}, in contrast to the linearly increasing maximum load F_{max}. Similar results were found during earlier examinations of carbon-black reinforced emulsion SBR vulcanizates (see [30]), while silica-filled natural rubber compounds show a continuous increase of crack toughness up to filler contents of 60 or 70 phr [16]. This means, the fracture resistance is strongly influenced by the filler as well as the polymeric matrix. Fig. 29 a shows selected load–extension diagrams of blend materials as examples. It can be seen that the addition of the filler increases the load and reduces the extension. This is reflected by the measuring values plotted in Fig. 29 b. Furthermore, the comparison of the results of instrumented notched-tensile impact tests with that of the quasi-static fracture mechanics tests indicates a partly strong effect of viscoelasticity since the test speeds are very different. Additionally, the comparably low level of J_d of the NR rubber vulcanizates denotes the absence of the strain-induced crystallization in the case of impact-like loading conditions as a reinforcing material effect. For this reason, investigations of the influence of test speed and test temperature were performed. Results are presented within the following section.

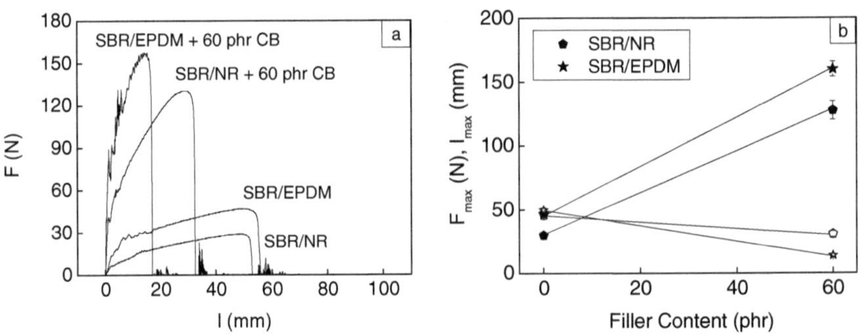

Fig. 29 Examples of load–extension diagrams (*F–l* diagrams) of vulcanizates basing on polymer blends (a) and belonging maximum load F_{max} (filled symbols) and extension at maximum load l_{max} (open symbols) (b)

3.2.2 Influence of Test Temperature and Test Speed on the Fracture Behaviour

Polymers are viscoelastic materials. For this reason, the mechanical and fracture mechanical behaviour is strongly dependent on the temperature and the rate of loading. Generally, for elastomers, experimental investigations of the fracture behaviour under impact-like loading depending on the test temperature are unknown. Seldom, the influence of the test speed in the range of quasi-static loading was investigated and discussed [31]. However, also for thermoplastics and duromers, the influence of test speed on fracture properties has been reported within literature relatively seldom. One reason may be experimental difficulties with the realization of test speed over a wide range. Often, more than one experimental method must be used. On the basis of linear-elastic fracture mechanics, investigations with various thermoplastic materials (POM, PEEK, PE-C modified PVC, pure and impact-modified PMMA, PP, PP/EPR, PEI, PBT, PEK-C, PA6.6/GF and PBI) were performed [32]. Usually, decreasing fracture toughness with increasing test speed was observed. Additionally, the application of concepts of elastic–plastic fracture mechanics for PP, PTFE, PC/ABS [33] delivered decreasing J values as measure for the resistance against unstable and stable crack propagation, respectively, with increasing test speed.

Summarizing, it can be stated that for thermoplastics and duromers the crack toughness parameters stress intensity factor K, energy release rate G and J-integral decrease with increasing test speed, while for elastomers a contrary behaviour was found [31, 34]. In Fig. 30, results of instrumented notched tensile-impact tests for 4 different filler-reinforced materials are shown illustrating the increase in fracture resistance.

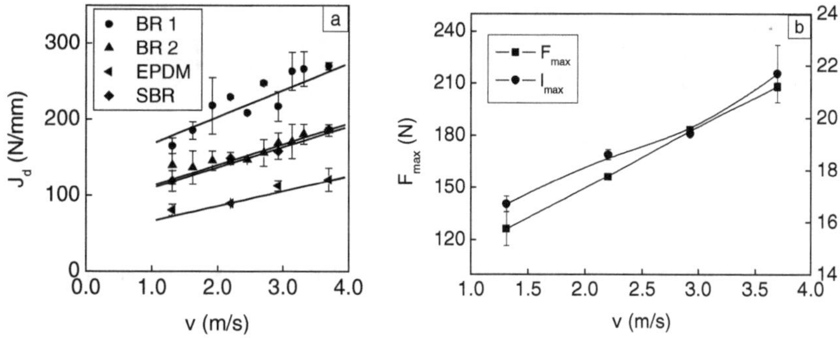

Fig. 30 Crack toughness J_d from instrumented tensile-impact tests (test temperature 23 °C) for two different filler reinforced polybutadiene rubber vulcanizates (BR), for EPDM and for S-SBR with 60 phr carbon black N550 [34] (a) as well as the measuring values F_{max} and l_{max} for SBR with 60 phr N550 (b) as a function of the test speed

For all tested materials an increase of the crack resistance was found under impact-like loading conditions. This result points out that concepts usually applied for explanation of toughness increase for polymers or metals do not work for elastomers in this form. Both, load and deformation component increase with increasing test speed (see Fig. 30 b). In literature [35], an increase of tearing energy T was described, but not discussed for a carbon-black reinforced natural rubber vulcanizate with increasing test speed in the quasi-static range. A possible explanation for this exceptional behaviour may be an additional energy dissipation mechanism working only at higher test speeds. Furthermore, the stress singularity at the crack tip may be reduced at lower test speeds due to strong blunting of the notch tip as it is known from quasi-static tests. Both effects have not been taken into account for fracture mechanical considerations yet, but investigations of *Marder* et al. [36] showed that they have to be regarded for quantitative analysis of fracture behaviour of elastomers. Unstable cracks may propagate within an elastomeric body with appearance of a diminished stress singularity. Restrictively, it is to note here, that the experimental procedure in [36] is not comparable to usual fracture mechanical tests.

Compared to the influence of the test speed, that of the testing temperature on the fracture behaviour is larger. In Fig. 31, crack toughness values J_d of SBR and NR vulcanizates determined by using the instrumented notched tensile-impact test are plotted as a function of the test temperature. The test speed was 3.7 m/s. Both the temperature and the material composition influence the fracture behaviour strongly. For the NR compounds with 40 and 60 phr CB, after a constant level of J_d up to 0 °C a decrease in J_d with rising temperature was observed. Unfilled NR and the compound with 20 phr CB show continuously decreasing J_d values.

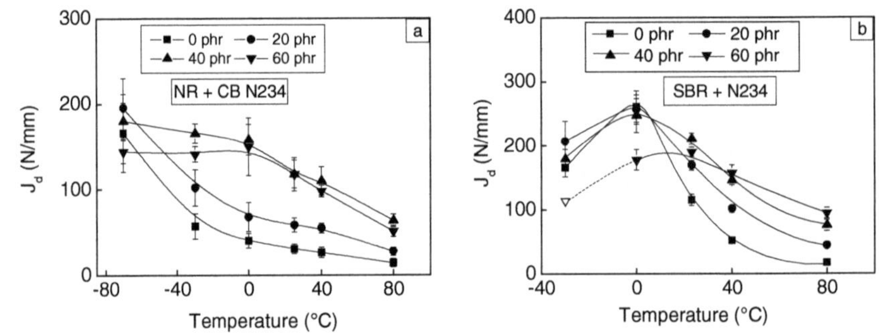

Fig. 31 Crack toughness J_d of carbon-black reinforced NR (a) and SBR (b) vulcanizates as a function of the test temperature [29, 37]

For SBR vulcanizates, at the lowest temperature of –70 °C, no reliable and assessable load–extension diagram could be obtained due to the very brittle deformation behaviour. Therefore, large oscillations in the signal appear and the loading curves cannot be correctly analysed. The fracture behaviour becomes more brittle with the increase in CB content. That is the reason for which 60 phr CB filled SBR cannot be correctly analysed already at –30 °C: only one valid value could be determined, which is marked by an open symbol in Fig. 31 b. Generally, in contrast to the NR compounds, the crack toughness of the SBR materials has a maximum at 0 °C, except from the vulcanizate with 60 phr CB. To explain this different behaviour it is useful to discuss also the measuring values maximum load F_{max} and the maximum extension l_{max}. Since J_d is an energy-determined fracture parameter, its trend is influenced by two contributions, that of F_{max} and of l_{max}. In Fig. 32 and Fig. 33, these values are plotted in dependence on the temperature for the SBR and NR materials. Results concerning EPDM compounds are not shown in this publication, because they are similar in trend (not in absolute values, of course) to those found for NR vulcanizates: F_{max}

decreases with temperature and l_{max} shows a maximum, whose position is once again dependent on CB content, even if it occurs at higher temperatures with respect to EPDM compounds.

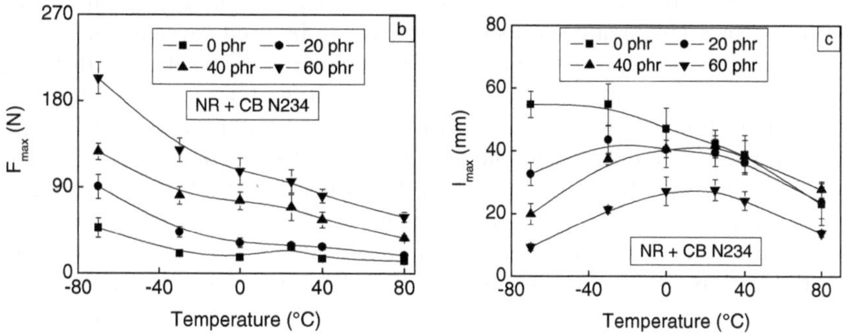

Fig. 32 Measuring values F_{max} (a) and l_{max} (b) for SBR reinforced with CB N234 as a function of the test temperature

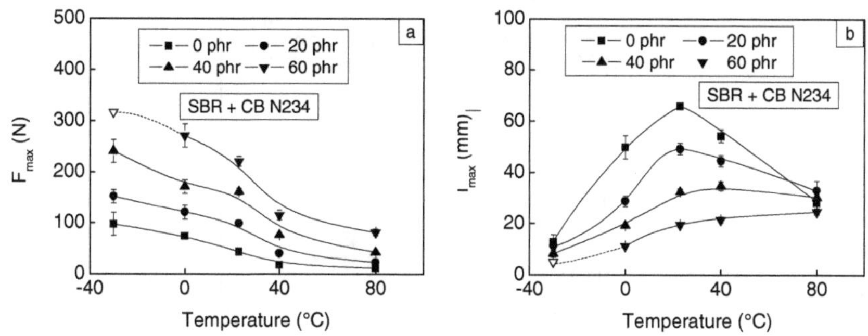

Fig. 33 Measuring values F_{max} (a) and l_{max} (b) for SBR reinforced with CB N234 as a function of the test temperature

As expected, the load is reduced with increasing temperature for all materials (Fig. 32 a and Fig. 33 a). However, the deformability of the vulcanizates expressed by l_{max} (Fig. 32 b and Fig. 33 b) shows a non-monotonous trend, with a maximum. Its position is influenced by the polymer type and the content of CB. Increasing filler content is connected with a reduction of the maximum extension, especially for the SBR elastomers, and the temperature of the maximum is shifted to higher temperatures. For SBR, the maximum of l_{max} appears at the comparably highest temperatures. This could be assumed to be influenced by the glass transition temperature T_g. From the maximum of the tan δ, determined by DMA tests, the T_g values of the materials were obtained (see Table 7). It can be assumed that the glass transition area of the SBR materials is in the range of the maximum of J_d and l_{max} (Fig. 31; Fig. 33). Similar behaviour was found for duromers, which are also crosslinked materials, and is explained by the Fluctuation–Dissipation

Theorem [38]. This theorem states that a correlation exists between the increase of the molecular mobility e.g. in the range of the glass transition and macroscopic energy dissipation during loading independently of the amplitude of the external load. This was proven for a number of thermoplastic materials [39, 40].

For fast determination of application limits, mapping of fracture mechanical parameters in a temperature–velocity diagram can be used. Because of the high experimental expense, often the time–temperature superposition principle is applied. The underlying time–velocity superposition laws are for elastomers mostly of WLF type. Unlimited validity of these laws also for high speeds can be assumed when all failure points plotted for different temperatures and velocities are in one line. Fig. 1 shows clearly that this is not the case for the investigated carbon-black reinforced EPDM and S-SBR materials. While the failure points for both materials decrease with the temperature, a contrary functionality is observed with the test speed. This means, the validity of the time–velocity superposition laws must be checked. A similar result was found by *Hoo Fatt* et. al [41] who used unnotched tensile specimens of SBR.

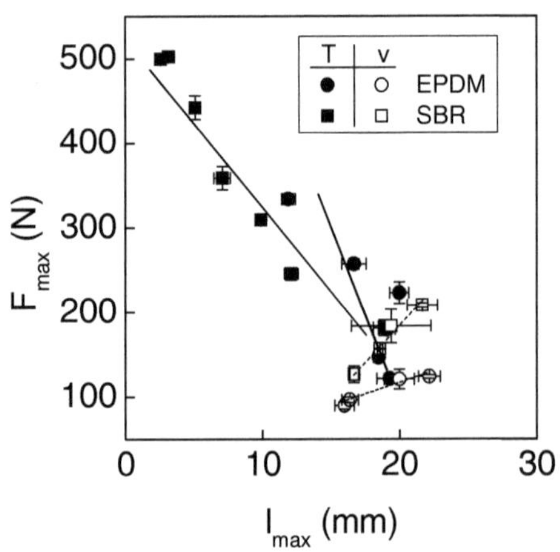

Fig. 34 Failure points of load–extension diagrams from instrumented notched tensile-impact tests for carbon-black filled S-SBR and EPDM vulcanizates; CB type N 550 [42]

3.2.3 Fracture Surface Characteristics

Former tests [19] led to the conclusion that a connection exists between fracture surface and crack toughness characteristics of elastomers. In Fig. 35, fracture surfaces from SBR specimens from the quasi-static fracture mechanics tests with single-specimen method are shown. On the left side, the notch can be seen and the crack propagation took place from left to right. As one can see, the content of the filler – in this case carbon black of the type N234 – determines the roughness of the fracture surface. Macroscopically, an increasing CB content leads to a smoothening of the fracture surface. This effect was quantitatively characterized for SBR with different amounts of CB N339. Fig. 36 summarizes results of different surface topography measurements of carbon-black filled SBR vulcanizates [18].

As one can see, the roughness depends strongly on the length scale. The mean macro roughness determined by tactile stylus method and MicroGlider® decreases with increasing carbon-black content. At the same time, the micro roughness from atomic force microscopy (AFM) measurements increases and shows a maximum at a filler content of 40 phr. This is in agreement with the results of the fracture mechanics experiments where for parameters describing the crack propagation, maximum values were found (Fig. 36 b). However, the tests showed that there is a need for systematic investigations with the aim of fixing the best methods as well as the optimal experimental parameters. Therefore, further investigations by AFM including a statistical assessment of the fracture surfaces were performed. Firstly, an important task was the experimental optimization of AFM measurements by systematic investigations. Here, the influence of scan size, scan frequency, amplitude and type of cantilever was investigated [43, 44]. As a result of these experiments optimized AFM parameter sets were defined for reliable and reproducible as well as economic determination of mean roughness values for filler-reinforced elastomers.

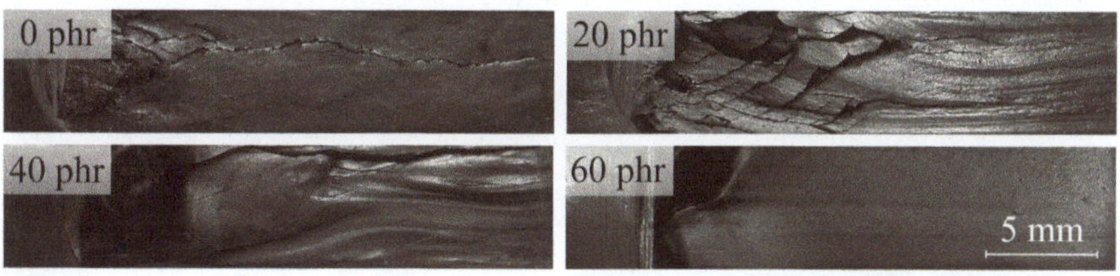

Fig. 35 Fracture surfaces of specimens from the quasi-static fracture mechanics tests with SSM (crack propagation direction from left to right); material SBR with CB N234 [10]

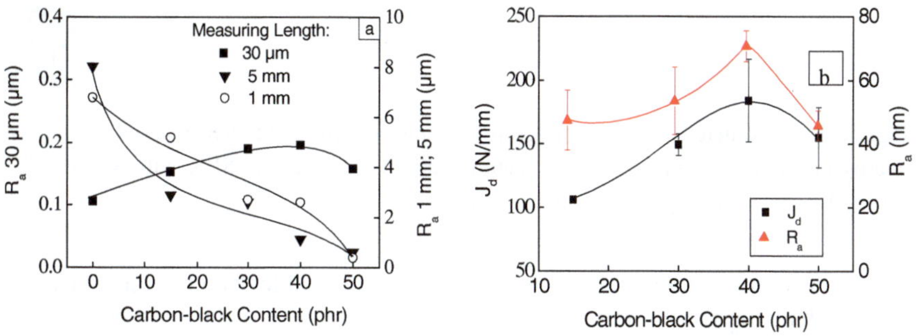

Fig. 36 Mean roughness value R_a of fracture surfaces from tactile stylus method (measuring length 5 mm), MicroGlider® experiment (measuring length 1 mm) and AFM measurements (measuring length 30 µm) (a) and correlation of mean roughness R_a and crack toughness J_d (b); material SBR + CB N339

To close the gap between the macroscopic and the microscopic characterization of the fracture surface by AFM tests, additional scanning electron microscopic investigations combined with 3-d reconstruction of the fracture surface topography were performed [10]. In Fig. 37 and Fig. 38, reconstructed 3d topography images of fracture surfaces of unfilled SBR and SBR with 40 phr CB N234 from SEM investigations are shown as examples.

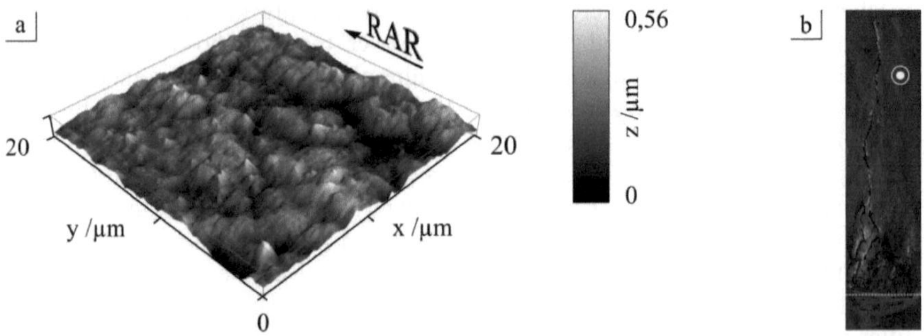

Fig. 37 Reconstructed fracture surface topography (a) from the area marked in the light microscopy image of the complete fracture surface (b); material unfilled SBR (series 2), RAR = crack propagation direction [10]

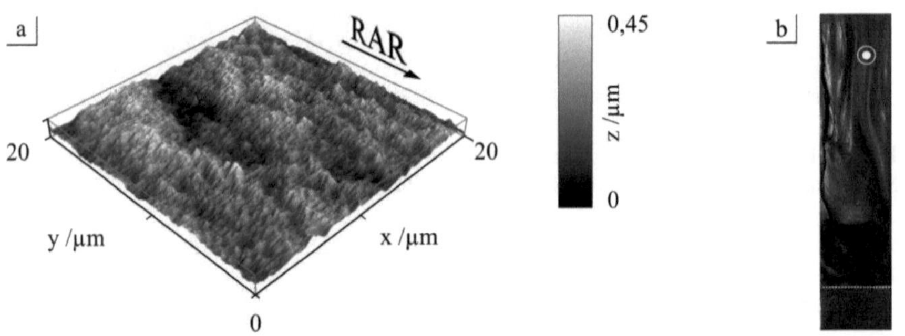

Fig. 38 Reconstructed fracture surface topography (a) from the area marked in the light microscopy image of the complete fracture surface (b); material SBR with 40 phr N234 (series 2) RAR = crack propagation direction [10]

The difference in the surface character is clearly visible. The fracture surface of the filled materials has much finer structure compared to the unfilled material. However, the main aim of these time-consuming investigations was not to determine R_a but to get reliable data which allow for quantitative analysis of fracture surfaces by means of statistical methods so that the size of the fracture process zone can be determined. Thus, it is possible to discuss how the microstructure influences the fracture process. In [45] and in chapter "Mesoscopic and macroscopic aspects of crack propagation and modelling at the interface between statistical mechanics and continuum mechanics", results of these investigations are presented and discussed.

4 Methodical Considerations

From the practical point of view, various aspects of materials testing and diagnostics are important. One the one hand, material parameters should be easily to determine, this means, the expense in time, material and costs should be as small as possible. On the other hand, tests should deliver reliable and reproducible results which possibly reflect of the inner structure and the morphology of the materials. Especially fracture mechanics parameters are sensible to the changes of the structure [2]. However, fracture mechanics testing of elastomers is still a field of research with some open questions. For example, one point is the geometry influence, which is often disregarded. Therefore, in own works [16, 19] the influence of the specimen thickness and the *a*/*W* ratio on the fracture mechanics parameters was investigated. Basing on these results, requirements for such experiments could be defined [25]. However, with proceeding experimental experience, an optimization of the test conditions for fracture mechanics tests is connected. As an example, the results indicate that not only for cyclic fracture mechanics tests (see chapter "Analysis of Dynamic Crack Propagation in Elastomers by Simultaneous Tensile- and Pure-Shear-Mode Testing") but also for quasi-static fracture mechanics tests, the use of pure shear specimens can be beneficial. In Fig. 39 a, R-curves from quasi-static fracture mechanics tests with aged pure shear specimens are shown. The thermo-oxidative ageing took place at a temperature of 70 °C and the ageing time was different.

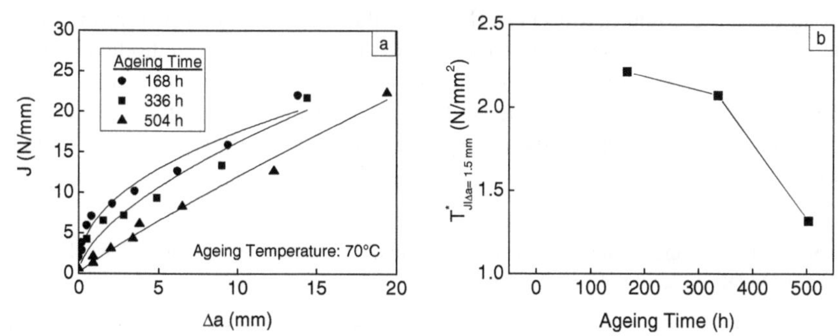

Fig. 39 Crack resistance curves of aged SBR with 40 phr CB N234 from quasi-static tests with pure shear specimens

The ageing resulted in a change in the crack propagation behaviour, especially with an ageing time of 3 weeks. The crack resistance is reduced by the ageing processes, which itself have not been investigated. The crack propagation values T_J* (see Fig. 39 b) show a decrease by 41 %. In Fig. 40, the tensile strength of SBR and NR with 40 phr CB N234 (series 2) is shown. It becomes clear that the thermo-oxidative loading over a period up to 3 weeks (504 h) does not reduce the tensile strength σ_{max} in the same way like the crack propagation value. In Table 10, for the purpose of comparison, the percentage change of the tensile strength σ_{max}, of strain at fracture ε_R, of tear resistance T_s and of the Shore A hardness due to the ageing is given.

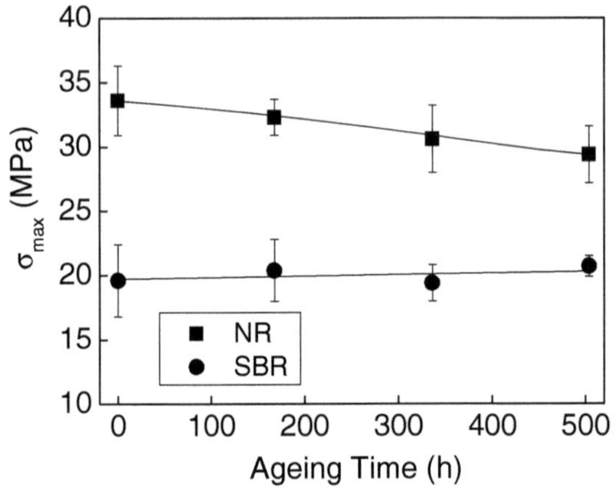

Fig. 40 Tensile strength of SBR and NR with 40 phr CB N234 as a function of the ageing time

For the both compounds, different ageing processes (chain scission, crosslinking) seem to appear. While for the SBR elastomer, except from the strain at fracture, the mechanical parameters stay relatively constant, for NR a partly strong decrease could be observed. Here, further investigations would help to find the reasons for the different behaviour. However, from the methodical point of view it is interesting to see that the fracture mechanics parameter is the only one which indicates structural changes. With this example, the importance of the selection of suitable methods of materials testing and diagnostics should be illustrated.

Table 10 Percentage changing of various mechanical parameters due to a 3 weeks lasting thermo-oxidative ageing of two elastomers

Elastomer	σ_{max}	ε_R	T_s	Shore A
SBR + 40 phr CB N234	+6	−13	+8	+3
NR + 40 phr CB N234	−12,5	−19	−37	+9

A second methodical aspect is the need for fast and possibly inexpensive testing methods which provide structure-sensitive material parameters. For example, cyclic fracture mechanics tests are mostly very time consuming and one needs expensive technical equipment and experience. Especially in material development and optimization for tyres, there is a certain interest to get fast first information on the crack growth behaviour. Therefore, the idea was to check if there is a correlation of fracture mechanics parameters from impact tests with that of the cyclic tests. Within the chapters "Analysis of Dynamic Crack Propagation in Elastomers by Simultaneous Tensile- and Pure-Shear-Mode Testing" and "Morphology and Micro-Mechanics of Filled Elastomer Blends" results of the cyclic fracture mechanics tests are represented and discussed in detail. The following Fig. 41 contains the exponent m of the course of da/dN–T for the SBR materials.

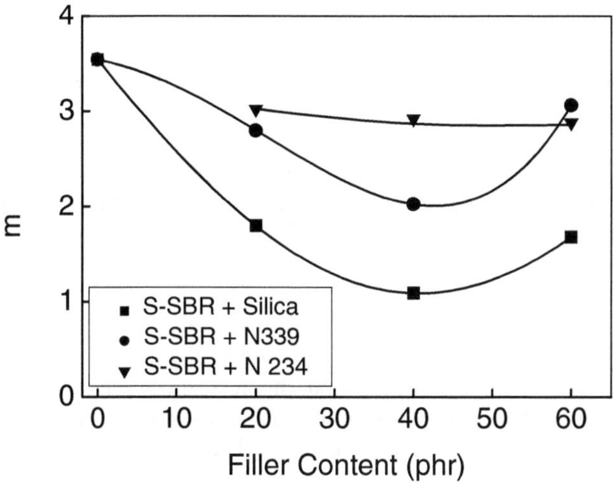

Fig. 41 Power law Exponent m of the power law from da/dN–T data resulting from cyclic fracture mechanics tests with SBR materials of series 1 and 2 [9]

It is assumed that a decreasing slope of the da/dN–T course is connected with an increase in crack resistance [9]. Similar to J_d (see Fig. 28 a) having a maximum at 40 phr filler content, the exponent m has a minimum at 40 phr content of silica and of CB N339. Adding CB N234, this effect is not as pronounced as for the other fillers. Furthermore, for NR and EPDM such a relationship could not be found. Thus, a general correlation between J_d and m seems to be not existent.

5 Conclusions and Outlook

This contribution contains results of comprehensive mechanical and fracture mechanics tests with various elastomeric materials. The compounds investigated based on different polymers and on polymer blends. Beside the polymer type, also the filler types and filler content were varied. Furthermore, investigations such as TEM or determination of the filler dispersion regarding the characterization of the structure were performed. For selected materials, the network density was additionally determined. The mechanical basic characterization with tensile tests, tear tests, DMA tests, Shore A hardness tests delivered a performance map with ultimate properties of the various materials.

The main part of the work was a large number of fracture mechanics tests under quasi-static and under impact-like loading conditions. The quasi-static tests included two different tests: with single-specimen method and with multiple-specimen method. Both methods are very suitable to characterize the stable crack propagation behaviour of elastomers and to assess the structural influence, e.g. the polymer type or the filler content. The advantage of the multiple-specimen test is the possibility to obtain J–Δa crack resistance curves and to separate the dissipated energy and the energy for creating a new crack surface from the elastic deformation energy. Latter has a great content in relation to the whole energy which is usually determined from the area under the load–deformation curve. The quasi-static fracture mechanics tests as well as the impact tests showed that the crack resistance has mostly a maximum value at a filler content of 40 phr. Furthermore, the type of polymer has an influence on the fracture mechanics parameters. Especially NR vulcanizates have a high resistance against stable crack propagation when the material has enough time to crystallize during loading. However, in the case of impact tests, NR behaves comparably to the other polymers.

For the first time, results of impact tests for elastomers with variation of the test temperature could be performed and also new are the results regarding the influence of the test speed in the range of impact loading. It was interesting and not expected how the temperature and the test speed variation influence the fracture resistance. The energy-determined J_d values show a maximum when they are plotted as a function of the temperature. The temperature of this maximum depends on the polymer type, but is not only determined by the glass transition temperature. Latter is nearly uninfluenced by the filler content, but the maximum of J_d in the temperature dependence is shifted with rising filler content. The maximum of J_d is can be explained by using the measuring values F_{max} and l_{max}.

Very successful was that part of the work dealing with the fracture surface characterization. Systematic investigations regarding the influence of test method and test parameters delivered useful information for the performance of such tests with the aim of a quantitative description of the surface topography as the basis for an analysis with statistical methods. Generally, the fracture surface characteristics are dependent on the carbon-black content. Increasing the amount of CB, a decrease in the fracture surface roughness at large length scales takes place, while at small length scales an increase was observed.

Under methodical aspects, the work could contribute again to the development of the fracture mechanics testing of elastomers. Further experience could be collected regarding the applicability, the reproducibility and the informational content of various material parameters coming from various methods of technical fracture mechanics. Moreover, the results were a source for the further development of the experimental methods. Thus, a solid basis is given for future investigations e.g. with the aim of describing the effect of ageing due to the influence of mechanical and/or media, heat or weathering.

Acknowledgements. The authors would like to thank the German Research Foundation (DFG) for the financial support of the sub-project 2 "Technical Material Diagnostics – Fracture Mechanics of Filled Elastomeric Blends" within the framework of the Research Unit FOR 597. Furthermore, we acknowledge Dr. Sybill Ilisch from the Martin-Luther-University Halle-Wittenberg, Germany, Dr.-Ing. Silvia Agnelli from the University of Brescia (Italy) and Dr.-Ing. Ralf Lach from the Martin-Luther-University of Halle-Wittenberg for their substantial experimental work and the helpful scientific discussions. We thank our former students Dipl.-Ing. Katja Oßwald, Dipl.-Ing. Nicole Kahnt, Dipl.-Wirtsch.-Ing. Marcus Tischer, Dipl.-Wirtsch.-Ing. Johannes März, Dipl.-Wirtsch.-Ing. Mirko Rennert for their contribution to this work through Diploma thesis or students' project works. Last but not least, we would like to thank all colleagues from the FOR 597 for the fruitful cooperation during 6 years, especially Dr.-Ing. Thomas Horst and Dr.-Ing. Radek Stoček for a very intensive and successful cooperation and scientific discussion.

References

1. Grellmann, W., Seidler, S. (eds.): Polymer Testing, 2nd edn. Hanser Verlag, München (2013)
2. Grellmann, W., Seidler, S. (eds.): Deformation and Fracture Behaviour of Polymers. Springer, Heidelberg (2001)
3. Anderson, T.L.: Fracture Mechanics – Fundamentals and Application, 3rd edn. CRC Taylor & Francis, Boca Raton (2005)
4. Standard Draft ESIS TC 4: A Testing Protocol for Conducting J-Crack Growth Resistance Curve Tests on Plastics (2001)
5. Blumenauer, H., Pusch, G.: Technische Bruchmechanik. Deutscher Verlag für Grundstoffindustrie, Leipzig (1993)
6. Rivlin, R.S., Thomas, A.G.: Rupture of Rubber. I. Characteristic Energy for Tearing. J. Polym. Sci. 10, 291–318 (1953)
7. Röthemeyer, F., Sommer, F.: Kautschuktechnologie. Carl Hanser Verlag, München (2001)
8. The Columbian Chemicals Company, Marietta (GA), USA: Data Sheet 08/09: Carbon Blacks for Rubber, `http://www.columbianchemicals.com` (assessed at January 16, 2013)
9. Reincke, K., Klüppel, M., Grellmann, W.: Investigation of Fracture Mechanical Properties of Filler-Reinforced Styrene–Butadiene Elastomers. Kautsch. Gummi Kunstst. 62, 246–251 (2009)
10. Rennert, M.: Untersuchungen zur quantitativen Bewertung von Bruchflächen. Diploma Thesis, Martin Luther University Halle-Wittenberg (2011)
11. ISO 37: Rubber, Vulcanized and Thermoplastic – Determination of Tensile Stress–Strain Properties (2011)
12. ISO 7619-1: Rubber, Vulcanized or Thermoplastic – Determination of Indentation Hardness – Part 1: Durometer Method (Shore Hardness) (2010)
13. ISO 34-1: Rubber, Vulcanized or Thermoplastic – Determination of Tear Strength – Part 1: Trouser, Angle and Crescent Test Pieces (2010)

14. Stoček, R., Reincke, K., Gehde, M., Grellmann, W., Heinrich, G.: Einfluss der Kerbeinbringung auf die Rissausbreitung in elastomeren Werkstoffen. Kautsch. Gummi Kunstst. 63, 364–370 (2010)
15. Stoček, R., Heinrich, G., Reincke, K., Grellmann, W., Gehde, M.: Rissausbreitung in elastomeren Werkstoffen unter dynamischer Beanspruchung – Einfluss der Kerbeinbringung. Kautsch. Gummi Kunstst. 64, 22–26 (2011)
16. Reincke, K., Heinrich, G., Grellmann, W.: Investigation of Mechanical and Fracture Mechanical Properties of Elastomers Filled with Precipitated Silica and Nanofiller Based upon Layered Silicates. Rubber Chemistry and Technology 77, 662–677 (2004)
17. Reincke, K., Grellmann, W., Heinrich, G.: Engineering Fracture Mechanics for Crack Toughness Characterisation of Elastomers. In: Proceedings of the European Conference of Fracture (ECF 16), Alexandroupolis, Greece, July 3–7 (2006)
18. Reincke, K., Grellmann, W., Heinrich, G.: Fracture Mechanical Investigations of Filler-reinforced Elastomers. In: Boukamel, A., Laiarinandrasana, L., Méo, V.E. (eds.) Constitutive Models for Rubber V, pp. 221–227. Taylor & Francis Group London (2008)
19. Reincke, K.: Bruchmechanische Bewertung von ungefüllten und gefüllten Elastomerwerkstoffen. Mensch & Buch Verlag Berlin (2005)
20. März, J.: Untersuchungen zur Weiterentwicklung bruchmechanischer Untersuchungsmethoden zur quantitativen Beschreibung des Deformations- und Bruchverhaltens von Elastomerwerkstoffen. Diploma Thesis, Martin Luther University Halle-Wittenberg (2010)
21. Reincke, K., Grellmann, W., Lach, R., Heinrich, G.: Toughness Optimization of SBR Elastomers – Use of Fracture Mechanics Methods for Characterization. Macromolecular Materials and Engineering 288, 181–189 (2003)
22. Gerber, G., Struve, J.: Einfluß der Mischungszusammensetzung und Belastungsart auf das Versagensverhalten von Elastomeren. Kautsch. Gummi Kunstst. 52, 400–404 (1999)
23. Reincke, K., Grellmann, W.: Fracture Mechanical Assessment of Elastomeric Materials. Lecture, University of Brescia, Italy (November 12, 2008)
24. DIN EN ISO 8256: Plastics – Determination of Tensile-impact Strength (2005)
25. `http://wiki.polymerservice-merseburg.de/index.php/Kerbschlagzugversuch#Instrumentierter_Kerbschlagzugversuch`, `http://www2.iw.uni-halle.de/ww/mpk/mpk_ISZV.pdf`
26. Reincke, K., Oßwald, K., Grellmann, W.: Experimental Investigations for Characterization of Crack Toughness of Filler-reinforced SBR Vulcanizates. In: 11. Tagung „Problemseminar: Deformation und Bruchverhalten von Kunststoffen“, Merseburg, Tagungsband, June 20-22, pp. S.131–S.141. CD-ROM (2007) ISBN 978-3-86010-918-2
27. Oßwald, K.: Untersuchungen zum Rissinitiierungs- und Rissausbreitungsverhalten von füllstoffverstärkten Elastomerwerkstoffen. Diploma Thesis, Martin Luther University Halle-Wittenberg (2007)
28. Toki, S., Fujimaki, T., Okuyama, M.: Strain-Induced Crystallization of Natural Rubber as Detected Real-Time by Wide-Angle X-ray Diffraction Technique. Polymer 41, 5423–5429 (2000)

29. Reincke, K., Grellmann, W.: Mechanische und bruchmechanische Eigenschaften von füllstoffverstärkten Elastomeren. In: 13. Tagung Problemseminar Deformation und Bruchverhalten von Kunstoffen, Merseburg, Tagungsband, June 29–July 1, pp. 33–45. CD-ROM (2011) ISBN 978-3-86829-400-2
30. Grellmann, W., Reincke, K., Lach, R., Heinrich, G.: Characterization of Crack Toughness Behaviour of Unfilled and Filled Elastomers. Kautsch. Gummi Kunstst. 54(7–8) (2001)
31. Ramorino, G., Agnelli, S., De Santis, R., Riccò, T.: Investigation of Fracture Resistance of Natural Rubber/Clay Nanocomposites by J-Testing. Eng. Fract. Mech. 77, 1527–1536 (2010)
32. Casiraghi, T.: Polym. Eng. Sci. 18, 833 (1978); Nezbedova, E.: Wissenschaftl. Zeitschrift TH Merseburg 28, 278 (1986); May, M., Hoffmann, H., Grellmann, W.: Plaste Kautschuk 31, 26 (1984); Gensler, R., Plummer, C.J.G., Grein, C., Kausch, H.-H.: Polymer 41, 3809 (2000); Goldbach, G.: Kunststoffe 64, 475 (1974); Goldbach, G.: Angew. Makromol. Chem. 14, 123 (1978); Beguelin, P., Kausch, H.-H.: J. Mater. Sci. 29, 91 (1994); Karger-Kocsis, J., Friedrich, K.: Polymer 27, 1753 (1986); Gensler, R. Beguelin, P., Plummer, C.J.G., Kausch, H.-H., Miinstedt, H.: Polym. Bulletin 37, 111 (1996); Hashemi, S., Mugan, J.: J. Mater. Sci. 28, 3983 (1993); Han, Y.C., Yanc, Y.M., Li, B.Y., Feng, Z.L.: J. Appl. Polym. Sci. 56, 979 (1995); Groves, J.F., Agrawal, C.M., Pearsall, G.W.: J. Mater. Sci. 27, 2335 (1992); Beguelin, P.: PhD Thesis, Ecole Polytechnique Fédérale de Lausanne, Lausanne (1996); Plummer, C.J.G., Beguelin, P., Kausch, H.-H.: Polym. Eng. Sci. 35, 1300 (1995)
33. Grellmann, W.: Habilitationsschrift. TH Leuna-Merseburg, Merseburg (1985); Joyce, J.A.: Polym. Eng. Sci. 43, 1702 (2003); Lu, M.-L., Chiou, K.-C., Chang, F.-C.: Polym. Eng. Sci. 36, 2289 (1996)
34. Reincke, K., Lach, R., Grellmann, W.: Influence of Viscoelastic Material Behavior on Fracture Properties of Elastomers. In: 12th International Conference on Fracture, Ottawa, Canada, July 12–17, T37.016, pp. 1–8 (2009)
35. Gdoutos, E.E., Daniel, I.M., Schubel, P.: Facta Universitatis – Ser. Mechanics, Automatic Control and Robotics 3, 497 (2003)
36. Marder, M.: Phys. Rev. Lett. 94, 048001 (2005); Pertersan, P.J., Deegan, R.D., Marder, M., Swinney, H.L.: Phys. Rev. Lett. 93, 0155504 (2004)
37. Reincke, K., Grellmann, W.: Mechanical and Fracture Mechanics Properties of Rubber Compositions with Reinforcing Components. In: Galimberti, M. (ed.) Rubber-Clay Nanocomposites: Science, Technology and Applications, 1st edn., pp. 305–342. John Wiley & Sons (2011)
38. Mizutan, I.K., Yoshii, M.: J. Mater. Sci. 23, 3501 (1988); Mizutani, K., Iwatsu, T.: J. Appl. Polym. Sci. 26, 3447 (1981); Araki, W., Adachi, T., Gamou, M., Yamaji, A.: Proc. Institution of Mechanical Engineers, Part L: J. Materials: Design and Applications 216, 79 (2002); Araki, W., Asahi, D., Adachi, T.,Yamaji, A.: J. Appl. Polym. Sci. 96, 51 (2005); Walter, H.: Dissertation, Martin-Luther-Universität Halle-Wittenberg, Halle/Saale (2003)
39. Lach, R., Grellmann, W.: Time- and Temperature-Dependent Fracture Mechanics of Polymers: General Aspects at Monotonic Quasi-Static and Impact Loading Conditions. Macromol. Mater. Eng. 293, 555–567 (2008)
40. Lach, R.: Habilitationsschrift, Martin-Luther-Universität Halle-Wittenberg, Halle/Saale (in preparation)

41. Hoo Fatt, M.S., Ouyang, X.: Intern. J. Solids and Structures 44, 6491 (2007); Hoo Fatt, M.S., Bekar, I.: J. Mater. Sci. 39, 6885 (2004); Al-Quraishi, A.A., Hoo Fatt, M.S.: Tire Sci. Technol. 35, 252 (2007)
42. Reincke, K., Lach, R., Grellmann, W.: Influence of Viscoelastic Material Behavior on Fracture Properties of Elastomers. In: 12th International Conference on Fracture, Ottawa, Canada, July 12–17, T37.016, pp. 1–8 (2009)
43. Tischer, M.: AFM-Untersuchungen von Bruchflächen füllstoffverstärkter Elastomerwerkstoffe, Martin Luther University Halle-Wittenberg (2008)
44. Tischer, M., Ilisch, S., Reincke, K., Grellmann, W., Radusch, H.-J.: AFM-Untersuchungen von Bruchflächen füllstoffverstärkter Elastomerwerkstoffe. In: 12. Tagung "Problemseminar Deformation und Bruchverhalten von Kunststoffen", Merseburg, Tagungsband, Juni 24–26, p. S. 394. CD-ROM (2009) ISBN 978-3-86829-170-4
45. Horst, T., Reincke, K., Ilisch, S., Heinrich, G., Grellmann, W.: Fracture Surface Statistics of Filled Elastomers. Physical Review E 80, 046120 (2009)

Analysis of Dynamic Crack Propagation in Elastomers by Simultaneous Tensile- and Pure-Shear-Mode Testing

Radek Stoček[1,2,3], Gert Heinrich[3,4], Michael Gehde[5], and Reinhard Kipscholl[6]

[1] PRL Polymer Research Lab., s.r.o., Nad Ovčírnou 3685, 760 01 Zlín, Czech Republic
[2] Centre of Polymer Systems, University Institute, Tomas Bata University in Zlin, Nad Ovcirnou 3685, 760 01 Zlin, Czech Republic
[3] Leibniz-Institut für Polymerforschung Dresden e. V., Hohe Straße 6, 01069 Dresden, Germany
[4] Technische Universität Dresden, Institut für Werkstoffwissenschaft, 01062 Dresden, Germany
[5] Technische Universität Chemnitz, Institut für Fördertechnik und Kunststoffe, 09107 Chemnitz, Germany
[6] Coesfeld GmbH & Co. KG, Tronjestr.8, 44319 Dortmund, Germany

Abstract. The present work proposes a new fracture mechanical testing concept for determination of dynamic crack propagation of rubber materials. This concept implements a method of simultaneous tensile- and pure-shear-mode testing. The present approach is based on an upgrade of the Tear Analyzer (Co. Coesfeld GmbH & Co. KG), on the fracture mechanics theory of dynamically loaded test specimens and on the definition of pure-shear states according to the test specimen's geometry ratio.

The main focus of this work can be divided into three parts. Firstly, it introduces the development of a method for analysis of dynamic crack propagation in filled rubber by simultaneous tensile- and pure shear mode testing. The servo-hydraulic machine with controlled temperature testing chamber is equipped with simultaneously operating two-mode test equipment that represents a new fracture testing method. This two-mode test allows the measurement of crack propagation on different rubber specimens simultaneously and under identical load. The data analysis allows a comparison between the two parallel running testing modes.

Secondly, this work deals with the development of a method for the defined creation of a notch in a rubber specimen. This method serves as a basis for the reproducible and reliable determination of fracture mechanical parameters for elastomers. After insertion of notches in a defined way, fracture tests under different loading conditions were performed. A significant influence on the notch geometry was observed in the test results. The results illustrated the importance of a defined and reproducible notching of elastomeric specimens.

Next, the analysis of crack propagation under dynamic loading conditions is practiced with this method. It is shown how the tearing energy and the crack

W. Grellmann et al. (Eds.): *Fracture Mechanics & Statistical Mech.*, LNACM 70, pp. 269–301.
DOI: 10.1007/978-3-642-37910-9_7

growth rate depend on the test specimen's geometry ratio and crack length. It is also demonstrated that the values for tearing energies and crack growth rates for short crack lengths in SENT, as well as in pure-shear test specimens, are identical. Another important aspect of the results is related to the different values of tearing energies and crack growth rates for cracks with short and large lengths in pure-shear test specimens. The results show the dependence of fracture behavior on the manufacture of the test specimens.

The new fracture mechanical testing concept offers a comparison between fracture behaviors of rubber materials independent of the test specimen's geometry.

Keywords: Rubber materials; Crack propagation; Tear Analyzer; Tearing energy; SENT; Pure-shear.

1 Introduction

The dynamic fracture behavior of elastomers has been intensively studied and published in a number of publications e.g. [1], [3], [3], [4], [5]. The mentioned works mainly deal with fracture mechanics of elastomers using different methods and testing equipment. Both methods allow an on-line evaluation of crack propagation, however, only one method is industrially used for the characterization of dynamic fracture behavior of elastomers: the tensile method. In the pure-shear method, the crack's propagation is caused due to uni-axial stress conditions in the crack tip.

The crack initiation and propagation mechanism of elastomeric materials is highly complex and always remains a subject of advanced research. An important criterion for characterization of dynamically loaded rubber materials is their resistance to fatigue crack growth. The problem of crack growth in rubber materials was first studied by Rivlin & Thomas [1]. In the publications of Lake & Lindley [2, 3], the authors first demonstrated fatigue crack growth dependent on the tearing energy of vulcanized rubber in the range of small as well as of critical strain. This dynamic crack growth behavior of vulcanized rubber was investigated in the SENT (Single Edge Notched Test) as well as in pure-shear test specimens, but the comparison of solutions for two different test specimen's geometry from the analysis is less well established.

The SENT test specimen is used in commercial studies of elasticity and fracture mechanics of rubber. However, in classical studies of fracture mechanics of rubber, the pure-shear test specimen featured prominently because of the amenability to a simple fracture mechanics analysis. The SENT test specimen consists of a thin, rectangular strip of rubber held by rigid clamps along its short edges and the definition is given by the geometry ratio L_0 (length or distance between the clamps in the un-deformed state): Q (width of test specimen) ≥ 1. Compared to a SENT test specimen, the pure-shear test specimen consists of a thin, rectangular strip of rubber, but held by rigid clamps along its long edges. The pure-shear test specimen is then defined by the geometry ratio $L_0 :Q << 1$. A simple FE-analysis of deformation of SENT as well as pure-shear test specimens represented by strain in the x-direction unambiguously showed the width of non-deformed range in the pure-shear compared to the SENT test specimen (s. Figure 1). However, when these test specimens are used to characterise the crack propagation of rubber, it is important to describe the influence of different test specimen areas as edge effect range (s. range *A* in the Figure 1-left) and pure-shear range (s. range *C*/2 in the Figure 1-right) on the crack length. Unfortunately, there is only limited information available in the literature (e.g. Yeoh [5]) on the influence of edge effects on the short crack length in the pure-shear test specimen, and limited information on the long crack in SENT test specimen.

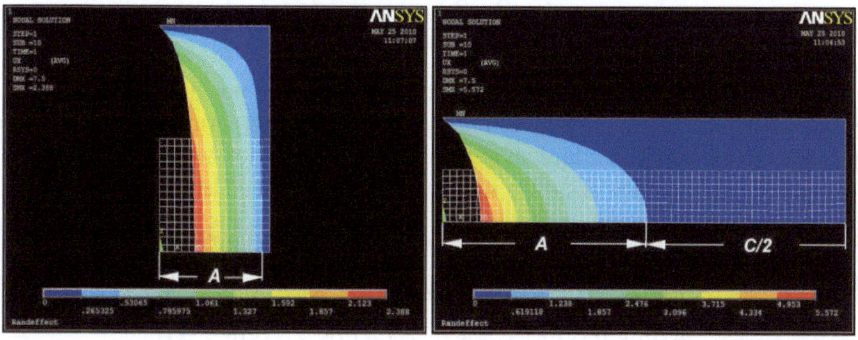

Fig. 1 FE - analysis of the deformation of quarter of SENT (left) as well as pure-shear (right) test specimens in the x-direction subjected to 100% tension

Considering the importance of the pure-shear and of the commonly used SENT test specimens, it is reasonable to consider the comparison of the crack propagation according to the different test specimen's geometry. Exact solution by traditional analysis of SENT and pure-shear test specimens is not possible, because of the use of different test machines and influence of test conditions. The new fracture mechanical testing concept based on a method of simultaneous SENT and pure-shear-mode testing was developed for use in the present work to describe the comparison of fracture behavior in SENT and pure-shear rubber test specimens.

Definition of the Test Specimen for Characterization of Fracture Behavior of Elastomers

The characterization of fracture behavior of elastomeric material is based on the analysis of notched test specimens. There is no information in the literature that deals with the definition of notch, or the exact definition of a test specimen's geometry. In the next paragraphs the test specimen's geometry and notch geometry will generally be defined for the characterization of fracture behavior in tensile as well as pure-shear methods.

The Figure 2 shows a schematic diagram of a test specimen's geometry as well as the geometry of a notch. The width of the test specimen is denoted as Q, the length is L_0 and the denotation of the thickness is B. The specification of the notch length is a_0, the actual crack's length a, and the radius of the notch tip is r_k.

The notch is defined as cut in the test specimen. The both of the opposed surfaces are defined as notch surfaces. The notch starts at the notch front edge and finishes at the notch tip.

The general definition of the notch's surface is:

$$\sin\alpha \cdot \cos\beta \cdot x - \cos\alpha \cdot \cos\beta \cdot y - \sin\beta \cdot z = 0\,, \tag{1}$$

where the boundary conditions are:

$$z \in \left\langle -\frac{B}{2}, \frac{B}{2} \right\rangle,\ y \in \left\langle -\frac{L_0}{2}, \frac{L_0}{2} \right\rangle.$$

The notch's tip line is created from the intersection of the notch's surface and tip's plane of notch. The tip's plane of notch is defined with a normal vector, which is oriented in the notch's surface and has an orthogonally orientation to the tip line of the notch. The equation of the tip's plane of notch is defined as follows:

$$\begin{aligned} &[\sin\alpha \cdot \sin\beta \cdot \sin\gamma - \cos\alpha \cdot (1-\cos\gamma)] \cdot x + \\ &+ [-\cos\alpha \cdot \sin\beta \cdot \sin\gamma - \sin\alpha \cdot (1-\cos\gamma)] \cdot y + \\ &+ (\cos\beta \cdot \sin\gamma) \cdot z + a_0 \cdot (1-\cos\gamma) - \sin\gamma = 0\,. \end{aligned} \tag{2}$$

The front plane belongs to the yz-plane and is expressed with following general equation:

$$x = 0\,. \tag{3}$$

The back plane is oriented parallel to the yz-plane. The next equation defines the back plane:

$$x - Q = 0\,. \tag{4}$$

The side planes are located parallel to the xy-plane, and they are described as follows:

$$z - \frac{B}{2} = 0;\ z + \frac{B}{2} = 0\,. \tag{5}$$

The first parallel plane to the xz-plane is denoted as the upper plane:

$$y - \frac{L_0}{2} = 0 \tag{6}$$

and the second one with the definition:

$$y + \frac{L_0}{2} = 0 \tag{7}$$

is denoted as the bottom plane.

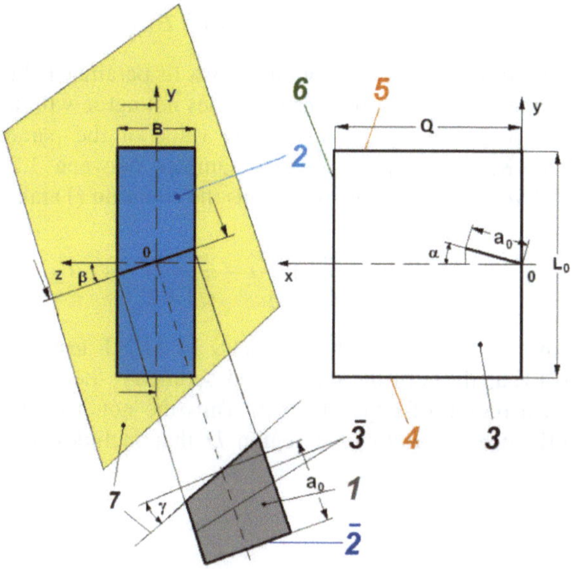

Fig. 2 Schematic diagram of the test specimen with: 1 – notch's surface, $\bar{2}$ – front edge of the notch, 2 – front plane, $\bar{3}$ – side edges of the notch, 3 – side plane 4 – bottom plane, 5 – upper plane, 6 – back plane, 7 – tip's plane of notch

Tearing Energy

An important quantity for fracture mechanical investigations is the tearing energy T, i.e. the energy released per unit area of crack surface growth. It was first introduced by Griffith [6] for metallic materials and Rivlin & Thomas [1] formulated the tearing energy for elastomers. It proposes that the strain energy release rate is the controlling parameter for crack growth and it is mathematically defined as,

$$T = -(\delta W/\delta A). \tag{8}$$

Where, T is tearing energy, W is the elastic strain energy, A is the interfacial area of crack and partial derivative denotes that no external work is done on the system.

In the language of geometry they defined the tearing energy for rubber material in SENT as well as in pure-shear test specimens. For SENT test specimens with a notch on one side, it is determined as follows:

$$T_t = 2 \cdot k \cdot w \cdot a, \tag{9}$$

Where w is the strain-energy-density stored in the un-notched test specimen; a - crack length; k = strain-dependent term. An approximate relation for k was determined by Lake [7]:

$$k = \frac{\alpha}{\sqrt{\lambda}} = \frac{\alpha}{\sqrt{1+\varepsilon}}, \tag{10}$$

where λ = extension ratio; ε = strain and α is a fit parameter. Furthermore, it was found by Klüppel [8] that the parameter α does not agree with the literature value $\alpha = \pi = 3.14$ proposed by Gent [9]. The value of the parameter α has been determined by Klüppel [8] for various samples between 1.2 and 3.1. In the presented work $\alpha = \pi = 3.14$ was used. Thus the equation (1) takes the form

$$T_t = \frac{2 \cdot \alpha}{\sqrt{\lambda}} \cdot w \cdot a = \frac{2 \cdot \alpha}{\sqrt{1+\varepsilon}} \cdot w \cdot a. \tag{11}$$

Figure 3 shows schematically the geometry of SENT test specimen, in which, according to FE-analysis of the un-notched geometry, the width of test specimen Q consists of a region of edge effect A. Thus the notch in the edge of the test specimen influences a width of a region D that includes a complicated strain around the crack tip.

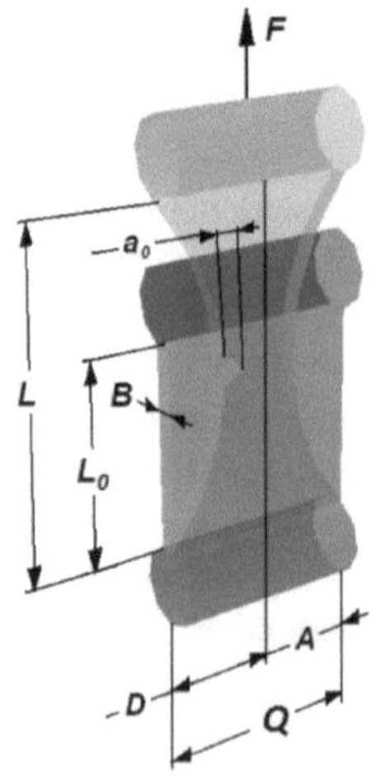

Fig. 3 Schematic diagram of SENT test specimen

The pure-shear test specimen first defined by Rivlin & Thomas [1] is schematically visualized in the Figure 4. Usually, the notch length a_0, which develops into the crack length a, is sufficiently long compared to the length of test specimen L_0. The test specimen can be divided into different regions. Region A_1 is unstrained and region C-D is in pure-shear state, whereas the region C is identical to the pure-shear state region C in the un-notched test specimen visualized in the Figure 1 (right) and is geometrically defined as follows:

$$\begin{aligned}\lambda(x) &= \frac{dx}{dx_0} = 1 \\ \lambda(y) &= \frac{dy}{dy_0} = \frac{y}{y_0} = \varepsilon(y)+1\end{aligned} \quad \Rightarrow \quad \begin{bmatrix} x \\ y \end{bmatrix} = \begin{bmatrix} 1 & 0 \\ 0 & \varepsilon(y)+1 \end{bmatrix} \cdot \begin{bmatrix} x_0 \\ y_0 \end{bmatrix}. \tag{12}$$

Furthermore, there is an area of complicated strain around the crack tip in region D, and a region of edge effect shown as A that is identical with the edge effect region A of the un-notched test specimen visualized in Figure 1 (right). If the maximum cyclic deformation remains constant and the crack length is increased by da, the region of complicated strain D moves along by da, but the pattern of strain, and hence the energy stored, remains unaltered. As a result, the net effect causes a decrease of region A and an increase of region C by a volume $L_0 \cdot t \cdot da$ and, thus, a decrease of the total elastic energy $w \cdot L_0 \cdot t \cdot da$, where w is the strain-energy-density in the pure-shear region C.

For a pure-shear test specimen's geometry, the tearing energy T_p is considered by Rivlin & Thomas [1] to be independent of the crack length:

$$T_p = w \cdot L_0 \tag{13}$$

The strain-energy-density w can be determined from the measurement of the tensile stress-strain curve of the un-notched test specimen.

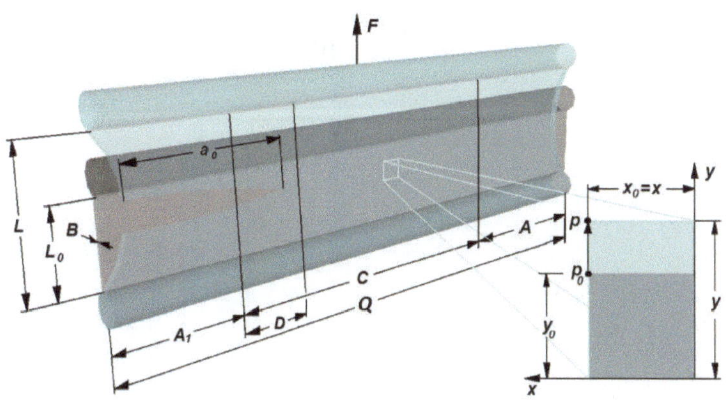

Fig. 4 Schematic diagram of a pure-shear test specimen with the visualisation of regions under different stress conditions

Fracture Mechanical Treatment of Failure

Gent, Lindley and Thomas [10] determined experimentally the crack growth rate *da/dn* in dependence on the tearing energy T for rubber materials.

Figure 5 shows the typical relationship for a rubber material on a double logarithmic plot. Lake & Lindley [3] divided this plot into 4 regions that characterize different tear behaviors. The crack growth rate *da/dn* depends on the tearing energy T in each of the 4 regions in a characteristic manner.

As long as the value of tearing energy T is lower than T_0, crack growth proceeds at a constant rate r and the crack growth is independent of the dynamical loading, but affected by the environmental attack.

$$T \le T_0 \Rightarrow \frac{da}{dn} = r \tag{14}$$

In region II between T_0 and T_I one finds a transition between a nucleation and propagation of crack growth:

$$T_0 \le T \le T_I \Rightarrow \frac{da}{dn} = A(T - T_0) + r \tag{15}$$

After this transient state the crack propagates in a region between T_I and T_C of stable crack growth which is denoted as region III. The relationship between fatigue crack growth rate *da/dn* and tearing energy is described by Paris & Erdogan [11] with the power-law:

$$T_I \le T < T_C \Rightarrow \frac{da}{dn} = b \cdot \Delta T^m, \tag{16}$$

where b and m are material constants.

In the last region, IV, the tearing energy T_C proceeds to the instable state of crack growth and the crack growth rate will become essentially infinite.

$$T \approx T_C \Rightarrow \frac{da}{dn} = \infty \tag{17}$$

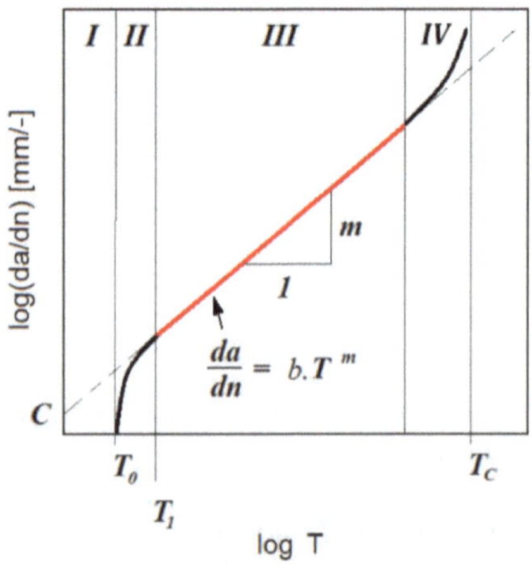

Fig. 5 Double logarithmic plot of crack growth rate *da/dn* vs. tearing energy T for rubber material [3]

Region III was utilised as the region that corresponds most closely to crack growth rates found in the engineering fatigue range. In this work, an experimental simultaneous analysis of SENT and pure-shear test specimens was performed according to this region to compare the stable crack growth in both of the test specimens depending on the notch and crack lengths.

2 Evaluation of Tearing Energy by Using Finite Element Analysis

The simplicity in shape of the pure-shear test specimen makes it suitable for an evaluation of the tearing energy in test specimens by using FEM, whereas the test specimen's geometry ratio $L_0:Q$ was taken into account. The first evaluations of tearing energy were performed by Yeoh [5], where the geometry ratio $L_0:Q = 1{:}5$ was studied and the different tearing energy influenced by crack length was observed.

The tearing energy has been evaluated using non-linear finite element analysis with software ANSYS by using the 6-parameters Ogden material model. Test specimens with three different geometry ratios $L_0:Q = 1{:}12$; 1:4 and 1:1.5 with constant length L_0 and modified width Q were modeled to study the effect of different geometry ratios. Taking advantage of symmetry, only one quarter of each test specimen was modeled, meaning that two crack growths in opposite edges were simulated. The simulation was performed by 10, 30 and 50 % strain applied on the long edges of modeled test specimen.

Figure 6 shows FEM results for the dependence of tearing energy on crack length for a test specimen based on styrene-butadiene rubber (SBR) with geometry ratios $L_0:Q = 1{:}12$ and 1:4, whereas the width Q was constant and the length L_0 was varied. The simulation was subjected to 10, 30 and 50% strain. The notation, which is used in the diagrams, is explained in Figure 7.

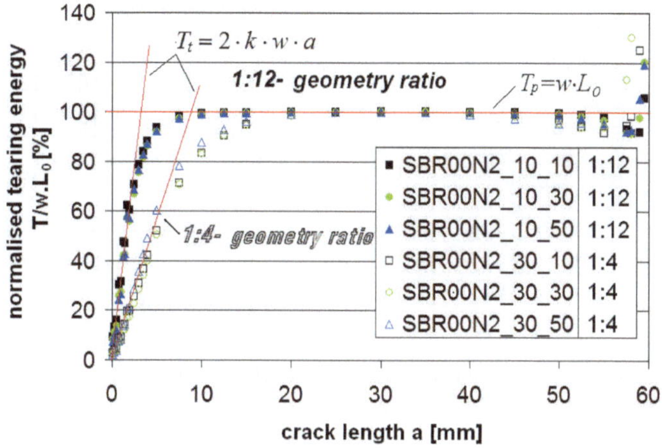

Fig. 6 Dependence of tearing energy on crack length for edge cracks in test specimens with geometry ratios 1:12 and 1:4 at various strains

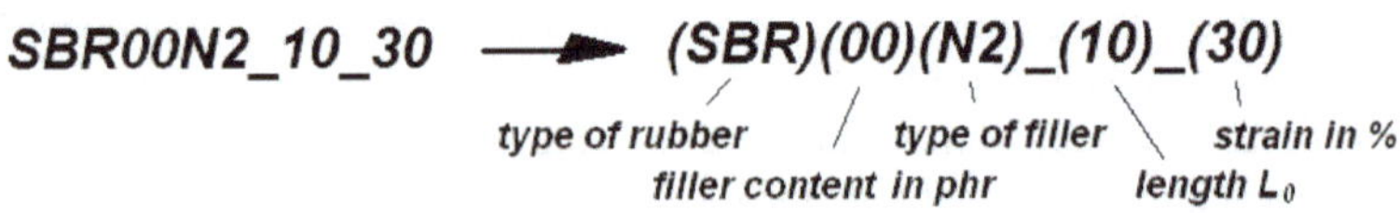

Fig. 7 Specification of the notation used in the diagrams

Figure 6 shows that independent of the geometry ratio, the tearing energy is approximately linearly dependent on a for short crack lengths according to the equation (1). This narrow range of the applicability of the Rivlin & Thomas theory changes into the non-defined transition of dependence of tearing energy on crack length. The calculation of tearing energy dependent on crack length shows in the center of the test specimen a region in which the tearing energy is independent of crack length. This region is deformed in the pure-shear state and formulated by the equation (5). The last region of the large crack lengths varied from the constant state of tearing energy by reason of the symmetric model of the test specimen. The comparison of the results obtained for the geometry ratios 1:12 and 1:4 shows a relatively small central region of the test specimen for the geometry ratio 1:4 deformed in the pure-shear state. On the other hand, it demonstrated the independence of various strains fromthe different geometry ratios.

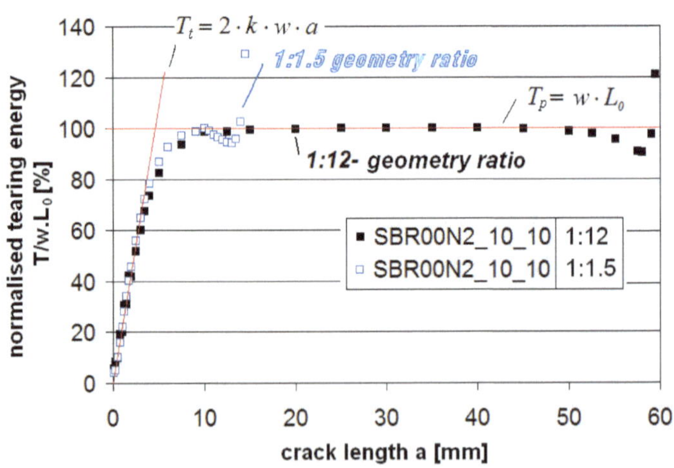

Fig. 8 Dependence of tearing energy on crack length for edge crack in test specimens with geometry ratios 1:12 and 1:1.5 at strain 10%

Figure 8 shows the FEM results of an edge crack in test specimens with geometry ratios 1:12 and 1:1.5 at strain 10%. The higher geometry ratio 1:1.5 was achieved by means of constant length L_0 and various widths Q. The geometry ratio 1:1.5 presents the geometry of the SENT test specimen. The simulation proved an identical slope of the linear dependence of tearing energy on crack length for small cracks a suggested by equation (1), in both test specimens with different geometry ratios. This documented that the identical crack growth in the region of availability of the equation (1) provided the same length L_0 of test specimens defined with different geometry ratios. There is no region of tearing energy independent of crack length in the center of the test specimen with the geometry ratio 1:1.5 according to the FEM simulation performed in the Figure 1 (left).

The FEM simulations were performed subjected to stable crack growth in rubber test specimens under dynamic loading. This basic fact according to the power-law described with equation (8) forms a real hypothesis for the constant crack growth velocity in the regions of availability of equations (1) and (5). Therefore, the insignificant dependence of the of equations (1) and (5) on the different rubber materials was established. The crack growth velocity could be an important indicator for the dependence of tearing energy on crack growth rate, which can be determined by experimental analysis.

3 Development of Testing Equipment for the Analysis of Dynamic Crack Propagation by Simultaneous Tensile and Pure-Shear Mode Testing

The new testing equipment for the analysis of dynamic crack propagation in filled rubber by simultaneous tensile and pure-shear mode testing is based on the redevelopment of the commercially and industrially used Tear Analyzer made by the company Coesfeld GmbH & Co. KG The Tear Analyzer available at the Technical University of Chemnitz was used for the development of this new testing method.

3.1 Design of Commercial Tear Analyzer

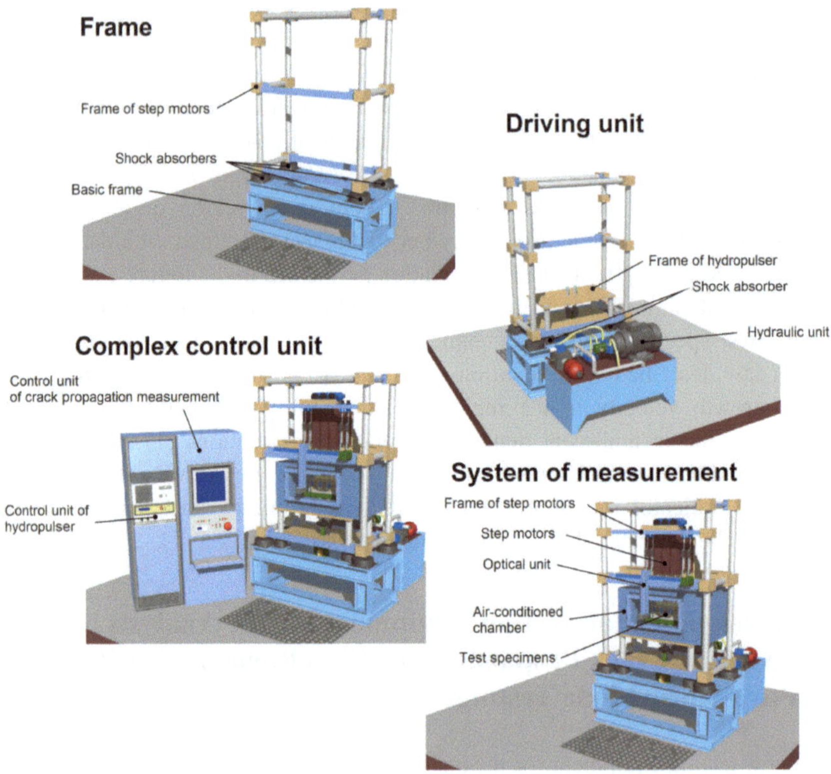

Fig. 9 Design of commercial Tear Analyzer

3.2 Basic Functionality of the Redeveloped Tear Analyzer

A schematic function diagram is shown in Figure 10 of the Tear Analyzer for investigating the fatigue crack growth of rubber materials by simultaneous tensile and pure-shear mode testing. It is possible to measure four tensile and three pure-shear test specimens simultaneously. Each bottom clamp attachment of the samples is fixed to the traverse. The traverse is connected to the piston, whereby it dynamically powers a hydropulzer. The hydropulzer is driven by a frequency sweep generator powered sinusoidal waveform, within the frequency range 0,1 - 50 Hz. The Amplitude of displacement is adjusted up to 20 mm in extension. Each upper clamp attachment of the samples is fixed to the load cell and its corresponding sample clamp attachment is connected to a separate computer-controlled stepping motor to ensure constant pre-stress during the whole time of analysis.

The crack growth of each rubber specimen is monitored through an image processing system with a CCD monochrome camera mounted on the linear motion axis system. The camera moves along the x-axis from specimen to specimen and takes a picture of the involved sample. The picture is then transferred to a frame grabber and stored. After the picture has been digitalized, the software localizes in situ the crack position and determines the contour length by following the black and white boundary line of the crack. The infinitesimal enlargement of crack surface area through dynamic loading can be determined according to Rivlin & Thomas criterion [1].

The separate connection of each stepping motor with control unit 2 additionally supplies the Tear Analyzer with a quasi-static testing method. This method is realized as follows: the hydropulzer is stopped and the stepping motor can move with the upper clamp attachments of the sample in the enabled range. The range of deformation is 300 mm. Each load cell transfers the stress-strain data to the computer and CCD camera system grabbed the crack propagations in the involved sample.

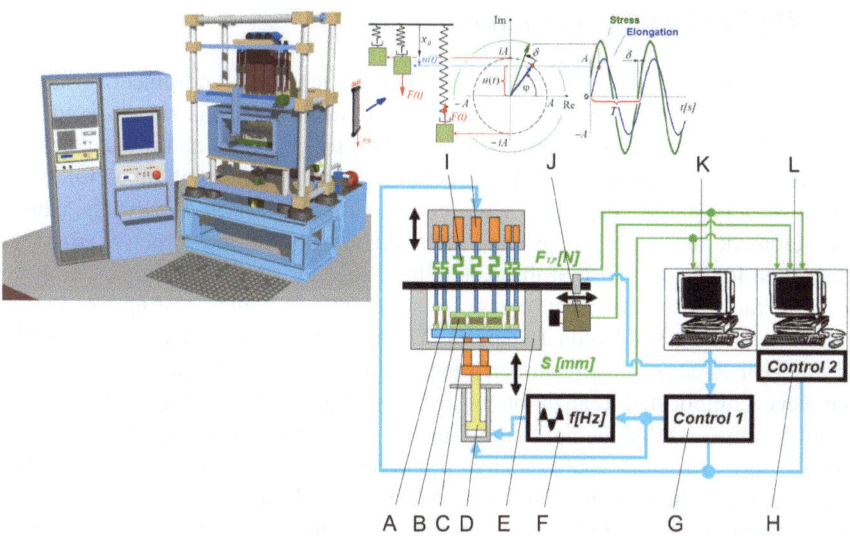

Fig. 10 Schematic and functional diagram of the test machine: A-tensile test piece; B-pure-shear test piece; C-traverse; D-hydropulzer; E-isolated chamber; F-frequency generator; G-control unit 1; H-control unit 2; I-load cells; J-CCD monochrome camera; K-PC1; L-PC2;

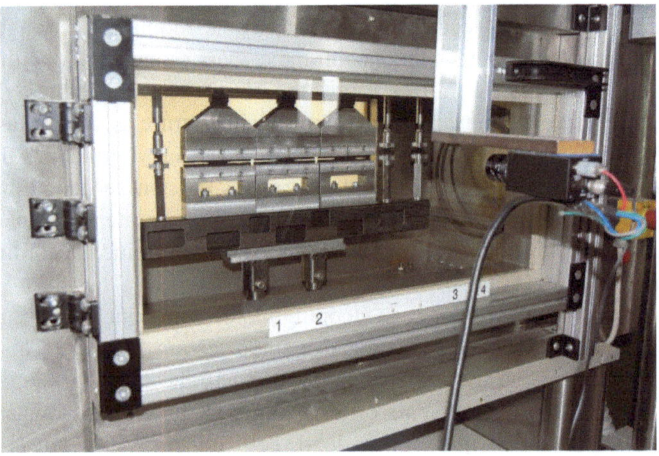

Fig. 11 The real visualization of the air-conditioned chamber with camera and SENT as well as pure-shear test specimens

3.3 Adjusting of the Tear Analyzer's Design for Measurement at Higher Loading Conditions

The Tear Analyzer is dynamically working mechanical equipment, which is characterized with natural mode shapes and frequencies dependent on the design body. Why is it important to measure the natural mode shapes and frequencies? At certain natural frequencies of the structure, a small amount of input force can cause a very large response. The redevelopment of Tear Analyzers caused measurement at higher forces and thus the design of the structure must be controlled for the response of the system in changed loading conditions. Certainly, two of the most important areas of structural dynamics testing is that of experimental and FEM modal analysis. The identification of dynamic properties of the Tear Analyzer has been performed using FEM modal analysis, because of the very complicated realization of experimental analysis. The FEM analysis simplifies the establishment of natural modal shapes because of the possibilities to modify the structure before any prototype is available for experimental measurement. The structure can be modified much cheaper, faster and easier than it could be redesigned for the real prototype. The building of the model with up to date hardware is the next advantage of the FEM method. The FEM model can predict a structure's behavior under real world dynamic operating conditions.

Discrete Model of the Tear Analyzer

For the creation of a discrete model of the Tear Analyzer, it is assumed that a solid material can be represented as a collection of particles corresponding to the material model interacting among themselves in different directions. The discrete model was created with respect to the weak points of the Tear Analyzer's structure. The most important weak point was expected in the interaction between the rigid clamps and rubber test specimens, because of the high difference of their mechanical behavior.

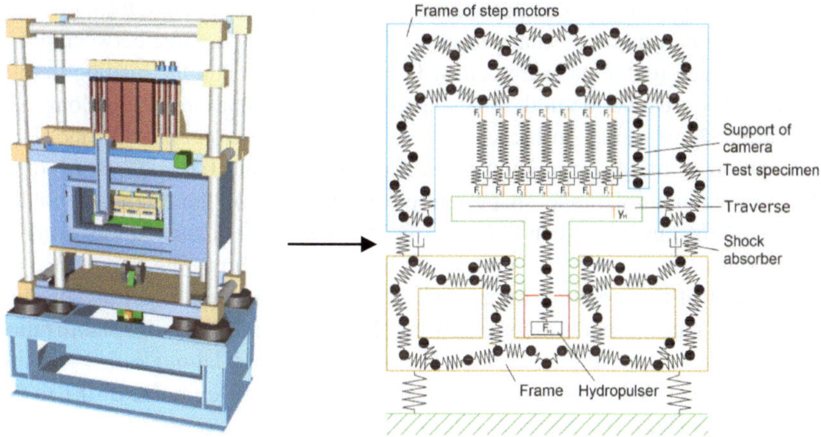

Fig. 12 Discrete model of the Tear Analyzer

Frequency Analysis and Optimizing of the Tear Analyzer's Discrete Model
The complex model of the Tear Analyzer has been calculated using the software ANSYS. The model of the structure consists of 62.981 elements, 118.152 nodes and 351.984 degrees of freedom. The calculation of natural mode shapes has been established in the range of frequency from 0.1 to 20 Hz.

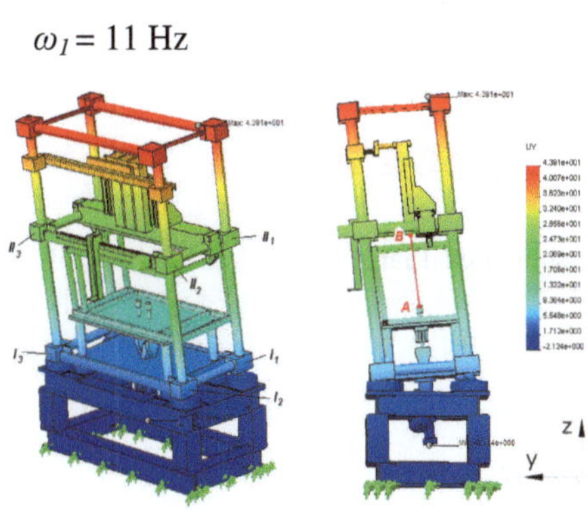

Fig. 13 The first modal mode shape

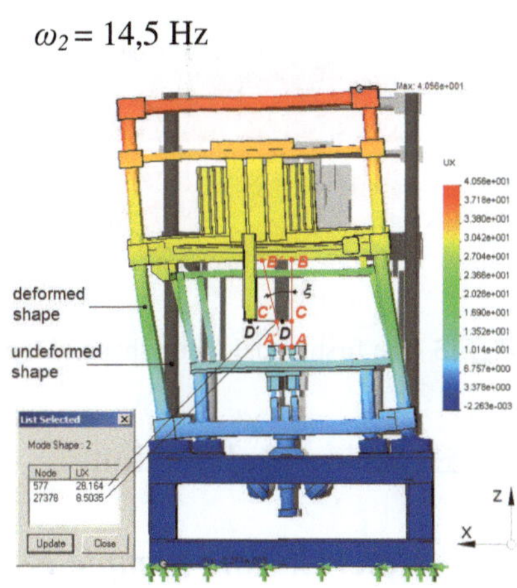

Fig. 14 The second modal mode shape

The calculation showed that the first two natural mode shapes are caused by the frequencies ω_1 = 11 Hz and ω_2 = 14.5 Hz. The first natural mode shape is caused by forwarding deformation, and the second one by longitudinal deformation.

The first natural mode shape has no influence on the distance between points A and B. Therefore, the measurement will be not influenced at the loading frequency 11 Hz. The second natural mode shape with deformation of the longitudinal direction significantly influences the distance between points A and B. The calculation of the distance showed a difference of about 0.0542% (s. Figure 14). The second natural mode shape caused the moving of the camera's holder (s. Figure 14, distance between C-C` and D-D´). The modal distance was established at the values $x_{rC\text{-}C'}$ = 8,5 and $x_{rD\text{-}D'}$ = 28,2. This result shows the importance of the attunement of the structure on natural mode shapes at frequencies higher than 20 Hz. As the natural mode shapes showed, this will be reached with stabilization or reinforcement of the frame. Figure 15 shows the first modal mode shape of the re-supported Tear Analyzer occurring outside of the frequency range of 0.1 to 20 Hz at ω_{1v} = 21,8 Hz.

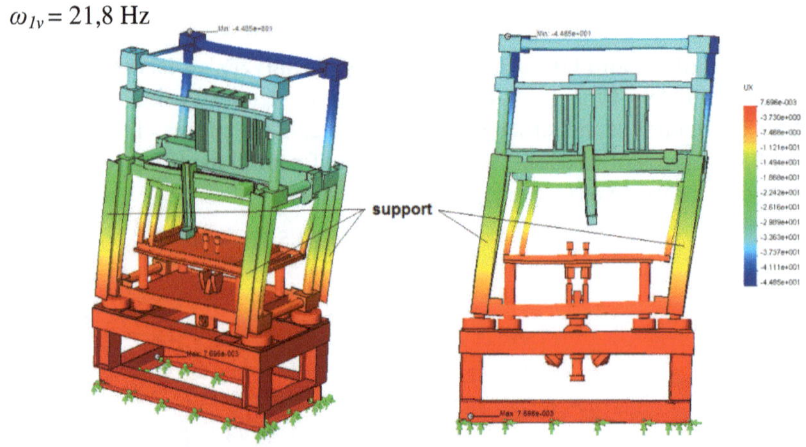

Fig. 15 The first modal mode shape of the re-supported Tear Analyzer

4 Experimental Section

4.1 Materials and Sample Preparation

For experimental investigations, unfilled and filled elastomers were compounded in an intermeshing mixer by DIK Hannover. Vulcanization of the samples was

Table 1 Composition of the materials investigated; data in phr

No	Label	NR	SBR	EPDM	CB N234	CB N234 Mas.bat.	Sulph.	ZnO	Stear. acid	ASM	T_{90} [min]
1	NR00	100	-	-	-	-	1	3	1	1,5	20,9
2	NR20N2	100	-	-	20	-	1	3	1	1,5	7,6
3	NR40N2	100	-	-	40	-	1	3	1	1,5	11,3
4	NR60N2	100	-	-	60	-	1	3	1	1,5	9,7
6	SBR00	-	100	-	-	-	1	3	1	1,5	5,1
7	SBR20N2	-	100	-	20	-	1	3	1	1,5	4,7
8	SBR40N2	-	100	-	40	-	1	3	1	1,5	3,7
9	SBR60N2	-	100	-	60	-	1	3	1	1,5	2,5
11	EPDM00	-	-	100	-	-	1	3	1	1,5	12,7
12	EPDM20N2	-	-	100	20	-	1	3	1	1,5	11,1
13	EPDM40N2	-	-	100	40	-	1	3	1	1,5	8,4
14	EPDM60N2	-	-	100	60	-	1	3	1	1,5	3,4
16	S-NR60N2	50	50		60	-	1	3	1	1,5	3,9
17	S-E60N2N	-	50	50	60	-	1	3	1	1,5	5,8
18	S-E60N2M	-	50	50	-	60	1	3	1	1,5	6,3

performed semi-efficiently with sulfur and accelerator (CBS-N-Cyelohexyl-2-benzothiazole sulfonamide and DPG-Dyphenylguanidine) in a heat press up to 90 % of the vulcameter torque maximum (T 90-time). As reinforcing filler, a constant amount (20, 40 and 60 phr) of carbon black type N234 was used. In the first phase the basic polymers were studied: natural rubber (NR), solution-styrene-butadien rubber (SBR) with 50% vol. vinyl and 25% vol. styrene and an amorphous ethylen-propylen-dien rubber (EPDM; Keltan 512). In the second phase, the blends consisting of combinations of NR, SBR or EPDM were investigated. All samples were compounded with the processing additives stearic acid and ZnO and protected against aging by IPPD. The ingredients are listed in Table 1. Standard SENT ($L_0 : Q$ = 1:1) and pure-shear ($L_0 : Q$ = 1:8) test specimens based on all types of compounds were cured and used for the fatigue experiments.

4.2 Influence of the Notch Procedure on Crack Propagation in Elastomers

The mechanisms of crack initiation and propagation in elastomeric materials are the subject of high scientific interest, because at present it is still not exactly known how these processes start and how they proceed under quasi-static as well as dynamic loading conditions. Therefore, the effect of the notch's geometry on crack propagation has been studied. A new method for the defined insertion of a notch into a rubber specimen was developed. This method serves as a basis for the reproducible and reliable determination of fracture mechanical parameters for elastomers. After the insertion of notches in a defined way into the test specimens with the geometry ratio $L_0 : Q = 1{:}2$, fracture tests under quasi-static loading conditions were performed.

The method of the notching is based on the blade razor insertion in the orthogonal direction to the supposed main strain of test specimen. The schematic diagram of the notching process and the created notch's surface are shown in the Figure 16. The principle of the defined insertion of the blade razor into the test specimen was realized with the redeveloped instrumented tensile test machine. Thus, the quantitative method of notching with exact positioning and length of the notch could be obtained.

For the investigation of the notch geometry and position influence on the crack propagation, four different notch's geometries have been realized using this new method. The Table 2 shows the realized notch geometries according to the definition of the test specimen and notch geometry (s. Figure 2). The geometry of the notch denoted 1 is created exactly orthogonally to the main stress. The second notch is defined with $\alpha = 0°$ and $\beta = \gamma = 20°$. The geometry of the notch denoted 3 has an orientation defined with $\alpha = \beta = \gamma = 10°$ and the last notch occurs with geometry $\alpha = 20°$, $\beta = \gamma = 0°$.

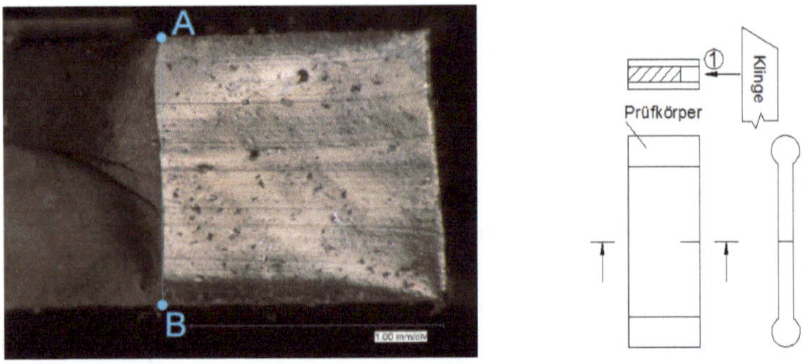

Fig. 16 Vizualization of the method of notching process and created notch's surface

Table 2 List of the notch geometries

	a_0 [mm]	α [°]	β [°]	γ [°]
1	2	0	0	0
2	2	0	20	20
3	2	10	10	10
4	2	20	0	0

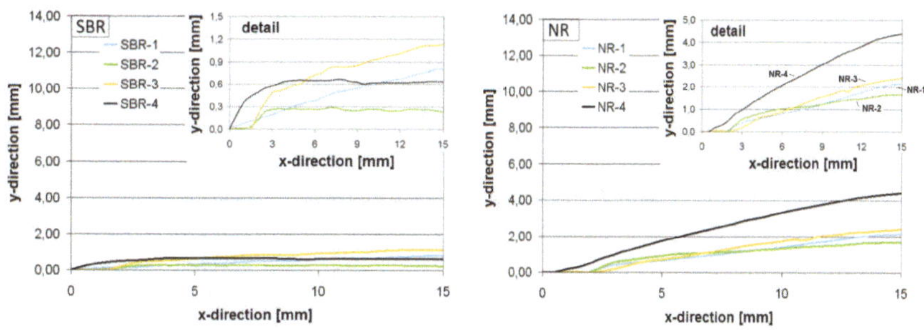

Fig. 17 The outline of the propagated crack in SBR (left) and NR (right), where: blue – notch's geometry 1 with $\alpha = \beta = \gamma = 0°$, green – notch's geometry 2 with $\alpha = 0°$, $\beta = \gamma = 20°$, orange – notch's geometry 3 with $\alpha = \beta = \gamma = 10°$ and black – notch's geometry 4 with $\alpha = 20°$, $\beta = \gamma = 0°$

Experimental data are plotted in Figure 17 and show the outlines of the propagated cracks in SBR (left) and NR (right) in the xy-direction. The influence of the notch geometry on the crack propagation is unambiguously evident. We found a higher deviation of crack growth outlines from the orthogonal direction to the main stress in the compound based on NR in comparison to the rubber based on SBR.

The slopes of the crack growth outlines measured in SBR demonstrate the parallel propagation of cracks in the case of a notch's geometry denoted 1: $\alpha = \beta = \gamma = 0°$. The other more complicated notches induce a substantial discrepancy when compared with the orthogonal direction to the main strain. The observed outlines of crack growth in rubber based on NR show an increase in the deviation from the orthogonal direction to the main strain over the whole effective test specimen's width. That effect occurs in strain-crystallizing rubber, such as analyzed natural rubber.

4.3 Investigations about Pure-Shear Width and Minimal Notch Length in Test Specimens

Because of the different strained regions in the pure-shear test specimen (s. Figure 4) it was necessary to investigate and to define the width of the pure-shear region and thus the minimum length of the notch for the initiation of crack propagation for the analysis of dynamic crack growth in the pure-shear region of a test specimen. The theory assumes that the pure-shear region is completely unstrained in the orthogonal direction to the main stress. The objective of the observation was to determine the width of this region. A schematic visualization of strained and unstrained regions in a stressed test specimen is shown in Figure 18. The region *C* visualizes the pure-shear region and the other regions denoted with *A* correspond with regions of deformed edges.

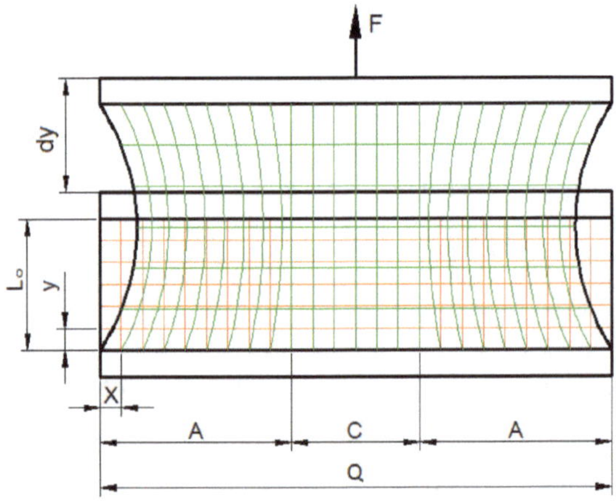

Fig. 18 Schematic visualization of determination of the width C of pure-shear area

Standard test specimens based on non-reinforced and reinforced SBR with amounts (20, 40 and 60 phr) of carbon black type N234, with different geometry ratios L_0:Q in the range of 1:12 to 1:4 and without initial cracks have been investigated first using FEM and second experimentally. The analyzed test specimen's geometries are listed in the Table 3.

Table 3 List of analyzed test specimen's geometries

No.	Length L_0 [mm]	Width Q [mm]	Geometry ratio L_0:Q
1.	10	120	1:12
2.	15	120	1:8
3.	20	120	1:6
4.	30	120	1:4

The width of the pure-shear region has been evaluated using non-linear finite element analysis with the software ANSYS by using the 6-parameter Ogden material model. Taking advantage of symmetry, only one quarter of each test specimen was modeled. The simulation was performed by different strains from 10 to 100 % of the test specimen's length L_0 applied on the long edges of the modeled test specimen.

The newly developed experimental method is based on tensile analysis of test specimens, wherein the surface of the test specimen is modified with a white squared layer. The stretched white squared layer is evaluated in an optical system that measures the displacement of the vertical layers in the points of intersection with the horizontal axis of the test specimens. The points with constant displacement over the whole stretched layers define the pure-shear region. The applied strain corresponds with the FEM calculation and was set at different strains from 10 to 100 % of test specimen's length L_0.

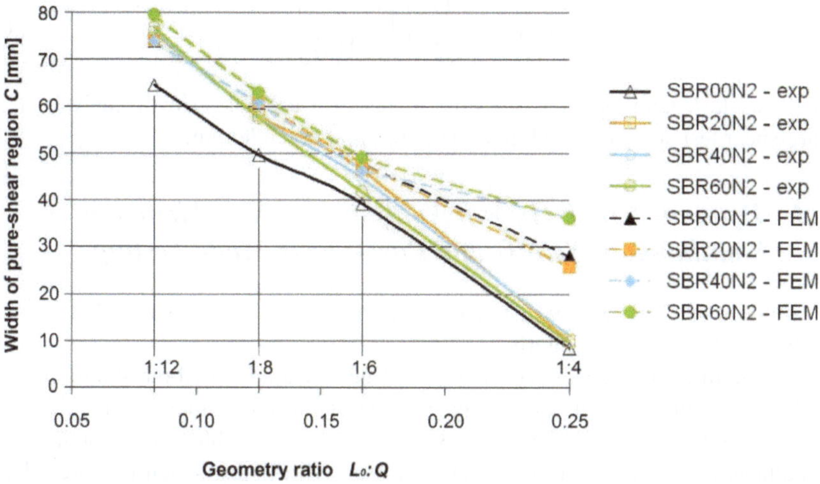

Fig. 19 Influence of the test specimen's geometry ratio on the width of pure-shear region *C*, evaluated experimentally as well as numerically at a maximum strain of 100% of the test specimen's length L_0

Figure 19 shows more detailed information about the coupling effect between the geometry ratio $L_0 : Q$ and the width of the pure-shear region C evaluated at a maximum strain of 100% of the test specimen's length L_0. A comparison of FEM results and experimental data is plotted in the diagram. By increasing the geometry ratio $L_0 : Q$ we determined a decrease in the width of the pure-shear region C. This effect was evaluated with FEM as well as with experimental methods. Clearly, both of the analysis methods show an accordant value of the pure-shear region's width C in the range of the geometry ratio from 1:12 to 1:6. The different values evaluated with FEM in comparison to the experimental method at the geometry ratio 1:4, are caused due to the optical method used for determination of the stretched white squared layer.

It is possible to correlate the observed dependence of the test specimen's geometry ratio on the width of pure-shear region C. We derived the following equation for C as a function of the strain in the range from 10 to 50% of the test specimen's length L_0:

$$C = \left[-3{,}115 \cdot \frac{L_0}{Q} + 0{,}979 - \kappa \cdot \varepsilon \right] \cdot 100, \qquad (18)$$

where: $\kappa = 0{,}0905$ [-] - factor described the influence of strain,
ε [-] - strain.

Then, making use of this definition, we can propose the equation of minimal efficient notch length for the analysis of crack propagation in the pure-shear region:

$$a_{s0\min} = \frac{Q}{2} \cdot \left[1 - \frac{C}{100} \right]. \tag{19}$$

With the last equation, it is possible to perform a crack growth analysis in the pure-shear test specimen with a defined minimal notch length.

4.4 Dynamic Crack Growth Analysis by a New Method of Simultaneous SENT- and Pure-Shear-Mode Testing

Evaluation of Crack Growth Velocity Dependent on Test Specimen's Geometry Ratio L_0:Q and Notch Length

This experimental analysis has been performed by a new method of simultaneous SENT- and pure-shear-mode testing. Four SENT test specimens with single edge notches and three pure-shear test specimens with double edge notches were simultaneously analyzed in the Tear Analyzer. All the test specimens were based on the compound SBR20N2. Each of the SENT test specimens was notched with a notch length of $a_0 = 2.0$ mm. Two of the three pure-shear test specimens were notched with a notch length of $a_0 = 21.0$ mm and the last pure-shear test specimen was notched with a notch length of $a_0 = 2.0$ mm, identical to the SENT test specimens. The quantity of test specimens enables the tester to obtain and describe four crack growth curves according to all of the analyzed SENT and pure-shear test specimens with the notch length $a_0 = 21.0$ mm and two crack growth curves according to the last pure-shear test specimen with the notch length $a_0 = 2.0$ mm. The fatigue test conditions were set at a frequency of 1Hz, loading amplitude 25% of L_0, preload of test specimens 1N and the tests were performed in a laboratory atmosphere.

The averaged values of crack growth curves for each of the analyzed test specimens and relevant calculated crack growth velocities are shown in the Figure 20. The relationship between the crack growth length and number of cycles is visualized in diagram (I). The averaged value of crack growth in the SENT test specimen is the curve denoted (a). The curve can be divided into two different regions. In the first region the crack growth curve linearly increases as long as the crack length $a < 3.5$ mm and the increase is given at an angle of ξ_{T1}. In the second region, the crack propagates by a non-defined function of growth. A similar average of crack growth curves was observed in the pure-shear test specimen notched with the notch length $a_0 = 2.0$ mm, denoted (c). However, compared to the crack growth in SENT, three different regions of crack growth in the pure-shear test specimen were established. The linearity of crack growth curve occurs until a maximum crack length of $a < 3.0$ mm and is characterized by the angle $\xi_{T2} \approx \xi_{T1}$. The next region, indicated as non-defined crack growth function, has a curve similar to the crack growth in the SENT test specimen. Because of the large width Q of the pure-shear test specimen, the crack growth proceeds in the last region of the second part of the linear increase of crack growth characterized by $\xi_{P2} > \xi_{T2}$. The second linearity of crack growth in the pure-shear test specimen (c) starts with a crack length $a < 22.0$ mm and continues over a broad range of crack lengths. The curve, denoted (b_1), represents the average value of crack growths in the pure-shear test specimens notched with the notch length $a_0 = 21$ mm. A linear relationship between crack length and number of cycles was observed in the complete range of analyzed cycles. The corresponding slope is ξ_{P1}. The horizontal translation of the curve (b_1) to the position (b_2) shows an increase similar to the second linear part of the crack growth curve (c), obtained in the pure-shear test specimen with the notch $a_0 = 2.0$ mm.

The following equation was used for the calculation of the crack growth velocities a' shown in the lower diagram (II) derived from the crack growth rate visualized in the upper diagram (I):

$$a'_{i+5} = \left(\frac{a_{i+5} - a_i}{t_{i+5} - t_i} \right); \; i \in \, <1, j>, \tag{20}$$

where t is a time dependent on cycles n, and j is the total number of measured values of crack lengths during the complete analysis. The calculations unambiguously show the constant crack growth velocity a' in the regions of crack growth linearity, and also proved the independence of the test specimen's geometry and notch length in this region. The non-defined crack growth curve in the range of crack length $3.0 \leq a \leq 22.0$ mm was characterized as a second derivation of crack length and thus represents an increasing acceleration of crack growth.

The experimental results suggest a common pattern of crack growth in rubber test specimens, demonstrate the complexity of the crack growth curve and describe the dependence of crack growth behavior on the test specimen's geometry ratio over the broad scale of the crack length.

The comparison of the crack growth curves obtained for different geometry ratios recognize the independence of the crack growth curve from the geometry ratio, whereas the variation of the ratio has been reached by the modification of a test specimen with width Q and invariable length L_0. The maximum possible crack length by the crack growth is then indirectly proportional to the geometry ratio. Thus, the description of a crack growth curve over a broad crack length scale can be obtained only in a test specimen with a low geometry ratio, as is demonstrated with the crack growth curve in the test specimen with the ratio $L_0:Q = 1:8$.

Three regions characterized with different crack growth behavior have been identified in the crack growth curve as dependent on the crack length. The first narrow region of short crack length, where the crack propagates under a constant crack growth velocity, is characterized by a stable state of crack propagation.

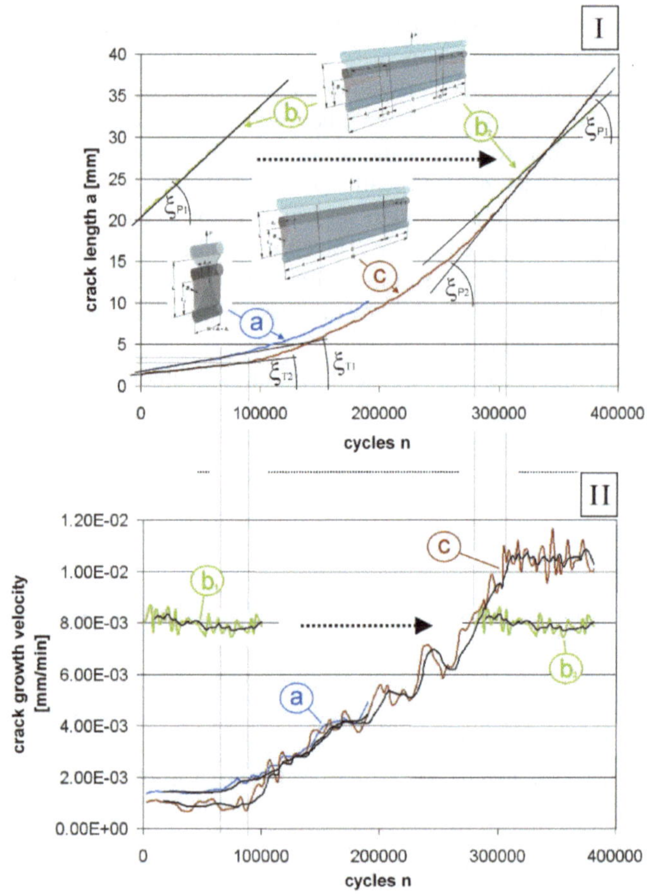

Fig. 20 Comparison of the crack growth rate (I) and calculated crack growth velocities (II) in the SENT and pure-shear test specimens for variable notch lengths

The crack growth curve occurs in this region independent from the geometry ratio. This approach was confirmed by FEM results (s. Figure 6), where the linear crack growth curve was confronted with tearing energy according to the expression in equation (9). The linear character of the crack growth allows for applying a very short notch length, which is a good premise to utilize the availability of the maximum length of the region of equation (9). However, on the other hand, the understanding of this approach guarantees the economic utilization of the test machine, represented by the termination of the analysis and leaving the stable state of crack propagation or availability of equation (9) for the analysis of the fracture mechanic in the SENT test specimen. This approach offers a hypothesis of the common usage of double notch test specimens (DENT) with a high geometry ratio and thus improves the quantity of measured data.

In the second narrow region of the crack growth curve, an accelerated propagation of crack has been determined. According to the FEM simulation, this crack growth curve shows complicated progression and represents a transition to stable crack growth in the pure-shear region.

This very important pure-shear region is situated over a broad crack length scale in the center of test specimen. However, it is defined by the low geometry ratio and occurs in the complete center only in the single edge notched test specimen. In this region, the characteristics of dynamic crack growth are given by stable crack growth in pure-shear region characterized by the equation (13) and are consistent with FEM simulation. This approach defines the utilization of the pure-shear test specimen and proves there is also a presence of the pure-shear state in the test specimen with the higher geometry ratio. For extremely high values of the geometry ratio approaching 1, there is no more pure-shear region in the test specimen. As the result of the FEM shows for the double-notched test specimen with the low geometry ratio, there is a region near the center of test specimen, where the double-notched test specimen does not show the pure-shear state. This region is characterized by the two crack tips, where the distance between them decreases, and thereby the stress fields around the crack tips influence each other and the fracture behavior changes.

Experimental Fatigue Crack Growth Analysis

The tests were carried out through a new method of simultaneous tensile and pure-shear mode testing. The fracture mechanical investigations under cyclic loading were performed with four tensile and three pure-shear test specimens. Four tensile test specimens were represented with three notched SENT and one not-notched tensile test specimen with the identical geometry ratios: $L_0 : Q$ = 1:1. The three simultaneously analyzed pure-shear test specimens were represented with two notched and one non-notched test specimen with the identical geometry ratios: $L_0 : Q$ = 1:8. Measurements were performed with preloading force 1N under sinusoidal loading conditions with the frequency 1 Hz at room temperature. The strain was varied between 5% and 45 %. The list of analyzed compounds or detailed testing and boundary conditions of the experimental fatigue crack growth analysis are summarized in Table 4.

For the evaluation of tearing energy T_t in the SENT test specimen, the equation (9) was used, and of tearing energy T_p in the pure-shear test specimen, the equation (13) was used. For the evaluation of crack growth rate *da/dn,* dependent on tearing energy T_t *and* $T_{p,,}$ the automatic evaluation of strain energy density *w* of each loading cycle was used in combination with an optical detection of crack length *a* implemented in the Tear Analyzer. The determined data were summarized to the double logarithmic plot of crack growth rate vs. tearing energy. This type of power law behavior is termed "Paris Plot" and expressed with the equation (16).

Table 4 Testing and boundary conditions for the fatigue crack growth analysis

Compound	
No.	Label
1.	NR00N2
2.	NR20N2
3.	NR40N2
4.	NR60N2
5.	SBR20N2
6.	SBR40N2
7.	SBR60N2
8.	EPDM20N
9.	EPDM40N
10.	EPDM60N
11.	S-NR60N2
12.	S-E60N2N
13.	S-E60N2M

Loading mode		Sinusoidal	
No. of frequencies		1	[Hz]
List of loading amplitudes		5; 10; 13; 16; 20; 25; 30; 35; 40; 45	[%]
Test specimen geometry (L_0 x Q x B)			No. of test specimens analyzed per amplitude
Tensile	15x15x1,5	[mm]	4
Pure-shear	15x120x1,5	[mm]	3
Notch length a_0			Type of notch
Tensile	2	[mm]	SEN (single edge notch)
Pure-shear	21	[mm]	DEN (double edge
Type of crack growth observation			Outline

In Figure 21, the influence of the volume of carbon black in NR on crack growth rate becomes apparent. We found that the crack growth rate at a given tearing energy decreases significantly with an increase in the volume of carbon black, whereas the crack growth exponent *m* slightly increases with a higher amount of carbon black in rubber test specimens based on NR. The increase of tear strength independent from carbon black concentration induces the higher crack growth rate. However, at the integration of an increasing concentration of carbon black, the tear resistance becomes higher. This indicates the positive reinforcement of NR by carbon black type N234. The increase of the crack growth exponent *m* pointed to the decreasing influence of the reinforcement with carbon black at higher tearing energies.

The simultaneous analyzation of two different test specimens determined the dependence of crack growth rate on tearing energy, by detecting the identical crack growth exponent *m* in the both of the test specimens. The comparison of the results determined in SENT as well as in pure-shear test specimens visualized in Figure 21, shows that the crack growth rate at a given tearing energy is higher in the pure-shear in comparison to the SENT test specimen, as observed in the non-reinforced and reinforced NR with a concentration of 20 phr of carbon black. The trends of the crack growth rate at the given tearing energy of NR reinforced by the concentration of 40 phr of carbon black are identical, as evaluated in SENT as well as in pure-shear test specimens.

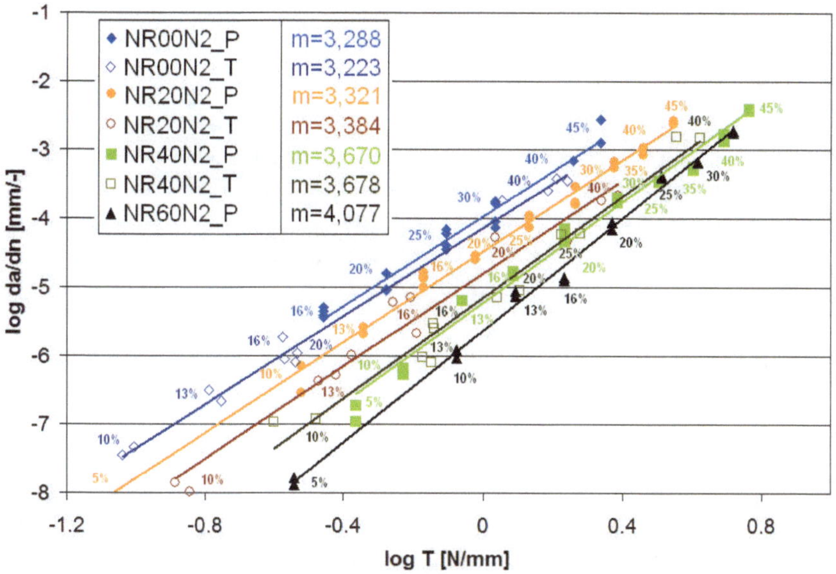

Fig. 21 Power law behavior of fatigue crack growth rate vs. tearing energy for the SENT (denoted T) as well as pure-shear (denoted P) test specimens based on carbon black filled NR

Figure 22 shows the effect of filler concentration in SBR on fatigue crack growth rates. Similar trends as in NR of dependence of fatigue crack growth rate on tearing energy are observed in SBR. With an increase of carbon black concentration, the decrease of the crack growth rate at a given tearing energy has been determined. It was found that the crack growth exponent *m* slightly decreases with a higher concentration of carbon black in rubber test specimens based on SBR.

The comparison of the results determined in SENT as well as in pure-shear test specimens proved a higher crack growth rate at a given tearing energy in the pure-shear as compared to the SENT test specimens. These trends have been observed in the all of the reinforced SBR test specimens independent from carbon black concentration. The result showed the identical crack growth exponent *m* independent from the test specimen's geometry.

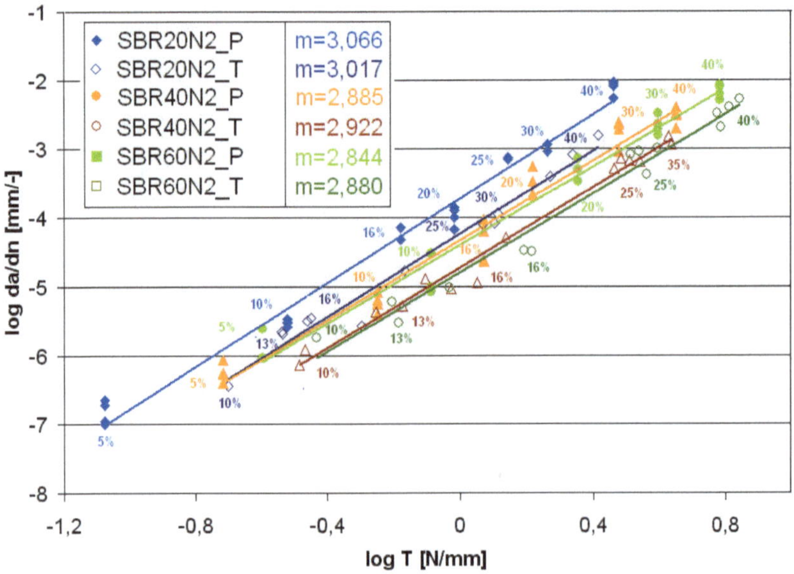

Fig. 22 Power law behavior of fatigue crack growth rate vs. tearing energy for the SENT (denoted T) as well as pure-shear (denoted P) test specimen based on carbon black filled SBR

Obviously, the crack growth exponent *m* observed in the reinforced EPDM (s. Figure 23) is significantly lower in comparison to the crack growth exponents determined in the reinforced NR and SBR. The crack growth exponent characterized by the dynamic fracture behavior of EPDM is not dependent on carbon black concentration. The increase of carbon black concentration has a decreasing influence on crack growth rate at a given tearing energy in SENT as well as in pure-shear test specimens based on EPDM. The comparison of the influence of a test specimen's geometry on dynamic fracture behavior in EPDM showed that the crack growth rate at a given tearing energy in the pure-shear test specimen is higher compared to the crack growth rate in the SENT test specimen.

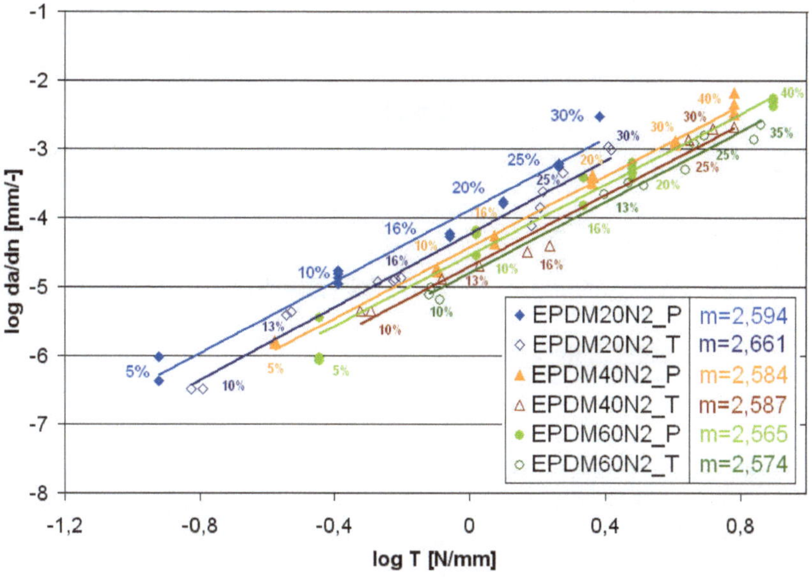

Fig. 23 Power law behavior of fatigue crack growth rate vs. tearing energy for the SENT (denoted T) as well as pure-shear (denoted P) test specimen based on carbon black filled EPDM

It was assumed that the behavior of the fatigue crack growth rate vs. the tearing energy of a blend based on two different rubbers depends on the fracture behavior of each type of rubber used. In Figure 24, we analyzed and plotted a comparison of power law behavior of fatigue crack growth rate vs. tearing energy for SBR, NR and blend SBR-NR based polymers, reinforced with 60 phr of carbon black type N234. We found that the crack growth rate at a given tearing energy of blend S-NR60N2 (based on 50phr of SBR and 50phr of NR) is derived from the fracture behavior of the individual components SBR and NR, because of the localization of the fracture values of blends between the fracture values of the both components. The crack growth rate at a given tearing energy of blend S-NR60N2 is closer to the fracture behavior of the NR component in comparison to the fracture data of the SBR component. This phenomenon shows the dominant influence of NR on the crack propagation in a blend based on SBR and NR.

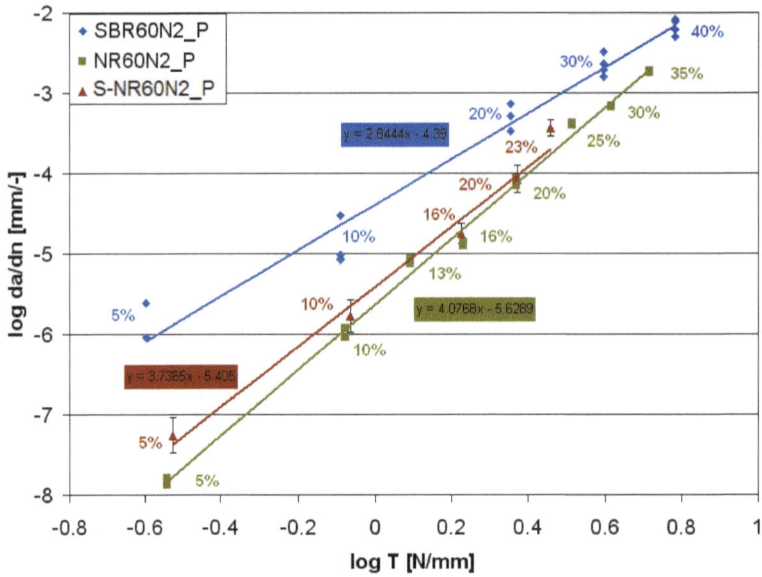

Fig. 24 Comparison of power law behavior of fatigue crack growth rate vs. tearing energy for SBR, NR and blend SBR-NR based polymers reinforced with 60 phr of carbon black type N234

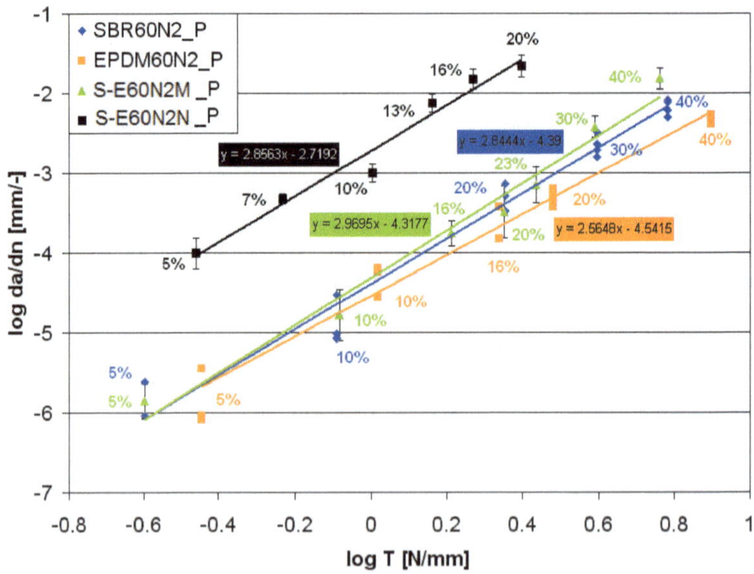

Fig. 25 Comparison of power law behavior of fatigue crack growth rate vs. tearing energy for SBR, EPDM and blend SBR-EPDM based polymers reinforced with 60 phr of carbon black type N234 normal or N234 master batch

In addition to this observation, the blend S-E60N2N, based on SBR and EPDM with 60phr of carbon black type N234 normal and the blend S-E60N2M, based on SBR and EPDM with 60phr of carbon black type N234 master batch were analysed dynamically in the Tear Analyser. The related results of these experiments are shown in Figure 25. The data of the crack growth rate at a given tearing energy for blend S-E60N2M are similar with the fracture behavior determined in the component SBR. Thus the component SBR is the main factor influencing the fracture mechanic of blends. The blend S-E60N2N has a lower dynamic crack growth resistance compared to the fracture behaviour of SBR, EPDM as well as blend S-E60N2M. The higher dynamic crack growth resistance of S-E60N2M could be reached by the increasing activity of master batch carbon black that leads to an increasing number of polymer-filler interactions and therefore to a higher level of fracture mechanical properties.

5 Conclusion

The aim of this research work was to develop a new method for the description of dynamic crack growth behavior of rubber by simultaneous tensile and pure shear mode testing, and to systematize efficiently the analysis of fracture behavior of elastomers. For a study of fracture material behavior, industrially used test methods were selected carefully. The material investigated were SBR, EPDM, NR, blends SBR-NR and SBR-NR vulcanizates reinforced with different filler content. The results of the analysis show different information regarding the various fracture mechanisms and methods for the efficient description of fracture behavior of rubber.

At first, the work deals with the theoretical description of the dependence of fracture behavior in rubber test specimens on its geometry, respective in its geometry factors L_0:Q. We assumed numerically that the deformation in the pure-shear test specimen is close to the pure-shear state in the center. However, the deformation in the regions near the free edges is complex. We demonstrated similar crack growth behavior in the free edges of SENT as well as of pure-shear test specimens.

The development of a new method for the analysis of dynamic crack propagation in filled rubbers by simultaneous tensile and pure shear mode testing is shown. The development is based upon the mechanical redevelopment of the test machine 'Tear Analyzer' (Co. Coesfeld GmbH), upon the equipment of the test machine with new hardware, and modification of the system for the recording of crack length. This machine is equipped with two modes operating simultaneously, which represents an extended fracture testing method. The redevelopment of the machine for the measurement of pure-shear test specimens was conditioned with the determination of test measurement accuracy. The FEM method was used for the calculation of the dynamic behavior of the machine's design in the range of used frequencies. The FEM calculation showed the weak points of the design and pointed out the necessary steps of redevelopment for measurement in the range of frequencies from 0.1 to 20 Hz.

The next objective was to investigate and to define the minimum length of a notch for the initiation of crack propagation in a pure-shear test specimen for the analysis of dynamic crack propagation in rubber. Test specimens with different geometry factors L_0:Q in the range of 1:12 to 1:4, with or without initial cracks have been investigated. The newly developed experimental method was supplemented by an appropriate FE–analysis. These methods firstly allowed a determination of the width of the pure-shear range, and thus a definition of the minimal notch length in the test specimen independent from its geometry. This definition forms the theoretical basis for an improvement in the efficiency of the analysis of dynamic crack propagation in rubbers.

The definition of the notch length had to be supported with the development of a method for the defined insertion of a notch into a rubber specimen. This method serves as a basis for the reproducible and reliable determination of fracture mechanical parameters for elastomers. After insertion of notches in a defined way, fracture tests under quasi-static loading conditions were performed. A significant influence on the notch geometry was observed in the test results. The variation of the angles of the notch showed a strong influence on the crack propagation. These results illustrated the importance of a defined and reproducible notching of elastomeric specimens.

In the experimental study of dynamic fracture behavior, experimental crack growth data of several rubber test specimens were obtained simultaneously with a new mechanical fracture testing concept. The test specimen used had different geometry ratios from those represented by commonly used SENT and pure-shear test specimens. This testing concept allowed a direct comparison of the dynamic crack growth behavior dependent on the test specimen's geometry ratios. We demonstrated the application of this testing method for the description of fracture behavior dependent on arbitrary crack lengths. The comparison of experimental data and FEM results show a dependence of the tearing energy on the crack growth curve and, thus, the correspondence between the regions of stable crack growth propagation obtained in experimental tests and regions of calculated tearing energy. The results identified the weak points of the dynamic crack growth analysis in such rubber test specimens and suggests steps for a follow-up research.

We found from the dynamic fracture investigation that the crack growth rate at a given tearing energy decreases significantly with increasing carbon black content, independent from the base of rubber. The results demonstrated the increase of the crack growth exponent m according to the strain. The decreasing influence of reinforcement with carbon black at higher tearing energies is shown. The dependence of crack growth rate vs. tearing energy simultaneously analysed in two different test specimens detected the identical crack growth exponent m in both of the test specimens. The comparison of the results determined in SENT as well as in pure-shear test specimens shows that the crack growth rate at a given tearing energy is higher in the pure-shear compared to the SENT test specimen, independent from the base of rubber. Therefore, the pure-shear test specimen represents a more critical geometry for the determination of fracture behavior of rubber.

The concept of simultaneously testing of SENT and pure-shear test specimens represents a new method for the improved description of the dynamic fracture behavior of rubber materials.

References

[1] Rivlin, R.S., Thomas, A.G.: Rupture of rubber. I. Characteristic energy for tearing. Journal of Polymer Science 10, 291–318 (1953)

[2] Lake, G.J., Lindley, P.B.: Cut Growth and Fatigue of Rubbers, II. Experiments on a Noncrystallizing Rubber. Journal of Applied Polymer Science 8, 707–721 (1964)

[3] Lake, G.J., Lindley, P.B.: The Mechanical Fatigue Limit for Rubber. Journal of Applied Polymer Science 9, 1233–1251 (1965)

[4] Stoček, R., Heinrich, G., Gehde, M., Rauschenbach, A.: Untersuchungen zur Kerbtiefe im Pure-shear Prüfkörper für präzise Analysen der dynamischen Rissausbreitung in Elastomeren. Zeitschrift Kunststofftechnik / Journal of Plastics Technology, 2–22 (January 2012)

[5] Yeoh, O.H.: Analysis of deformation and fracture of 'pure shear' rubber testpiece. Plastics, Rubber and Composites 30, 391–397 (2001)

[6] Griffith, A.A.: The phenomena of rupture and flow in solids. Philos. Trans. R. Soc. Lond., Ser. A 221, 163 (1920)

[7] Lake, G.J.: Fatigue and Fracture of Elastomers. Rubber Chemistry and Technology 68, 435–460 (1995)

[8] Klüppel, M., Huang, G., Bandow, B.: Evaluation of Tearing Energy of Elastomer Materials. Kautschuk-Gummi-Kunststoffe 61, 656–659 (2008)

[9] Gent, A.N.: Engineering with Rubber, 2nd edn. Hanser Publishers, Munich (2001)

[10] Gent, A.N., Lindley, P.B., Thomas, A.G.: Cut Growth and Fatigue of Rubbers. I. The Relationship between Cut Growth and Fatigue. Journal of Applied Polymer Science, 455–466 (1964)

[11] Paris, P., Erdogan, F.: A critical analysis of crack propagation laws. Journal of Basic Engineering, Transactions of the American Society of Mechanical Engineers, 528–534 (1963)

[12] Stoček, R., Gehde, M., Heinrich, G.: Analyse des dynamischen Risswachstums von Elastomeren im simultanen Zug-(Tensile-) und Pure-Shear-Prüfmodus bei optimierter Risserfassung. Kautschuk-Gummi-Kunststoffe 62, 170–176 (2009)

Author Index

MIX
Papier aus verantwortungsvollen Quellen
Paper from responsible sources
FSC® C105338

If you have any concerns about our products,
you can contact us on
ProductSafety@springernature.com

In case Publisher is established outside the EU,
the EU authorized representative is:
Springer Nature Customer Service Center GmbH
Europaplatz 3, 69115 Heidelberg, Germany

Printed by Libri Plureos GmbH
in Hamburg, Germany